The Inquisitive Problem Solver

Library of Congress Catalog Card Number 2001097391
ISBN 0-88385-806-1
Printed in the United States of America
Current Printing (last digit):
10 9 8 7 6 5 4 3 2 1

The Inquisitive Problem Solver

by

Paul Vaderlind, Richard K. Guy, and Loren C. Larson

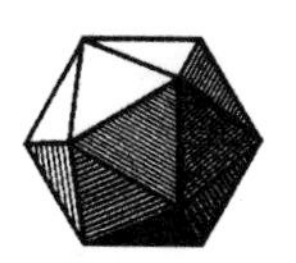

Published and distributed by
The Mathematical Association of America

MAA PROBLEM BOOKS SERIES

Problem Books is a series of the Mathematical Association of America consisting of collections of problems and solutions from annual mathematical competitions; compilations of problems (including unsolved problems) specific to particular branches of mathematics; books on the art and practice of problem solving, etc.

The Inquisitive Problem Solver, by Paul Vaderlind, Richard K. Guy, and Loren C. Larson

Mathematical Olympiads 1998–1999: Problems and Solutions From Around the World, edited by Titu Andreescu and Zuming Feng

Mathematical Olympiads 1999–2000: Problems and Solutions From Around the World, edited by Titu Andreescu and Zuming Feng

USA and International Mathematical Olympiads 2000, edited by Titu Andreescu and Zuming Feng

USA and International Mathematical Olympiads 2001, edited by Titu Andreescu and Zuming Feng

MAA Service Center
P. O. Box 91112
Washington, DC 20090-1112
1-800-331-1622 fax: 1-301-206-9789
www.maa.org

Preface

Parvis e glandibus quercus
Large streams from little fountains flow,
Tall oaks from little acorns grow.

David Everett (1769–1813)
Lines Written for a School Declamation

Not just another problem book! To see why you'll have either to look fairly carefully through the book, or to read here about its development. For reviewers and others with the patience for neither, we just say that the core of the book is a collection of 256 entertaining miniatures: the original problems and solutions were published in Swedish by PV, translated into English by LL and augmented and embellished by RG.

But the book's long history, spanning many years of problem posing with contributions from cultures around the globe, can lead the reader to a deeper benefit than might otherwise be derived. The first version was published by PV (*Fler Matematiska Tankenötter*, Svenska Dagbladt, 1996). PV is a Polish mathematician who immigrated to Sweden, and is Professor of Mathematics at the University of Stockholm. He is a product of the mathematical training common in many east European countries, where gifted students are coached in mathematical problem solving from an early age, inspired by the mathematics of Paul Erdős.

PV had a base of experience to bring to mathematics education in Sweden. One manifestation of his energy was the writing of several mathematics problems books, aimed at a general audience. They feature problems that one might find in the puzzle section of a newspaper, so it was most important to write problems in a compelling way: not so much to instruct as to provide stimulation and entertainment; solutions were often abbreviated.

LL, who has long service on the Putnam Competition committee, and who has published two problem books, met PV on a trip to Sweden, knew that he was the coach of the Swedish Mathematics Olympiad Team and had an international reputation as a teacher. PV gave a copy of his latest problem book to LL, who decided to translate it as a way of learning Swedish. LL soon found that reading mathematics in this way provided a double joy: first, the mystery of each sentence, the discovery of the puzzle, word by word from the dictionary: second, the urge to solve and compare solutions. This enforced deliberation allows the essense to sink in and allows time for reflection and appreciation of the underlying mathematics.

LL was continually impressed by the quality of the problems, rich in mathematical ideas but not so bone-crunchingly difficult as to intimidate or to stifle curiosity. All were immediately understandable and approachable with virtually no mathematical prerequisites. Most folks with modest persistence could experience success and feel the satisfaction of accomplishment and empowerment: sometimes surprise and wonder.

LL thought that the collection should be available in English, and although many of the problems were not quite at a level suitable for the MAA, whose book list already features numerous excellent problem books, he sent the translation to them. It arrived at the Spectrum committee, then chaired by Arthur Benjamin, where RG, with a quarter-century's experience in the Unsolved Problems section of the *Monthly* and two books of unsolved problems, recognized its potential. He suggested that it be rewritten to bring out the mathematics behind the solutions. So we have a new book with a different purpose.

Originally it was in three sections of difficulty, but difficulty is often in the eye of the beholder, and with the cross-sectioning of Hints and Solutions, the book became too subdivided. It is still true that earlier problems are easier than later ones, but you will find some surprises—different surprises for different people.

Several vis-à-vis collaborations and much email correspondence between LL and RG saw a series of suggestions for the title. "Yes or No?" reflected our changes of wording to make this the short answer to each problem, and to emphasize the openness of a true mathematical investigation. "Why or Why Not?" expressed our hope that readers would provide much more than a simple 'yes' or 'no'. But we wanted more: at least a quarter of the original problems suggest additional problems of a more demanding nature, and we've inserted several new problems that we came across while working on the book. So our eventual title, "TIPS", encapsulates our main fascination, which is to explore the mathematics which lies behind often quite innocently simple little puzzles and problems. In this way we wish to show, by example, that solving a problem is not the end of the process: a good problem usually suggests other good problems that can lead to active involvement in mathematics and mathematical research.

In **P9**, making a wire frame leads to partitions and Euler circuits; in **P21** a simple number-arranging problem leads to a labeling of the Fano configuration and several other topics in combinatorics; several problems, including a classical pouring problem, **P41**, provide an introduction to linear diophantine equations; the placing of a gazebo in **P48** introduces the Fermat-Torricelli point, "Napoleon's" theorem and Steiner trees; the touching circles of **P111** lead to Platonic and Archimedean solids, Euler's formula and tilings of the plane; and the distances of **P219** to difference sets and projective geometries.

Many other examples are shown in the Treasury, which serves as Index, Glossary, and as a list of Terms, Techniques and Tricks of the Trade. Of course, problem solving is much more than a bag of tricks. A trick is a simple device for solving a particular problem; a trick used more than once is a method. Where a problem invites deeper inquiry we've added one or more further Queries, about a hundred in all, and given Responses to them in a later section. On a score of occasions (identified by an asterisk in the Response column in the Contents) we have left further investigation to **Rikki-Tikki-Tavi**, the mongoose in

Kipling's *The Jungle Book*, whose motto "Run and find out" inspired at least one of us from an early age. We're sure that the reader will be similarly inspired.

We occasionally give references and we encourage the reader to chase them down and to ask further questions themselves. As a well known mathematician has said, mathematics often owes more to those who ask questions than to those who answer them.

In order to accommodate a number of problems which came to our notice while we were working on the book, we have combined some of the original ones, and omitted a few which were variants of ideas which have already appeared in numerous problem collections. We specifically thank Barry Cipra, John Conway, Eugene Luks, Marc Paulhus, Marc Roth and David Savitt & Russell Mann for allowing us to publish their problems, and Cedric Smith for giving us the beautiful 13-coin problem which Herbert Wright sent to him around 1960, and which might otherwise have been lost to the world. Fred Galvin, Dean Hickerson, Michael Kleber, Dan Velleman, David Wilson and many others have also helped us, consciously or otherwise. Where we've added problems we've given their source, as far as we know it. The same problem often occurs to different people in different places, sometimes centuries apart. Any serious historian of problem ideas should have access to David Singmaster's *Sources in Recreational Mathematics* `zingmast@sbu.ac.uk`.

During the long gestation period, the MAA has started yet another series of books, in which we are proud to appear. We are grateful not only to Art Benjamin and members of the Spectrum committee, but also to Roger Nelsen and members of the Problem Books committee. The cheerful faith of Don Albers has been a constant encouragement, and the technical expertise of Beverly Ruedi continues to maintain the high standard of MAA books.

Have fun reading the book; we doubt if you'll have more than we had writing it!

Richard Guy, Calgary, Alberta
Loren Larson, Northfield, Minnesota
Paul Vaderlind, Stockholm, Sweden

September 2001

Contents

* An asterisk indicates that there's something for Rikki-Tikki-Tavi, who was introduced in the Preface.

1
Problems

P1. Bad calibration

My bathroom scale is set incorrectly but otherwise it works fine. It shows 10 kilograms when Dan stands on it, 14 kilograms when Sarah stands on it, but 22.5 kilograms when Dan and Sarah are both on it. Is the scale set too high?

P2. Getting inside a brick

You have three identical rectangular bricks and a ruler. Must you use a formula, such as the Pythagorean Theorem, to find the length of the brick's diagonal?

P3. In the running

In a cross-country run, Sven placed exactly in the middle among all participants. Dan placed lower, in tenth place, and Lars placed sixteenth. Is it possible to figure out how many runners took part in the race?

P4. The domino effect

Can each of these arrays of 1×1 squares be covered with 1×2 dominoes, with all dominoes inside the boundaries?

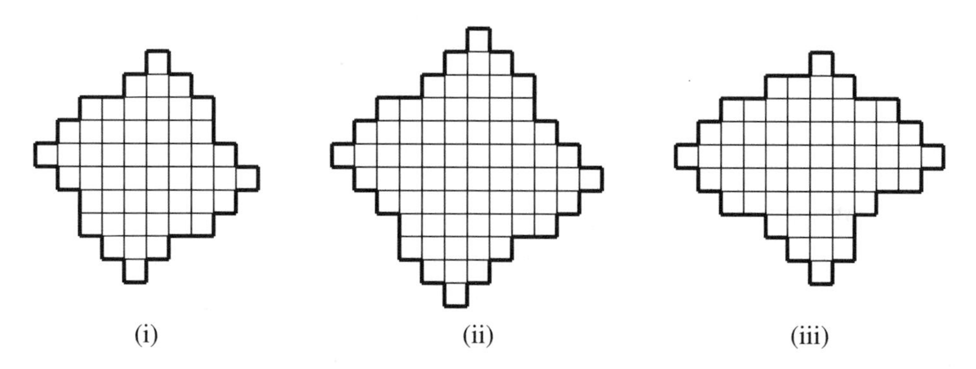

(i) (ii) (iii)

P5. What a mesh!

The figure below represents a system of gears, where the number of teeth on each gear is indicated by the number within the wheel. If A makes one revolution, will B make more than one revolution?

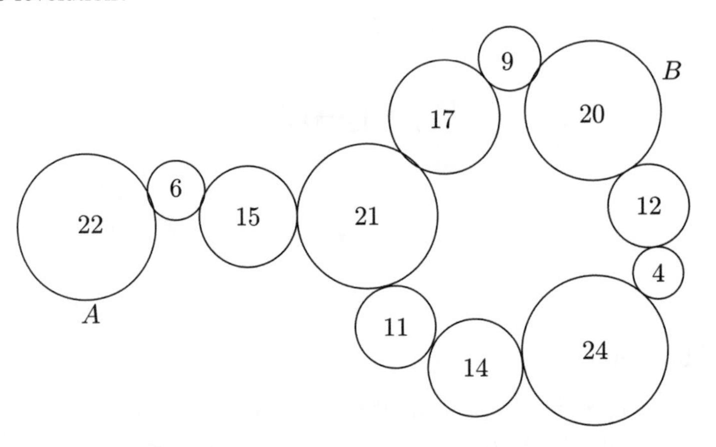

P6. The sands of time

Maia bought two unusual sandglasses. One measures a nine-minute interval, and the other measures a thirteen-minute interval. A certain love potion needs to boil for exactly thirty minutes. Is it possible to measure such a time interval with these sandglasses under the additional stipulation that you turn over the glass(es) for the *first time* just as the potion starts to boil? (Q)

P7. Anyone for tennis?

Each child in a school plays either soccer or tennis. One-seventh of the soccer players also play tennis, and one-ninth of the tennis players also play soccer. Do more than half the children play tennis?

P8. Tiresome parrot

We recently bought a parrot. The first day it said "O" and the second day "OK". The third day the parrot said "OKKO" and the day after that "OKKOKOOK". If this doubling pattern continues and the parrot squawks every second, will it get through squawking on the sixteenth day? (Q)

P9. A wire cube

A 12 inch long wire is to be divided into a number of parts and from these we want to construct the frame (that is, the edges) of a cube 1 inch on a side. Can you do it with three pieces? (Q)

P10. A cube of many colors

Each face of a cube is divided into four squares. Each of the 24 squares is to be painted with one of three colors in such a way that any two squares with a common edge have different colors. Can you arrange to have nine squares of the same color?

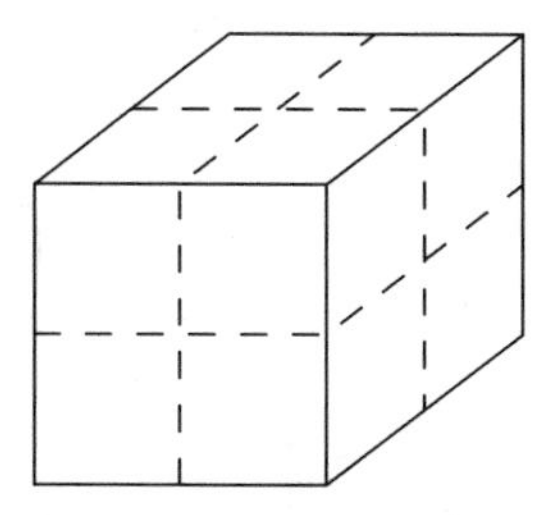

P11. Paper folding

You have a strip of paper that is two-thirds of a meter long. However, you need a strip exactly half a meter long. Must you have a ruler to cut off such a length? (Q)

P12. Colored cubes

You have nine red, nine yellow, and nine blue $1 \times 1 \times 1$ blocks. Can you build a $3 \times 3 \times 3$ cube such that no row of three small cubes will have two of the same color? (A row is a line of bricks parallel to one of the edges of the $3 \times 3 \times 3$ cube.)

P13. No-one twice as rich

One hundred children in a school are counting their money. Each child has between 1 and 100 cents, and no child has the same amount as any other child. Is it possible to divide the children into two groups so that no child in either group will have twice as much money as any other child in the same group?

P14. Find the dud

We know that one of 30 coins, all of which look alike, is counterfeit and weighs a little less than a true coin. With the help of a balance scale (two pans; no weights), can the counterfeit coin always be found in three weighings?

P15. Eternal triangle

In a round-robin volleyball tournament (everyone plays everyone once) the team from Lund had the same number of wins as the team from Uppsala. Must there be three teams in the tournament, A, B, C, such that A beat B, B beat C, and C beat A? (A tie is not possible in volleyball.)

P16. Painting models

Consider the three constructions shown below, each of which is made by gluing together 16 identical cubes. Does any one have less surface area than the others?

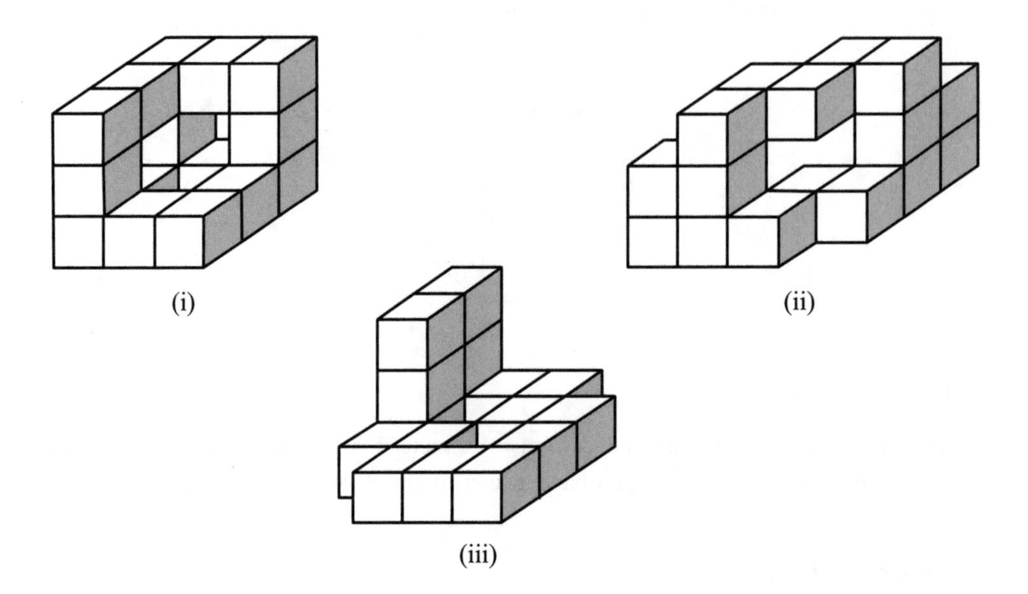

(i) (ii) (iii)

P17. Sums of consecutive numbers

The number 42 can be written in three different ways as the sum of two or more consecutive positive numbers:

$$42 = 13 + 14 + 15 = 9 + 10 + 11 + 12 = 3 + 4 + 5 + 6 + 7 + 8 + 9.$$

Can you write 105 in eight different ways as the sum of two or more consecutive positive numbers? (Q)

P18. Hand in hand

The women from Venus have five hands and the men have three. Their delegation to Varmland numbered 17 individuals. They decided to invite some Varmlanders to a dance. In their dance routine, everyone holds hands and together they move to the rhythm of Venusian music. For the dance to be done properly, every hand must be held by someone else's hand. Assuming all Varmlanders have two hands, will it be possible to do the dance with 13 Varmlanders? (Q)

P19. Unicursal

Can each of the following figures be traced without lifting the pencil? No part of the figure, except for the intersection points, should be traced more than once. (H)

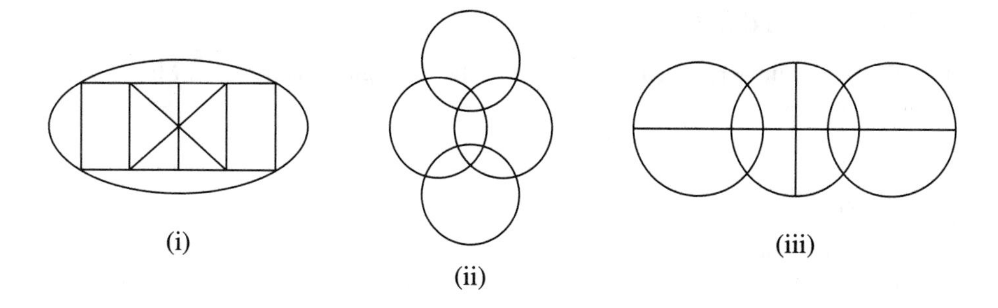

P20. Do-it-yourself

Is it possible to draw a figure in the plane with a number of lines (or arcs such as those in the preceding figure) that cannot be traced by a single path (without lifting the pencil), but can be drawn by tracing each line exactly twice?

P21. Number the regions

The three lines in the figure divide the circle into seven regions. Can you arrange the numbers 1 to 7 in these regions so that for each of the lines, the sums of the numbers on either side are all the same? (Q)

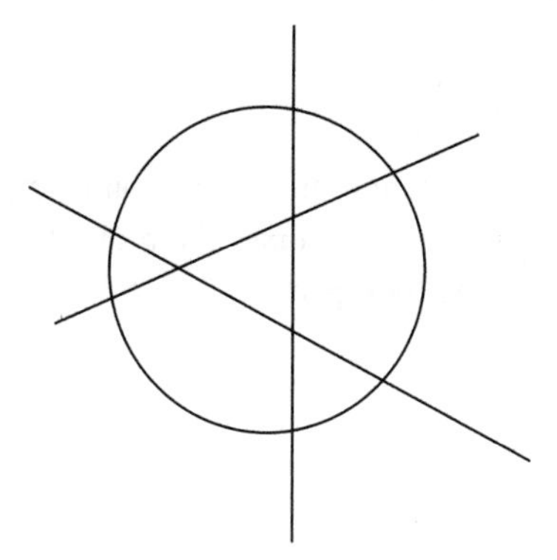

P22. Fair proportion

Someone said that the percentage of women employees in the office is more than 60% but less than 65%. Can there be as few as nine employees? (Q)

P23. More or less

(a) Are there more two-digit numbers with the first digit larger than the second than there are with the second digit larger than the first?

(b) Partition the numbers $0, 1, 2, \ldots, 123456$, into two sets in the following way. In the first set, place all numbers whose digits add to an even number, and in the second set, place all numbers whose digits add to an odd number. Are there more numbers in the first set than there are in the second set?

P24. Not much difference

Two five-digit numbers together use each of the ten digits exactly once. Must they differ by more than 250?

P25. Leapfrog

Four frogs sit in a row. Every five seconds two neighboring frogs hop into each other's places. Sometime after 80 seconds, but before 100 seconds, the frogs are seen to be in their original order. Is it possible to determine on which jump that occurred?

P26. Slicing a cube

A $4 \times 4 \times 4$ cube can be divided into 64 small $1 \times 1 \times 1$ cubes by making nine cuts. Can it be done in less? (H)

P27. Sums around a circle

I have written 50 integers, not necessarily different, around a circle, and their sum is 1 (so they're not all positive). A sequence of one or more consecutively placed integers is called *positive* if its sum is positive; otherwise it is called *nonpositive*. Are there more positive sequences than nonpositive?

P28. How many voted?

Among those taking part in an election, the proportion of men to women was 17:15. Had 90 fewer men and 80 fewer women taken part the proportion would have been 8:7. Does this determine how many people took part in the election?

P29. Shelf space

Soren works in a music store in the CD recordings department. He is faced with the problem of having to put all the opera recordings in a display case consisting of six shelves, each 1 meter long. The CD's are packaged in 150 small boxes, each of which is either 3 centimeters or 6 centimeters wide. Can Soren solve this problem if

(a) exactly 51 of the boxes are 6 cm wide?
(b) exactly 50 of the boxes are 6 cm wide?
(c) exactly 49 of the boxes are 6 cm wide?
(d) exactly 48 of the boxes are 6 cm wide?

P30. What's the difference?

Is it possible to write the numbers from 1 to 100

(a) along a line,
(b) around a circle,

so that the difference between any two adjacent numbers is at least 50?

P31. Shortest path

The following grid is made from 24 line segments, each 1 centimeter long. Is the figure traceable without lifting the pencil with a path of length less than 28 cm? (Some segments may have to be traversed twice.)

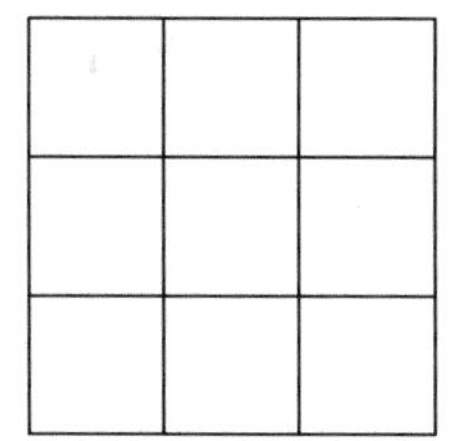

P32. Unneighborly circuit

Place a rook on an arbitrary square of a 4×4 chessboard. The rook is allowed to move horizontally or vertically but not to an adjacent square (that is, it must move at least two squares). Is there a starting square and a sequence of 16 moves that will take the rook to each of the squares exactly once and return to the starting square? (H, Q)

P33. Emptying heaps

You have three heaps of beans. There are 12 beans in the first heap, 123 beans in the second heap, and 1234 beans in the third heap. In a single move you may either take the same number of beans from each of the three heaps, or you may divide a heap into two equal parts (assuming there are an even number of beans in the heap) and move one of these parts to one of the other heaps. Can the moves be organized so as to obtain a position with

(a) no beans in the first and second heaps?
(b) no beans in the first and third heaps?
(c) no beans in the second and third heap?
(d) no beans in any of the heaps?

P34. The cookie monster

You have 15 cookie jars containing $1, 2, 3, \ldots, 15$ cookies, respectively. The cookie monster may choose any subset of jars and take the same number of cookies from each jar. Can he empty the jars in fewer than five moves? (Q)

P35. Bull's-eye

Anne, Bjorn and Camilla competed in shooting at a target. Each had six shots and each of their shots hit the target (see figure). Bjorn scored 22 points in his first few shots, and Camilla scored 3 points on her first shot. At the end, they each had the same number of points. Can you tell who hit the "bull's-eye" (50-pointer)?

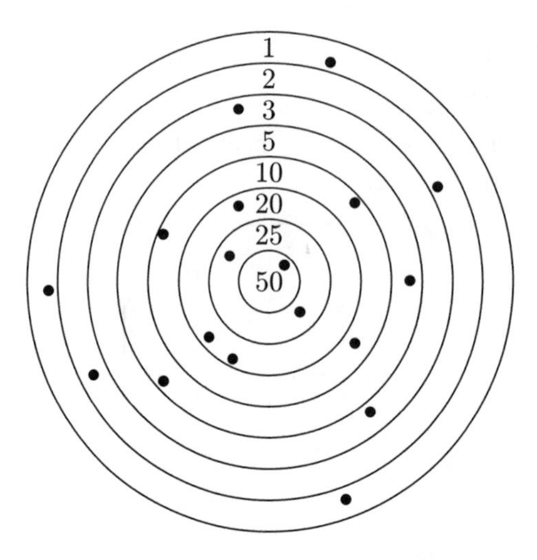

P36. Off-off-off-price

A sale item was marked down by the same percentage for three years in a row. After two years the item was 51% off the original price. Was the price more than two-thirds off the original price after three years?

P37. Sums and differences

Is there an arrangement of the ten numbers 1 1 2 2 3 3 4 4 5 5 in a row so that each number, except the first and last, is the sum or difference of its two adjacent neighbors? (H, Q)

P38. Fair shares

Can you divvy up 13 presents, of values 1, 2, . . . , 13 dollars, among Alice, Bertha, and Carol, so that each receive the same total value? What about 14 presents of values 1, 2, 3, . . . , 14 dollars? (Q)

P39. Turning $0.88 into $1.60

Sixteen coins are arranged in a row on a table: first a penny, then a dime, then a penny and a dime, and so on. In a single move you are permitted to select an arbitrary sequence of adjacent coins and to replace each penny in this sequence with a dime and each dime with a penny. Can you obtain a row of sixteen dimes in less than 8 moves? (H)

P40. Musical squares

(a) Children have drawn a large 6×6 square grid in their schoolyard. On each of the 36 small squares there stands a child. At the teacher's signal each child jumps to a neighboring square (two squares are neighbors if they share a common side). Can their jumps be organized so that after their jump there is once again a child on every square?

(b) One of the children gets a stomachache and has to drop out of the game, so they draw a 5×7 rectangular grid. They now wish to play the same game. Can their jumps be organized so that after their jump there is once again a child on every square?

P41. Pouring down

We have two buckets which hold 11 and 7 liters respectively. Both buckets are empty but we have unlimited access to water. We can fill either of the buckets from a pump, pour it from one bucket to the other until it is empty or until the other is full, or empty full buckets down the drain. Is it possible to obtain 1 liter of water? Two liters of water? (H, Q)

P42. Equal but different

A number of people throw darts at the target shown below. Each person throws four darts, and each person scores 62 points. Assuming that no two people had the same set of scores on all their throws, could there have been as many as nine people playing the game? (H)

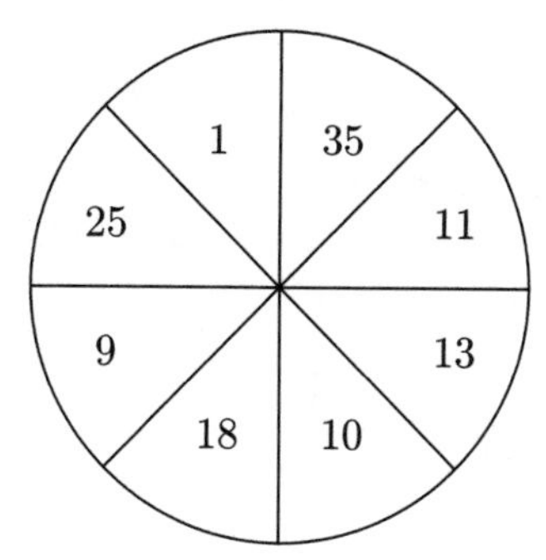

P43. Mitosis

At the beginning of an experiment, on the evening of May 1, there were a number of cells in an incubator. Every 24 hours each cell split into two. On the evenings of three different dates after May 1 an extra cell was added to the culture. By the evening of May 17 there were exactly a million cells. Is it possible to determine how many cells were originally put into the incubator on May 1? (H)

P44. Free-throw contest

For the opening of the new high school gym, five administrators agreed to compete in a free-throw contest. Each took 100 shots, with one point for each one made. Their scores were all different and totalled 200 points. The second and fourth finishers made more than 90 free-throws between them. Could the spread between the top and bottom scores be as little as 30 points? (H)

P45. Restore these multiplications

Here are two reconstruction problems. Can the stars in the following multiplication problems be replaced by digits so that the computations are correct? (H, Q)

$$
\begin{array}{cccccc}
 & & & * & * & 4 \\
 & & \times & 2 & 3 & * \\
\hline
 & & * & * & 2 & 4 \\
 & 1 & * & * & * & \\
1 & * & * & * & & \\
\hline
* & 1 & * & * & * & *
\end{array}
\qquad
\begin{array}{ccccccccc}
 & & & 2 & 3 & * & * & 8 & 5 \\
 & & & \times & & * & * & * & 5 \\
\hline
 & & * & * & * & * & * & * & * \\
 & & * & * & * & 7 & * & * & \\
 & * & * & 9 & 5 & 7 & 0 & & \\
* & * & 4 & * & * & * & & & \\
\hline
7 & * & * & * & * & * & * & * & *
\end{array}
$$

P46. Seven-eleven

The figure shows five lines that make five triangular regions (and a pentagon and ten infinite regions). Is there an arrangement of seven lines that make eleven triangular regions? (Q)

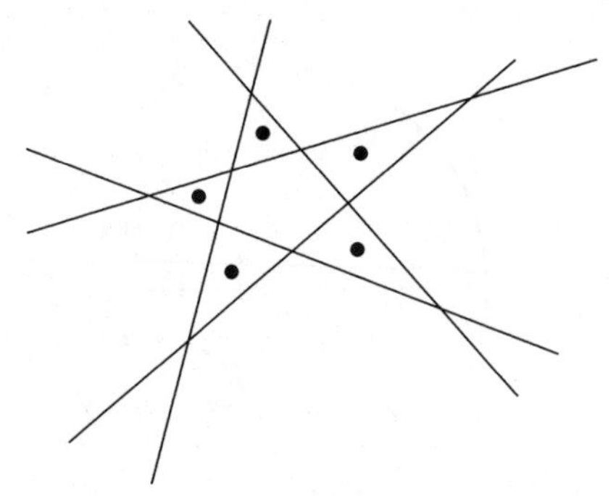

P47. All ten digits

Are there two three-digit numbers whose sum is a four-digit number where all the digits in the three numbers are different? (Q)

P48. Minimize the walking

Four houses are located at the corners of a quadrilateral. The occupants wish to share a gazebo. Is there a unique place to build it so that the sum of the distances to their houses is as small as possible? (Q)

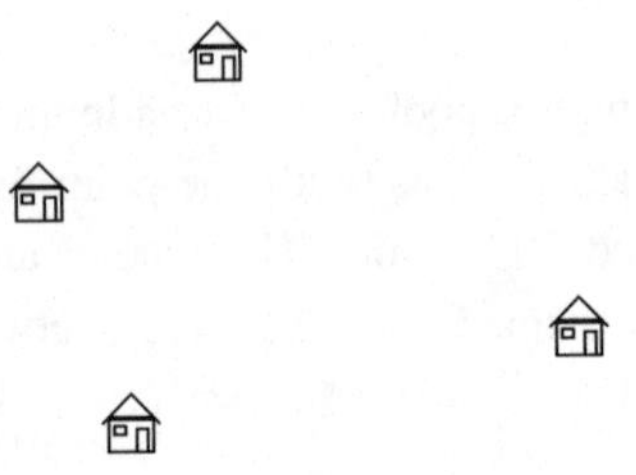

P49. Equally productive

Can the numbers from 1 to 25 be partitioned into two or more subsets so that the product of the numbers in each subset is the same?

P50. Join the dots

A rectangular grid consists of 19 rows, each of 21 squares. There is a dot in the middle of each of the 399 squares, except for the center square. Is there a path through each of these midpoints that only goes horizontally or vertically but doesn't pass through the middle square? The path is not to visit any midpoint more than once and doesn't have to start and finish in any prescribed squares. (H)

P51. Three truths, two lies

Consider the following conditions on the whole number A: $3A$ is larger than 35, $7A$ is at least 43, $2A$ is at most 99, A is at least 21, and $5A$ is at least 51. Can you determine the value of A if only three of these conditions are true?

P52. Diminishing returns

On July 31, a pound of chocolates cost \$2.56. The price went up on the next day (August 1). As a result, sales for the month of August were down 23.2% from the preceding month, and total income from sales was down by 12.4%. Was the cost of a pound of chocolates in August more than \$3.00?

P53. Fidgits with digits

Is there a number A with three decimal digits such that when you increase the first digit by d (d is a single-digit number), and simultaneously decrease the second and third digits by d, the result is a three-digit number equal to $A \times d$?

P54. Plus or minus

Are there more than seven different ways to replace the stars in the expression $1 * 2 * 3 * 4 * 5 * 6 * 7 * 8 * 9 * 10$ with either a plus sign ($+$) or a minus sign ($-$) so that the resulting sum equals 29? (Q)

P55. Truth or consequences

Two communities live on a remote island. In the one community, everyone has green hair, and in the other, everyone has blue hair. In the Lonely Planet's Tourist Guide it says that those in one of the communities always tell the truth, while those in the other community always tell lies. On my visit to the island I met with two representatives from the island, one from each community. They had their heads covered with scarfs so that I could not see their hair color. I asked each of them the same question: Is your hair green? They both answered "no." From this information, can I determine which community tells the truth? What if they had both answered "yes?" What if one had answered "yes" and the other "no?"

P56. Lying about lying

At another time on the same island (see the preceding problem) I overheard the following quarrel between four residents A, B, C and D. "You are a liar!" shouted B to A. "It is you who is a liar!" exclaimed C to B. "They're both liars," interjected D to C, "and furthermore, you are as well!" I could not see their hair but from this little dispute I was able to deduce the color of each person's hair. Can you? (H)

P57. Roads of destiny

On another occasion (see the preceding problem) I was on my way to see a doctor when I came to a junction in the road. There were three choices: the first led toward the woods, the second toward the mountains, the third toward the marshlands. Five inhabitants standing at the crossroads came forward to help me. "You should go through the woods," declared the first, whom we shall call A. "No, you should go through the mountains," said the second, B. "It is not true that both of these two have green hair," proclaimed C. The fourth, D, trying to astonish everyone, said "I will add that either A's hair is green or B's is blue." The last person, E, not to be outdone, added, "Either my hair is green, or C and D have the same color hair." Because their heads were covered with scarfs I was unable to see their hair. I sat on a stone and thought about which way I should go to reach the doctor. Can I determine the right choice? (H, Q)

P58. Twixt truth and falsehood

A year ago I visited another remote island. Here there were three towns: Trueton, Falseton, and Twixton. Those from the first were known to always tell the truth, whereas those from the second always lied. Those living in Twixton alternated between truth-telling and lying, every other statement, but you didn't know which came first. During my visit I heard a rumor that the mayor in one of these towns was quite sick. One evening a man approached me and asked, "Can you help us, our mayor is sick!" "In which community?" I asked. "In Twixton!" he exclaimed and then rushed off. Assuming the mayor of one of the towns really is sick, is there enough information to decide which town has the sick mayor?

P59. Stringing along

On a paper I have written a string of 0's and 1's, and you can operate on this string in three different ways:

(i) The substring 01 can be replaced by 100, and vice-versa;
(ii) The substring 10 can be replaced by 111, and vice-versa;
(iii) The substring 11 can be replaced by 000, and vice-versa.

Using these rules, can you transform

(a) 11001 into 0001000?

(b) 0110 into 0110010? (Q)

P60. Magic square

A number is placed in every square of a 3×3 grid. The sum of the numbers in each row and column and each diagonal is 33. Is the number in the center square determined? (H)

P61. Count up and count down

Margaret and Bruce start counting at the same time and at the same speed. Margaret counts forward by 3's starting with $70: \ 70, 73, 76, \ldots$, while Bruce counts backwards by 7's from $1996: \ 1996, 1989, 1982, \ldots$. Will there be two numbers they will count simultaneously that are less than four apart?

P62. Seventh-fold deletion

Is there a number A with three decimal digits which is equal to seven times the two-digit number obtained by deleting one of the digits of A? (H)

P63. A binary cryptarithm

The only digits in the binary (positional) system are 0 and 1. The operations of addition and multiplication are defined by $0+0=0$, $0+1=1+0=1$, $1+1=10$, $0 \times 0 = 0$, $0 \times 1 = 1 \times 0 = 0$, and $1 \times 1 = 1$. Can the asterisks in the following binary multiplication (done left to right), be replaced by 0's and 1's so that the computations are done correctly in the binary system?

$$\begin{array}{ccccccc}
 & & & * & * & * & * \\
 & & \times & & * & * & * \\
\cline{3-7}
 & * & * & * & * & & \\
 & & & * & * & * & * \\
\hline
* & * & * & 1 & * & * & *
\end{array}$$

P64. First service

The first set in the tennis match between Agassi and Becker in the U.S. Open was won by Agassi 6-3. There were five service breaks in the match. Did Becker serve the first game? (The service alternates between the players with each game. There were nine games in the match. A service break is a game won by the non-server.) (H)

P65. Eight into six won't go

Is there a dissection of the regular hexagon into eight congruent quadrilaterals? (Q)

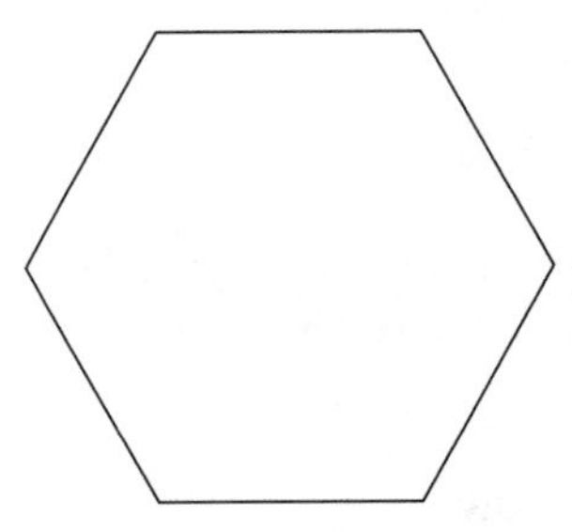

P66. Times are odd!

Let A, B, C, D, E, F, G, H, J denote any nine whole numbers, and let a, b, c, d, e, f, g, h, j denote a permutation (rearrangement) of these same nine numbers. Can the numbers be chosen so that the product

$$(A-a)\times(B-b)\times(C-c)\times(D-d)\times(E-e)\times(F-f)\times(G-g)\times(H-h)\times(J-j)$$

is an odd number? (H)

P67. Round-robin wrestling

Yesterday there was an arm-wrestling tournament in the mathematics classroom. Each student competed against each of the others exactly once. Each pair of competitors had 30 seconds to win the match. Each participant scored 1 point for a win, -1 point for a loss, and 0 points for a tie. Jack finished with 9 points and Jill finished with 12. Was there necessarily a tie? (H)

P68. Knockout tennis

Sixty-four players took part in a knockout tennis tournament; that is, there were 32 matches in the first round, 16 matches in the second round between the 32 winners of the first round, 8 matches in the third round, and so on. The winner of a match is the first person to win 3 sets. The number of matches that ended at 3–1 was double the number that ended 3–0 and half the number that ended 3–2. Were more than 256 sets played in the tournament?

P69. Rummy

John and Jane spent the afternoon playing rummy. John won 36 games, Jane won 24, and there were no draws. Each game took 5 minutes. Was there a time before the mid-afternoon break when the number of wins by John up to that time was equal to the number of games Jane would yet win in the remaining time? (H)

P70. Nine cubits

I have glued together three cubes into a block (a cubit), shown in the following figure. Can I build a $3 \times 3 \times 3$ cube from nine such identical blocks? (Q)

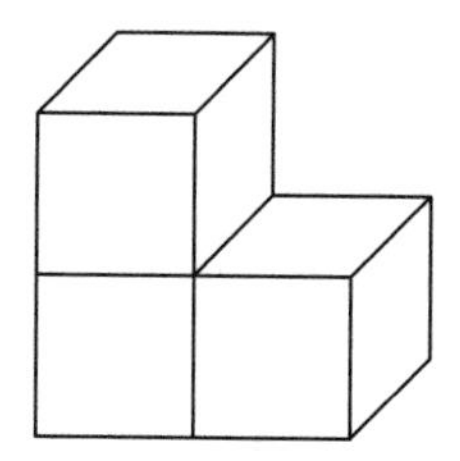

P71. Nine-fold insert

Is there a number A with at least two decimal digits such that if we insert an extra digit between the last two digits, the result is nine times A? (H)

P72. Magic rectangle

Can the numbers from 1 to 30 be placed in the squares of a 5×6 rectangular grid so that (two independent questions)

(a) the six column sums are all the same?
(b) the five row sums are all the same? (Q)

P73. Dunrovian coinage

In Dunrovia they have a curious coinage system which includes $4 and $7 coins.

(a) Can you determine the smallest amount of money that can be made up in two different ways, using only $4 and $7 coins? In three different ways?
(b) Can you determine the smallest amount of money having the property that it *and all greater amounts* can be made up in two different ways, using only $4 and $7 coins? In three different ways? (H)

P74. An elevating experience

The elevator in my apartment building (66 floors) has only two buttons, U and D. When you push the U-button the elevator goes up 8 floors (or not at all if there aren't 8 floors available); when you push the D-button it goes down 11 floors (or not at all). Is it possible to take the elevator from any given floor to any other given floor? (H)

P75. Another elevating experience

The apartment building next to mine has an elevator exactly like the one in my building, with an 8-floor up-button and an 11-floor down-button, as in the previous problem, but the number of floors is not the same. It is possible to get from any one floor to any other floor. However, if the building had one fewer floor it would be impossible to do this. Can you determine how many floors are in this building?

P76. Squaring the triangle

The two legs of a right-angled triangle are 1 inch and 2 inches long. Can you make a square with twenty such congruent triangles?

P77. Plus-minus grid

Place a 1 or a -1 in each square of an 11×7 rectangular grid. Compute the product of the numbers in each row (11 row-products) and each column (7 column-products). Can the numbers be placed in the grid so that the sum of the 18 products is 0? (H)

P78. Staffing problem

A travel agency employs a number of people. Among these, 15 are American and 24 are women. Only eight of the employees are neither American nor Swedish, and half of these are women.

(a) Are there more American male employees than Swedish female employees?
(b) If it is known that at least one American man works for the company, could there be fewer than 30 employees in the agency?

P79. Difference of squares

A square sheet of graph paper is ruled (by horizontal and vertical lines) into a number of small squares 1/4 inch on a side. Along the lines of this grid we cut out a square. There are 148 small squares in the remaining piece. Was the original sheet more than 10 inches on a side? (Q)

P80. Making a soccer ball

A soccer ball is made from pieces of black and yellow leather. The black pieces are regular pentagons (five sides) and the yellow ones are regular hexagons (six sides). Each pentagon is adjacent to five hexagons and each hexagon is adjacent to three pentagons and three hexagons. The ball has 20 yellow hexagons. Does the ball have more than 10 black pentagons?

P81. Clicker, the robot

The physics students have created a robot, Clicker, that can wheel through the rooms of the school and turn the lights on or off. So far, however, they've only managed to program it to flip the switch when it enters a room. If the lights are on, Clicker turns them off; if they're off, Clicker turns them on. The floor plan of the school resembles a checkerboard, with 64 identical rooms arranged in eight rows with doorways between rooms that share a wall. Clicker is stored in the southwest corner room. At 5:00 P. M. the lights are on in every room, and Clicker goes to work. Can you plot a course for Clicker, visiting individual rooms as often as needed, so that at the end, the lights will be off in Clicker's room (back into storage) and in alternate rooms as on a checkerboard? (H, Q)

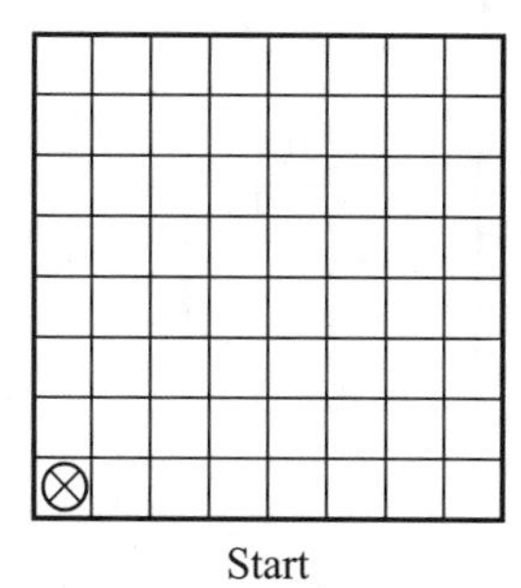

Start

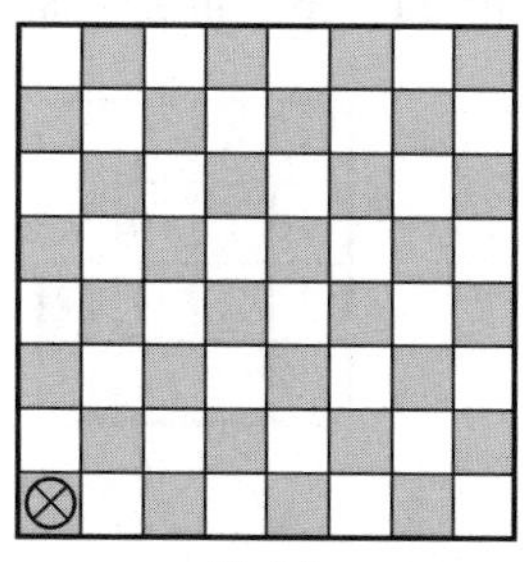

Finish

P82. Tricolored triangles

We have three steel wires of different colors, each 1 meter long. You take two of these wires and divide each of them into three pieces at your discretion, and I take the third wire and arbitrarily divide it into three lengths. Can you divide your wires so that regardless of how I divide my wire, you will be able to make three triangles with the nine pieces, each with different colored edges? (H)

P83. A game of Snap

A teacher and each of the 15 children have a shuffled deck of 17 cards numbered 1 to 17. Everyone turns their cards over one by one. When a child's card matches the teacher's, the child shouts "Snap" and scores a point. At the end, no two children have the same score and no child scored 0 or 15. Mark's score was next to the highest in the class. Can you say how many points Mark scored?

P84. Variegated products

The product $14 \times 59 = 826$ uses 7 different digits (all but 0, 3, and 7). Are there two numbers with at least two digits apiece so that the digits of these numbers and their product are

(a) $1, 2, \ldots, 9$ (with no repetition)?

(b) $0, 1, 2, \ldots, 9$ (with no repetition)? (H, Q)

P85. Complete satisfaction

Out of 100 people, 78, 70, and 63, answered "yes" to "yes-no" questions A, B, and C, respectively, while three people answered "no" to all three questions. Is it possible to determine the largest and smallest number of people who could have answered "yes" to all three questions? (H)

P86. Numberworm

A numberworm is made up of sixteen sections numbered consecutively from 1 to 16. It can fit onto a 4×4 grid in many different ways: one possibility is shown on the left below. Can a numberworm be placed onto the grid at the right so that sections 7, 10, and 11 will rest on the corresponding squares marked on the grid? (H, Q)

5	6	7	8
4	1	10	9
3	2	11	12
16	15	14	13

	11	10	
	7		

P87. Longer worms

We now have two longer numberworms to place onto larger grids; one has 25 sections and the other 36. In the same way as in the preceding problem, can numberworms be placed so their numbered sections match the numbers marked on the grids as shown? (H)

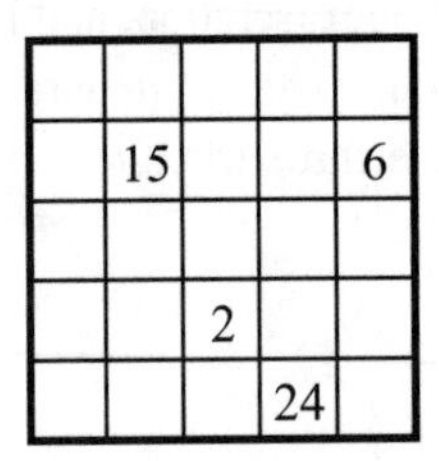

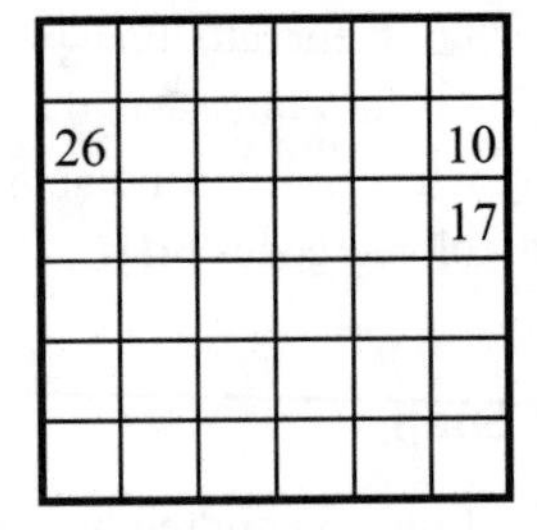

P88. Guideposts (Marc Paulhus)

The 11×11 grid shown below has guideposts set as indicated. Is there a path of horizontal and vertical segments that avoids the guidepost squares but passes through each of the other squares exactly once and returns to the starting square? (H, Q)

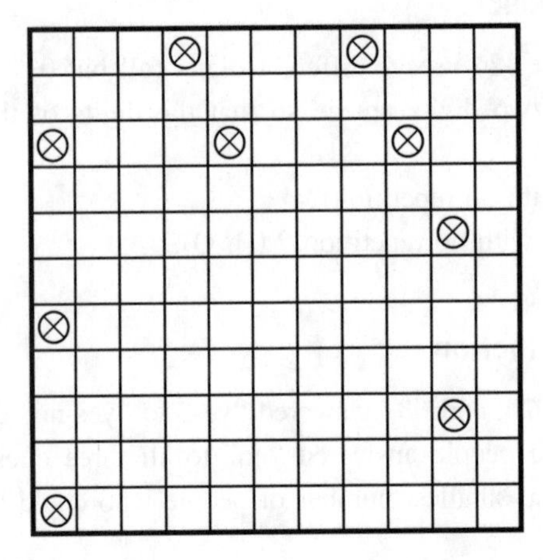

P89. Straight strokes

Is there a path of alternating horizontal and vertical segments on the 8×8 chessboard that passes through each square exactly once and returns to the starting position that is made up of

(a) exactly 25 segments?

(b) exactly 26 segments? (H, Q)

P90. Heads I win, tails you lose

Seven pennies lie heads-up on a table. In a single move you are allowed to turn over any four coins at the same time. Using a sequence of such moves, can you get all seven coins to lie tails-up on the table? Can it be done if you are allowed to turn over any five coins in a single move? (Q)

P91. Rewards for work well done

My three children and I have agreed on a reward-system for helping with household chores. Each week I distribute the same three numbers of points (distinct positive integers); the largest to whoever has helped the most, the next to the second best, and the smallest to the person who helped the least. After distributing last week's points, I noted that Sarah, who had the most points in the first week, has a total of nine points for all the weeks so far, Daniella also has nine points, and Camilla has 22 points. Can you say who came in second place in the second week? (H)

P92. The school dance

Two of Sarah's classes at school organized a dance. More than half of the 44 young people came to the dance, but Sarah was sick and had to stay home. Afterwards, she was told that during the evening each boy had danced (once each) with four girls, and each girl had danced (once each) with seven boys. Assuming this accounts for all the dances, is it possible to say how many of Sarah's classmates attended the dance? (H)

P93. Well-balanced neighbors

Is it possible to place 0, 1, 2, 3, 4, 5 in the circles so that

$$0 \to 4, \quad 1 \to 12, \quad 2 \to 7, \quad 3 \to 8, \quad 4 \to 3, \quad \text{and} \quad 5 \to 4?$$

By $a \to b$ we mean that the circles connected to a have numbers which add to b. (H)

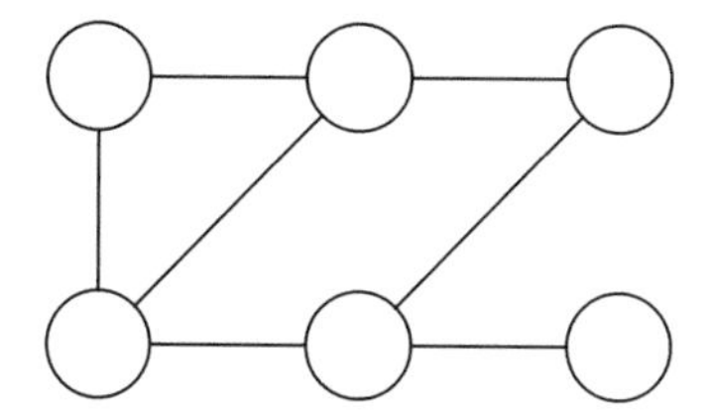

P94. A 9-hole and an 11-hole course

Now that you have trained yourself on the preceding problem, is it possible to fill in the numbers (a) $1, 2, \ldots, 9$ and (b) $0, 1, 2, \ldots, 10$, into the respective figures shown below, so that

(a)
$$1 \to 26, \quad 2 \to 8, \quad 3 \to 19, \quad 4 \to 13, \quad 5 \to 17,$$
$$6 \to 10, \quad 7 \to 4, \quad 8 \to 19, \quad \text{and} \quad 9 \to 24;$$

(b)
$$0 \to 5, \quad 1 \to 17, \quad 2 \to 7, \quad 3 \to 18, \quad 4 \to 8, \quad 5 \to 38, \quad 6 \to 10,$$
$$7 \to 23, \quad 8 \to 9, \quad 9 \to 6, \quad \text{and} \quad 10 \to 12. \quad \text{(H)}$$

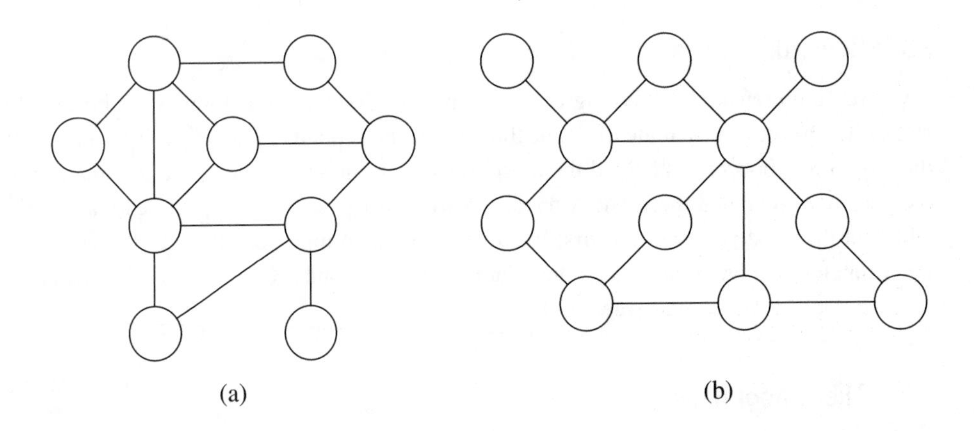

P95. Elephant waltz

With great dignity, two elegantly groomed elephants, Romeo and Juliet, entered the center ring of the big top. They paraded around the ring to ballet music, each going in a different direction. Each time they met they waltzed around each other, and this took them 10 seconds to accomplish. They finished their act when they met each other at the entrance, by waltzing around each other one last time. Romeo took 36 seconds to walk around the ring, disregarding the dances when they met, and Juliet took 45 seconds (disregarding dances). Did they finish in less than five minutes? (H)

P96. Equal sums

The letters $A, B, C, D, E, F, G, H, J$ denote 1 to 9 in some order. Assuming that

$$A + B + C = C + D + E = E + F + G = G + H + J,$$

is the value of E determined if it is known that the sum of these four equal numbers is as large as possible? (H, Q)

P97. Equal products

The letters A, B, C, D, E, F, G denote different digits. We know that

$$A \times B \times C = C \times D \times E = E \times F \times G.$$

Is there more than one possible value for D? (H)

P98. Four-letter words

If our alphabet has only two letters, say A and B, then we can only make 10 different four letter words if we're not allowed to have three consecutive letters the same. Can these ten words be arranged in a word square, with a different word in each of the four rows, four columns, and two diagonals? (H)

P99. Lost galoshes

During a recent heavy rainstorm, sixteen members of a men's club arrived for their annual banquet at a restaurant, each wearing galoshes. They all had different sizes. As they left the restaurant one by one after their meal, each took the first pair of galoshes whose size was not less than their own. If all the remaining galoshes were smaller than their own size, they left without galoshes. If three men left with their own galoshes, could there be seven men who left without galoshes? (Q)

P100. Six sums the same

Can the numbers $0, 1, 2, \ldots, 9$ be placed in the ten circles so that the sums around triangles A, B, C, D, E, F are all equal? (H)

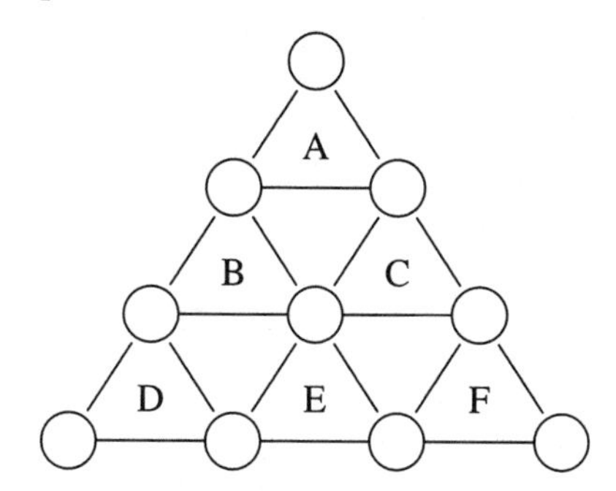

P101. Collective bargaining

Four collectors each own a single, but different, piece of a four-piece place setting (saucer, cup, small plate, large plate). Ye Olde Antique Shoppe sells the individual pieces, each for a whole number dollar amount. The collectors' costs for completing their sets are four consecutive numbers, one of which is $46. Can you find the cost of the saucer, assuming that it is the least expensive item?

P102. A confusion of coins

Peppi has 96 coins in her backpack, from several different countries. Whenever she takes out eleven coins she always finds that at least three of them are from the same country. Assuming that this will always necessarily happen, she concludes that she must have at least 20 coins in her backpack from the same country. Is Peppi right about this? (H, Q)

P103. Completing the square

Can the numbers $1, 2, 3, \ldots, 16$ (except for those already filled in) be placed in the squares so as to make a magic square; that is, so that the sums of the numbers in each row, column, and diagonal are equal? (H)

16			
		11	8
	15	14	

P104. Countdown

I have filled the squares in the following grid with numbers. You may choose any two adjacent squares (squares sharing a common side) and increase or decrease the two numbers by the same amount. Is it possible, with a sequence of such operations, to obtain a grid with

(a) 0 in each square?
(b) 1 in one square and 0 in all other squares?
(c) 1 in each square of one of the diagonals and 0 in all other squares?
(d) 1 in each square? (H, Q)

4	6	0
8	7	2
5	0	3

P105. Give or take

(a) Each star in the expression $1 * 2 * 3 * \cdots * 2001$ represents a plus or minus sign. Can the signs be chosen to make a positive number less than 5?
(b) Same question for the expression $1^2 * 2^2 * 3^2 * \cdots * 2001^2$.

P106. Skipping around the track

Harry and his son Skip agreed to run around the stadium each morning. Their first run was on Monday. They started at the same time from the same place and ran in the same direction, each at constant speeds. Skip passed his father twice and caught him for the third time just as his father was finishing his first lap. On Tuesday they each ran at the same constant speeds as they had the day before, starting at the same time and at the same place, but this time they ran in opposite directions. Did they meet more than four times in Harry's first lap? (H)

P107. Fifty-fifty

In the first row of a 10×10 grid, write the numbers $0, 1, 2, \ldots, 9$, from left to right, in the second row write the numbers $10, 11, 12, \ldots, 19$ in the same way, and so forth, in the bottom row write the numbers $90, 91, 92, \ldots, 99$ from left to right. Now attach 50 minus-signs to these numbers in such a way that exactly five numbers in each row and each column are negative. Let S denote the sum of all the numbers in the resulting grid. Can S be more than 10? (H)

P108. Casting out the threes

Begin with the sequence of numbers $0, 1, 2, 3, 4, 5, \ldots$ and cross out those numbers which contain any of the digits 3, 6, or 9. Does the 587th remaining number contain the digit 2? (Note that we are counting 0 as the first term of the sequence.) (H, Q)

P109. Location, location, location

The lines in the figure represent the streets (east/west) and avenues (north/south) of a city. The dots at the intersections denote the homes of eleven friends. They decide to meet one evening on a corner chosen so as to minimze the total walking distance (for all the friends taken together). Should they meet on 6th and 6th? (H)

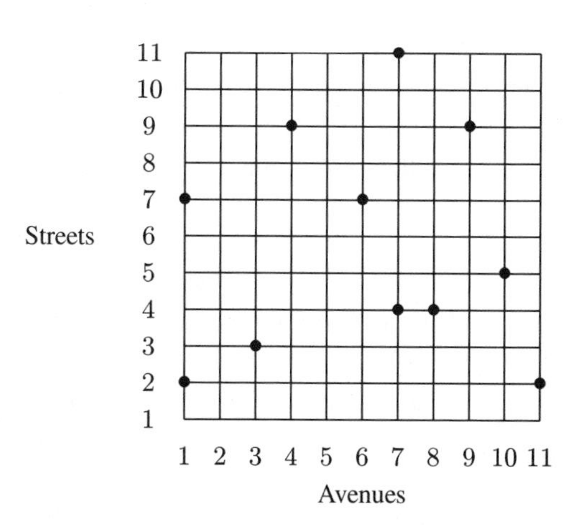

P110. Growth spurt

On the evening of May 1, a number of bottles containing cells were placed in an incubator. None of the containers were empty, and all together they contained 100 cells. Sometime in May (but after May 1 and before May 30), and for three consecutive evenings, a new container containing one cell was added to the incubator. Throughout the month, the number of cells in each container increased by one new cell every 24 hours. When the experiment ended on the evening of June 1 (after the new cells had appeared), there were a total of 653 cells in the containers. Is it possible to determine on which day the first new container was added to the incubator? (H)

P111. Wheels within wheels

(a) Can you make an arrangement of circles (not necessarily of the same radius) so that each circle is tangent to exactly four others and no three of them pass through the same point?
(b) Can you make an arrangement of circles (not necessarily of the same radius) so that each circle is tangent to exactly five others and no three circles pass through the same point? (H, Q)

P112. Crossed polygons

A closed polygon consists of a number of line segments (called edges), where each edge has two endpoints, call them the initial and terminal endpoints, and the terminal endpoint of one edge coincides with the initial endpoint of the next edge, and the terminal endpoint of the last edge coincides with the initial endpoint of the first edge. Here is an example of a closed polygon with five edges.

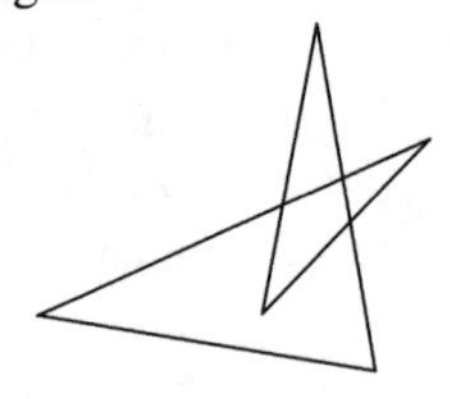

(a) Is there a closed polygon of six edges so that every edge intersects exactly one other edge (at a point other than an endpoint)? Exactly two edges?
(b) Is there a closed polygon of seven edges so that every edge intersects exactly one other edge (at a point other than an endpoint)? Exactly two edges? (H)

P113. Turning points

The figure opposite shows a 4×4 grid and on the right there is a closed polygon whose edges do not intersect each other (except at their endpoints) and whose corners lie on these gridpoints. No point on the grid is used more than once as part of the polygon. Is there such a polygon with 16 corners? The example in the figure has only seven corners. (Q)

P114. Two pointed questions

Is there a closed polygon with a corner at every point in either one of the following grids? As in the preceding problem, the edges of the polygon are not to intersect (except at the endpoints) and no point on the grid is to be visited more than once. (Q)

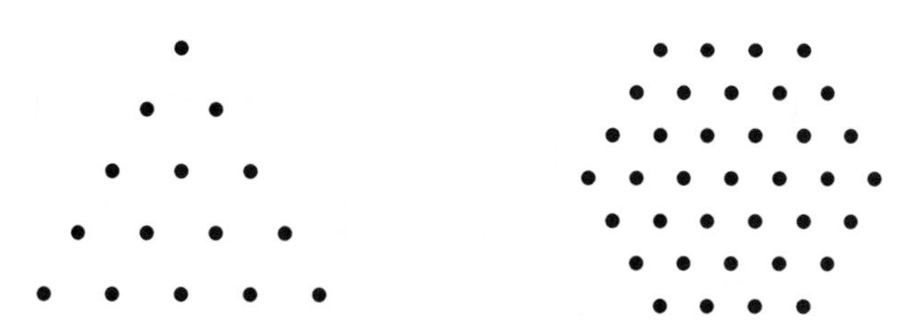

P115. Oddly even

Can 25 whole numbers, not necessarily all different, be arranged so that the sum of any three successive terms is even but the sum of all 25 is odd? What is the answer if you replace "three" with "five"?

P116. Latin squares with diverse diagonals

Is it possible to complete each grid with the digits $1, 2, 3, 4, 5$ so that that each digit occurs once in each row, each column, and both diagonal? (H)

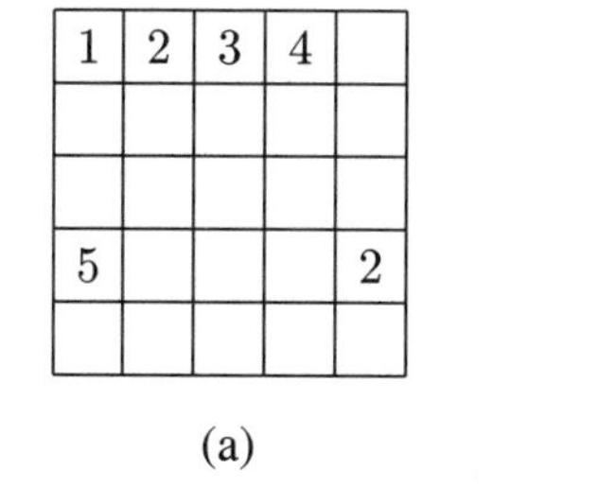

1	2	3	4	
5				2

(a)

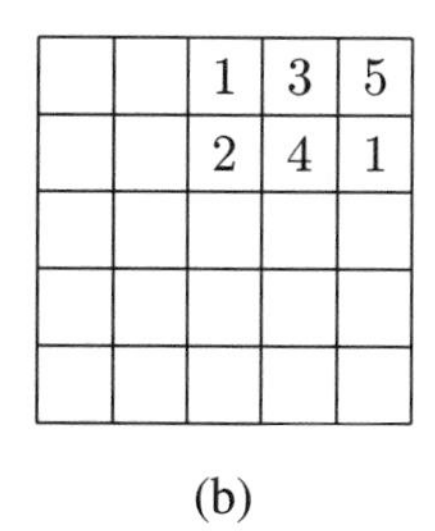

		1	3	5
		2	4	1

(b)

P117. Seeing double

The numbers 0 and 2 have the property that their squares have the same digits as their double. Is there another number whose square has the same decimal digits as its double (although possibly in a different order)? (H, Q)

P118. Empty nest

Can you build a $3 \times 3 \times 3$ cube with thirteen $1 \times 1 \times 2$ blocks if the $1 \times 1 \times 1$ space in the center is empty?

P119. Tetrominoes

Four congruent squares can be put together in five essentially different ways to form "tetrominoes" as shown. Will two sets of them fit into a 5×8 rectangle? (It is permissible to turn the pieces over.) Will three sets of them fit into a 5×12 rectangle? (Q)

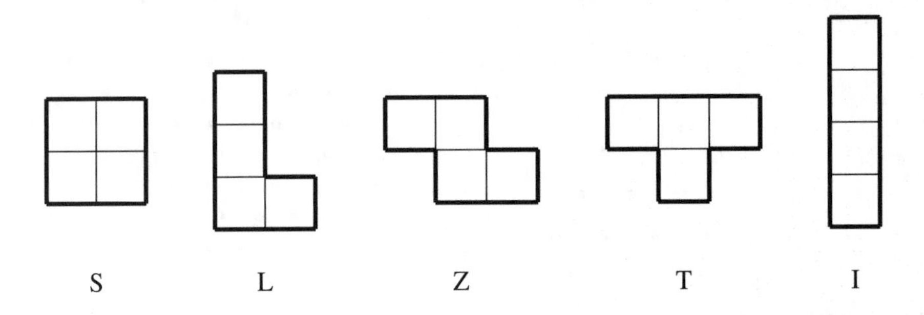

P120. Intersecting lines

Nine unbounded straight lines are drawn in the plane. There are twenty-one intersection points. Nineteen of these points have only two lines passing through them, one point has four lines through it, and one point has five. Are there any pairs of parallel lines? (H, Q)

P121. Domino cover-up

We have deleted five squares from the 10×10 board as shown below. Can you cover more than 90 of the remaining 95 squares with 1×2 dominoes? (Each domino covers exactly two squares of the deleted board.)

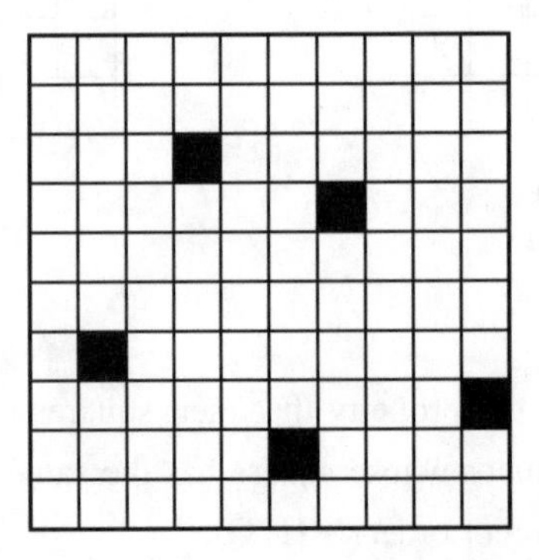

P122. Secret lives

Elizabeth has a secret hiding place where she presently has 35 coins, 38 baseball cards, and 39 pieces of candy. Her younger brother John has his own secret hiding place, but that does not prevent him from taking things from Elizabeth's (her place isn't as secret as she thinks). John always takes two items at a time, each of a different kind so that the number won't decrease too quickly, and he always adds one item of the third kind. One day, some time later, Elizabeth was shocked to find that all the items in her hiding place were of the same kind. Was it all candy? (H, Q)

P123. Reverse ordering

You have fifteen chips numbered from 1 to 15 and I have sixteen chips numbered from 1 to 16. Our chips are each arranged in increasing order. If we are allowed to exchange the positions of any two adjacent chips, you will need $14 + 13 + 12 + \cdots + 1 = 105$ exchanges to reverse the order of your chips (chip 1 must make 14 exchanges, chip 2 requires 13 exchanges, and so forth). I need $15 + 14 + 13 + \cdots + 1 = 120$ exchanges to reverse the order of my chips. Suppose we change the rule so we are allowed to exchange two chips provided exactly one chip lies between them. Can you reverse your original ordering in less than 100 moves? Can I? (H)

P124. Different sums for different lines

(a) Is it possible to place the numbers $-1, 0, 1$ on the squares of a 4×4 grid so that the eight row and column sums will all be different?
(b) The same problem as above except now we also want one of the two diagonal sums to be different from the previous sums. (Q)

P125. More diverse line sums

Is it possible to place the numbers $-1, 0, 1$ on the squares of a 4×5 grid so that the row and column sums will all be different? (Q)

P126. A dollar a year

Anna and Peter receive weekly allowances (in dollars) equal to their respective ages. Anna is older than Peter. If I should double Anna's allowance their combined weekly allowance would be less than 32 dollars, but if I should double Peter's allowance, their combined weekly allowance would be more than 28 dollars. Can you determine the ages of the children?

P127. The Sol LeWitt puzzle (Barry Cipra)

Sol LeWitt, an American conceptual artist who uses mathematical motives as inspiration, created an etching, *Straight Lines in Four Directions and All Their Possible Combinations*

(1973), which suggested a mathematical problem to Barry Cipra, a mathematician and mathematics writer. Notice that some of the lines in the figure below (adapted from LeWitt's design) continue from one square to the next, but others don't. Is it possible to rearrange the 16 squares, keeping them in a 4×4 grid and not rotating any of them, so that all the lines go all the way from one edge of the grid to the other? (H)

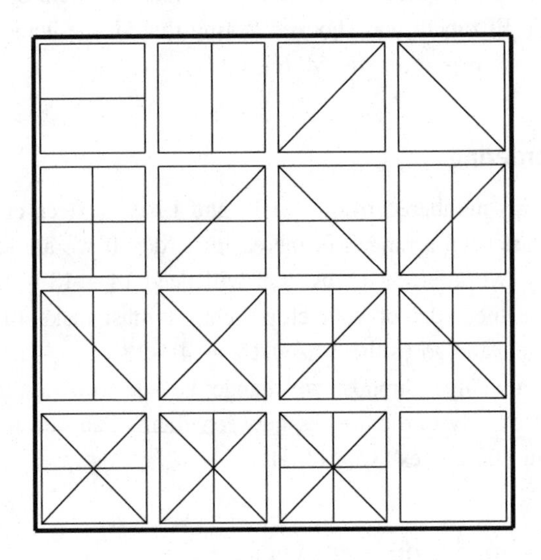

P128. Leveling up

(a) Disks are placed in the spaces of a wheel as shown in (a). You are allowed to add two disks to the wheel, placing one on each of two adjacent spaces. Is it possible, after a number of such moves, to get the same number of disks on each space?

(b) The same question for (b).

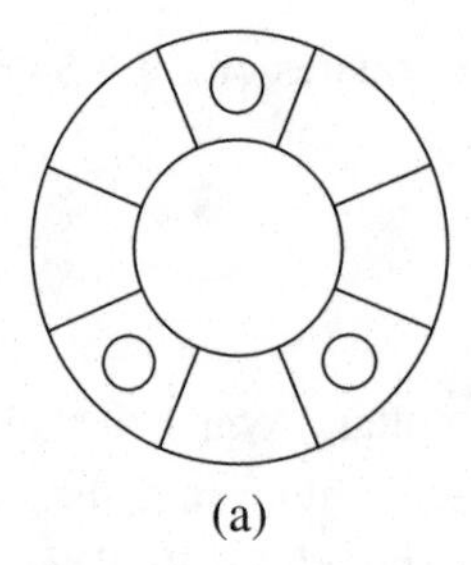

(a)

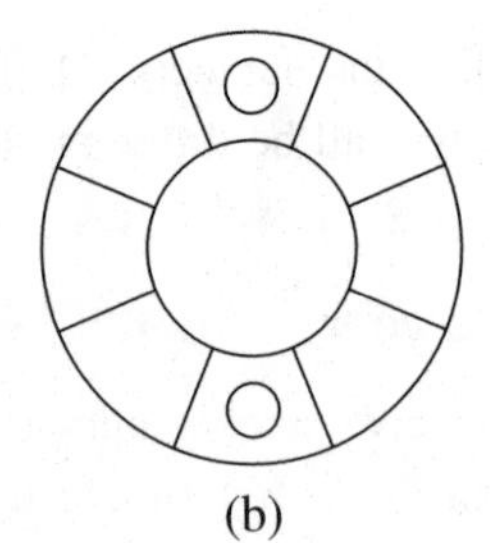

(b)

P129. Dial a pile

Disks are placed in the spaces of a wheel as shown. In each move you are to choose any two disks (they may be on the same space) and move each of them one space (independently, to the right or the left). For example, if you choose the disks on spaces 2 and 4, you could move the first to space 1 or 3 and the second to 3 or 5. With these moves, is it possible to reach a position where all the disks are on the same space? (H)

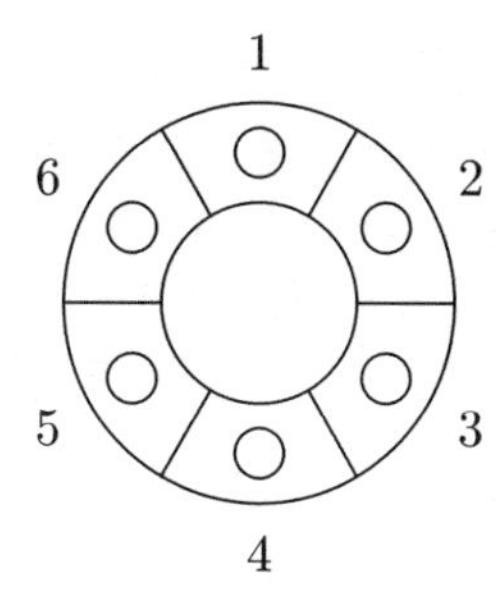

P130. Fencing a field

We have forty-three pieces of fencing whose lengths are $1, 2, 3, \ldots, 43$ meters. Is it possible to fence a square field using *all* of these pieces? (The pieces are not to be bent to form a corner.) A rectangle? (Q)

P131. Partitioned latin squares

Each of the squares is to be filled with a digit from 1 to 9. Some of the digits are already in place. Can both grids be completed so that every digit appears one time in each row, in each column, and in each of the nine 3×3 grids identified by the thick boundary lines? (H)

<table>
<tr><td></td><td></td><td>5</td><td></td><td>7</td><td></td><td>4</td><td></td><td></td></tr>
<tr><td></td><td>6</td><td></td><td>2</td><td></td><td>1</td><td></td><td>3</td><td></td></tr>
<tr><td>8</td><td></td><td>3</td><td></td><td>6</td><td></td><td></td><td></td><td>9</td></tr>
<tr><td></td><td>5</td><td></td><td>6</td><td></td><td>8</td><td></td><td>7</td><td></td></tr>
<tr><td>6</td><td></td><td>2</td><td></td><td></td><td></td><td>3</td><td></td><td>5</td></tr>
<tr><td></td><td>8</td><td></td><td>5</td><td></td><td>3</td><td></td><td>1</td><td></td></tr>
<tr><td>5</td><td></td><td></td><td></td><td>3</td><td></td><td>6</td><td></td><td>1</td></tr>
<tr><td></td><td>1</td><td></td><td>8</td><td></td><td>4</td><td></td><td>5</td><td></td></tr>
<tr><td></td><td></td><td>7</td><td></td><td>5</td><td></td><td>8</td><td></td><td></td></tr>
</table>

(a)

<table>
<tr><td></td><td></td><td>5</td><td>8</td><td></td><td>6</td><td>9</td><td></td><td></td></tr>
<tr><td></td><td></td><td>1</td><td></td><td>4</td><td></td><td>3</td><td></td><td></td></tr>
<tr><td>9</td><td>4</td><td></td><td></td><td>3</td><td></td><td></td><td>7</td><td>5</td></tr>
<tr><td></td><td></td><td></td><td></td><td>5</td><td></td><td></td><td></td><td></td></tr>
<tr><td></td><td>6</td><td>4</td><td>9</td><td></td><td>7</td><td>5</td><td>8</td><td></td></tr>
<tr><td></td><td></td><td></td><td></td><td>1</td><td></td><td></td><td></td><td></td></tr>
<tr><td>5</td><td>2</td><td></td><td></td><td>8</td><td></td><td></td><td>1</td><td>6</td></tr>
<tr><td></td><td></td><td>8</td><td></td><td>6</td><td></td><td>7</td><td></td><td></td></tr>
<tr><td></td><td></td><td>3</td><td>4</td><td></td><td>1</td><td>2</td><td></td><td></td></tr>
</table>

(b)

P132. Balanced numbers

Call a number *balanced* if some sequence of its beginning digits also occurs as its ending digits. For example, the numbers $\underline{92}0\underline{92}$, $\underline{5}370\underline{5}$, $\underline{31}\,\underline{31}$, and $\underline{270}43\underline{270}$ are balanced. Is there a balanced number that remains balanced when you adjoin any of the digits $1, 2, 3$ at the end of the number? Is there a balanced number that remains balanced when you adjoin any of the digits $1, 2, 3, 4, 5$ at the end of the number? (H)

P133. Divisible digits

(a) Write a digit in each square of a 2×2 grid. We can read the digits in the rows (from left to right) and the columns (from top to bottom) and think of them as four two-digit numbers. Suppose that the three two-digit numbers found in the rows and the first column are divisible by K. Will the number in the second column also be divisible by K?

(b) The same question as above for a 3×3 grid. In this case, suppose that the three three-digit numbers in the rows and the two three-digit numbers in the first two columns are divisible by K. Will the three-digit number in the last column be divisible by K?

(c) Once again, consider a 3×3 grid with a digit in each square. Suppose that five of the six numbers, formed by reading the digits in the rows and columns as three-digit numbers, are divisible by K, where K is a prime number larger than 5. Will the sixth number also be divisible by K? (H)

P134. Leg work

Sirens have twice as many heads as Neptunes. The Neptunes each have three legs. Several Sirens and some Neptunes were seen near a meadow by The Great Volcano; there were 23 heads and 134 legs in all. Does a Siren have more than 15 legs? (H)

P135. Contending contestants

Thirty-five young people were invited to take part in a mathematics contest. Unfortunately several of them were delayed and could not be present at the appointed time. Each problem in the contest counted for one point. If the women had each solved five problems and the men had each solved four problems, the total score of all contestants would have been 4 percent more than if the men had each solved five problems and the women had each solved four problems. Did more than 10 women take part?

P136. Magic products

In the 3×3 grid shown below, the product of the numbers in each row and column is 216.

1	6	36	216
12	9	2	216
18	4	3	216
216	216	216	

Can nine different positive integers be put into the squares of a 3×3 grid so that the products of the numbers in each row and column will all be the same but less than 150? (H)

P137. Goggles retrieval

John was swimming in a river, against the current, when he lost his goggles. He continued for 10 minutes, then decided to turn around and retrieve them. He caught up with the lost goggles one-half mile from the point where he lost them. Is the river flowing faster than one mile per hour? (You can assume that John is swimming at the same strength throughout.)

P138. Paper folding sequences

A 2×3 sheet of paper is divided into six squares numbered as shown.

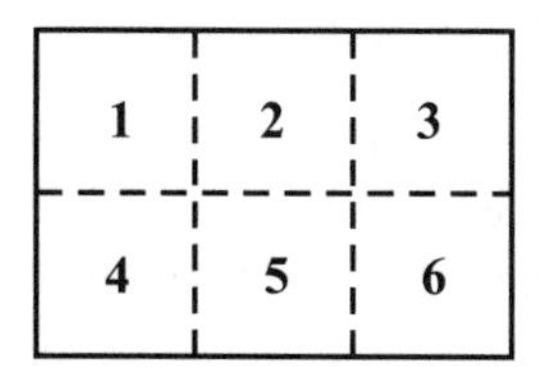

By folding the left third onto the middle third and then folding the right third on top of that, and finally the bottom half on top, we have the numbers arranged from the top in the order 5, 4, 6, 3, 1, 2.

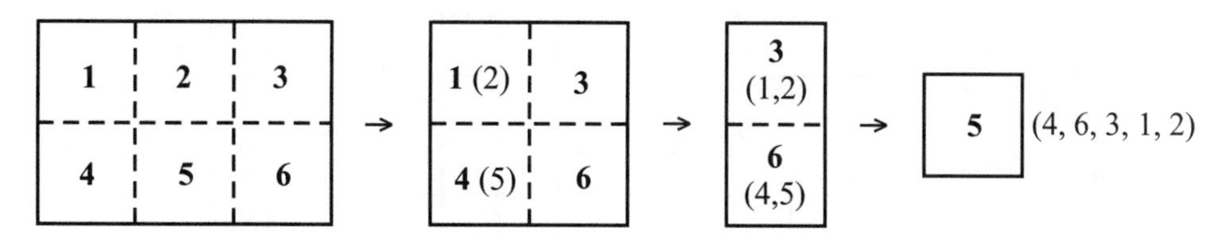

Can three folds about these three dotted lines arrange the squares in the order 4, 3, 6, 5, 2, 1? (Q)

P139. Green acres

A farm is made up of 15 rectangular fields plus a rectangular acreage on which the farmhouse is built. The acreage of seven of the fields is known, and is shown (but not to scale) below. Is it possible to determine the acreage of the field on which the house is located? (H)

	20	14	
12			
8		15	
	25		21

P140. Keeping in the nines

Among the first 1000 positive integers, are there more numbers with a 9 as one of its digits (in its usual decimal representation) than there are without a 9 among its digits? Same question for the first ten million positive integers. (Q)

P141. Seven divisions

Is there a seven-digit number, with seven different digits, that is divisible by each of these digits? Is there an eight-digit number with this corresponding property? (H, Q)

P142. Elementary, my dear Watson!

(a) Can positive integers (not necessarily different) be put into the empty squares in (a) so that the equations are true? The operations $+$, $-$, $\times$ and $\div$ are to be read in the order in which they appear, reading from left to right, and from top to bottom. For instance, if we put 6 in the third square of the third row then that row would read $(9+6)/3 = 5$ and not $9+(6/3) = 11$, and the third column would read $(12+6)/9 = 2$ and not $12+(6/9)$.

(b) Can nonzero digits be put into the empty squares in (b) so that the equations are true? (H)

	×	12	÷		=	12
+		+		+		×
9	+		÷	3	=	
÷		÷		−		÷
	+	9	÷		=	
=		=		=		=
4	×		+		=	16

(a)

	+		÷		=	7
+		+		+		−
	×		−	9	=	
÷		−		÷		×
	×	3	−		=	
=		=		=		=
8	+		−		=	

(b)

P143. Winners in, losers out

(a) Christopher, Victor, and Dan took turns playing a two-person game. They decided that the loser of a single game should sit out the next game while the other two played. Each game resulted in a winner and a loser (that is, there were no ties). They played for an afternoon, and at the end, Victor had played 17 times while Dan had played 35 times. Can you say who won the fifteenth game and how many games Christopher played?

(b) The next day the boys continued to play the same game. This time, Christopher played 22 times and Dan played 25 times. Can you say who lost the sixth game and who played in the antepenultimate game (the next to the next-to-the-last game)? (H, Q)

P144. Four square

Can the numbers $1, 2, 3, \ldots, 16$ be placed in the 4×4 grid shown below so that the product of the numbers in each of the four 2×2 squares is divisible by 16?

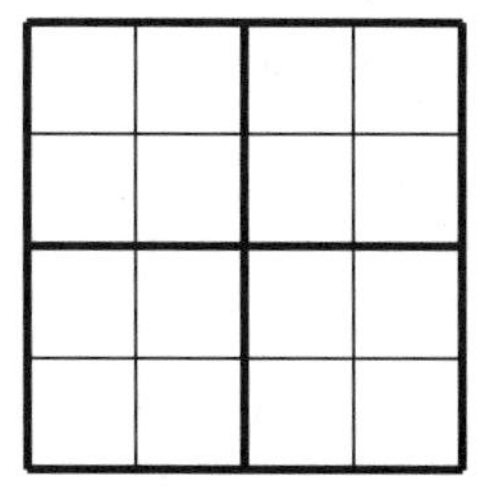

P145. Less is more

The sum of seven different positive integers is 101. Must three of these add up to more than the other four? (H)

P146. Take 'em away

I have placed a dime in nine of the 36 squares of a 6×6 grid. You may choose three rows and three columns and take all the coins you find in them. Can you always get all nine coins? (H, Q)

P147. Trivalence

Can the numbers $1, 2, 3, \ldots, 18$ be placed on the line segments (one on each segment) so that the sum of the numbers on the three edges meeting at each node is divisible by 3? (H)

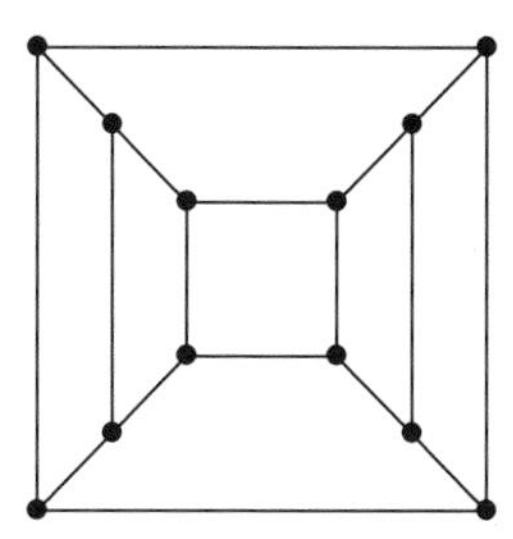

P148. Half and half

Place the thirty-two chess pieces on the ordinary 8×8 chessboard so that there is an equal number on each half of the board. Is the following statement always true?

> If there are more white pieces on one-half of the board than black pieces on the other half then some row or column contains five pieces.

P149. Go for gold!

Twenty-seven points are evenly distributed around a circle and these form the vertices of a regular 27-gon. Some of these points are colored green and the others are colored gold, but the coloring is done so that between any two green points there are always at least two gold points. Must there be three gold points which form the vertices of an equilateral triangle?

P150. Destructive Dan

(a) My son Dan has torn out 25 sheets (not necessarily consecutive) from a medical journal. Is it possible that the sum of the torn-out page numbers is 1273?
(b) Can the sum of the torn-out page numbers be 2450?
(c) On another occasion Dan tore out 24 sheets (not necessarily consecutive) from a journal. Is it possible that the sum of the torn-out page numbers is 2450? (H)

P151. Let's face it

Can the edges of a cube be labeled from 1 to 12 so that the sums of the four numbers around each face are each the same? (H)

P152. Packing calissons

We have 60 of those small diamond-shaped almond confections that come from Aix-en-Provence. Each covers exactly two triangles in the candy boxes pictured below, where we've placed one of them. Can any of the boxes be filled with these sweetmeats (without overlap) without moving the one that is already placed? What if you are allowed to move the one that's shown?

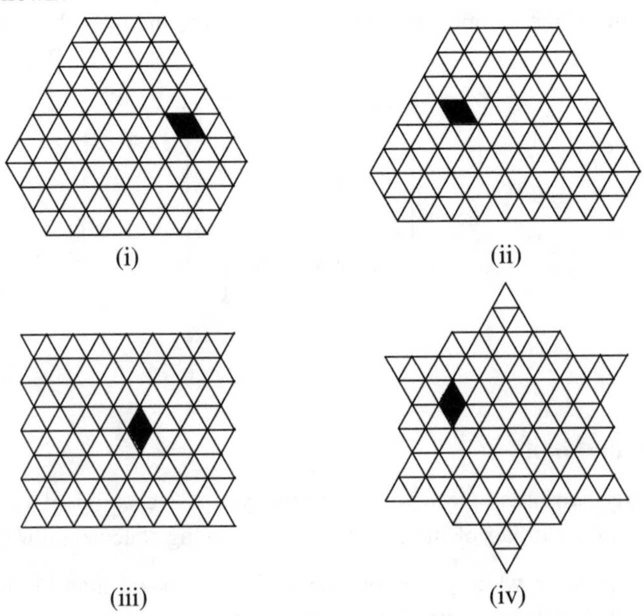

P153. Ballroom dance-cards

The King has invited n knights and n ladies to a ball at the castle. But he has given a peculiar order: At the end of the evening no two knights shall have danced with the same number of ladies, and no two ladies shall have danced with the same number of knights. At the same time, however, each person must dance at least once. Can the dance cards be organized to comply with the King's wish?

P154. Scrambled shoebox

Asa has 240 shoes in a box in his shoe store, in three different colors, 80 of each color. All the shoes are of the same size and style; 120 are for the left foot and 120 for the right foot. Can Asa always find 40 pairs in the box? (A pair, naturally, consists of a left and right shoe of the same color.) (H)

P155. Covering with trominoes

Delete a corner square from the ordinary 8×8 chessboard. Can this deleted chessboard be tiled using

(a) pieces such as those shown below (using at least one piece of each type)?
(b) only pieces of the type (i)?
(c) only pieces of the type (ii)?

Each piece covers three squares of the chessboard.

P156. Four for four

A rectangular grid can be tiled with some 2×2 squares and some 1×4 rectangles (each piece covers four small squares of the grid). Can the rectangle be tiled using one fewer 2×2 piece and one more 1×4 piece? (H)

P157. Tables for four

The committee members are sitting around the boardroom table in the conference center. The number of pairs of people of the same gender who sit next to each other is equal to the number of pairs of people of different gender who sit next to each other. Will it be possible to place these folks in the dining hall with four to a table? (H)

P158. Dot to dot

Can the squares shown overleaf be arranged in a 4×4 tray so that the lines will fit together to form a continuous loop which doesn't cross itself? You are allowed to rotate the pieces 180° about the center, but not 90°, nor are you permitted to turn the pieces over. (Think of the pieces as wooden blocks with horizontal grain; the requirement is that the grain of the wood must remain horizontal and the blocks can't be turned over. This is a variation of the Sol LeWitt puzzle, **P127**.) (H)

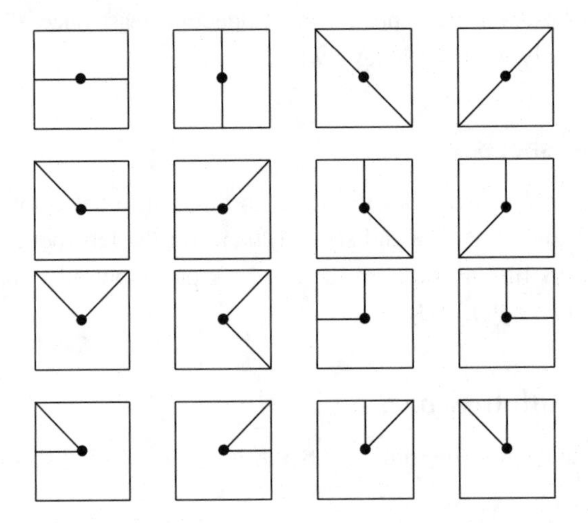

P159. Counterfeit coins

We have five coins that look alike; however, one is lighter and one is heavier than the other three (true) coins. Is it possible to find these counterfeit coins, and distinguish which is which, in three weighings on a balance scale without weights? (You do not know if the heavy coin is too heavy by the same amount as the light coin is too light.) (H, Q)

P160. Delegate mixer

Film studios were nominated for Oscars in this year's Academy Award Ceremony. They competed in three categories: best barber, best tatooist, and best plastic surgeon. Members of the jury voted for at most one studio in each category. The result was that each nominated studio received nineteen votes (distributed over the three categories). After the vote, the jury delegates were invited to a dinner in a hall in which 55 tables were set. Is it possible to seat the delegates at these tables so that no two people at the same table voted for the same studio in any of the categories? (You may assume that the tables can be extended to accommodate as many people as necessary.)

P161. Round-robin volleyball

Eight teams take part in a round-robin volleyball tournament (each team plays each other team once and there are no ties). Must there be four teams A, B, C, D such that A defeated B, C, D; B defeated C, D; and C defeated D? (H, Q)

P162. Encoded pentomino tiling

From five small squares, one can build twelve different connected (along their edges) shapes which Sol Golomb has called pentominoes, three of which are shown below. Will the other nine different pentominoes cover the 5×9 rectangle in such a way that the sums of the numbers covered by each pentomino are all the same (each pentomino covers five small squares)? (Two pentominoes are considered to be the same if one of them can be rotated or reflected and moved so as to coincide with the other.) (H, Q)

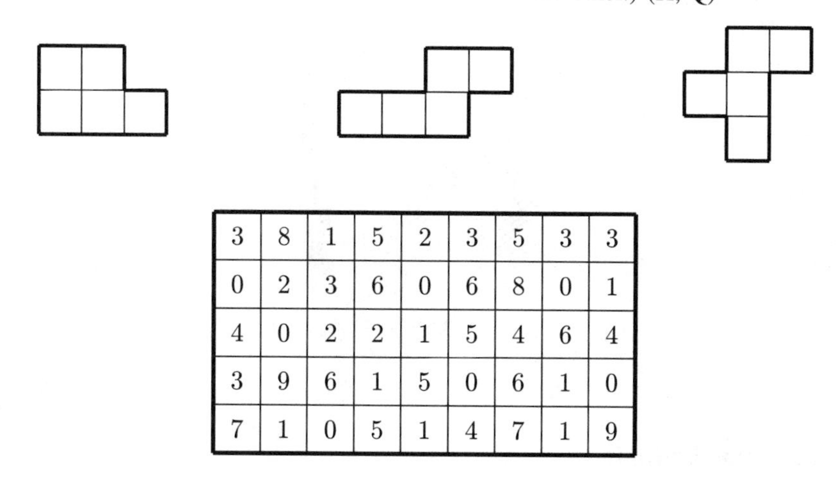

3	8	1	5	2	3	5	3	3
0	2	3	6	0	6	8	0	1
4	0	2	2	1	5	4	6	4
3	9	6	1	5	0	6	1	0
7	1	0	5	1	4	7	1	9

P163. More pentomino rectification

Place a dot in the center of each of the five squares that make up each pentomino. Consider the eight pentominoes shown below: U, V, W, Z, I, L, P, N. (What distinguishes these from the other four (T, X, Y, F) is that the five centers of the squares that make up these pentominoes can be connected by a continuous path of length 4.) Will these 8 unicursal pentominoes tile a 5×8 rectangle? (Q)

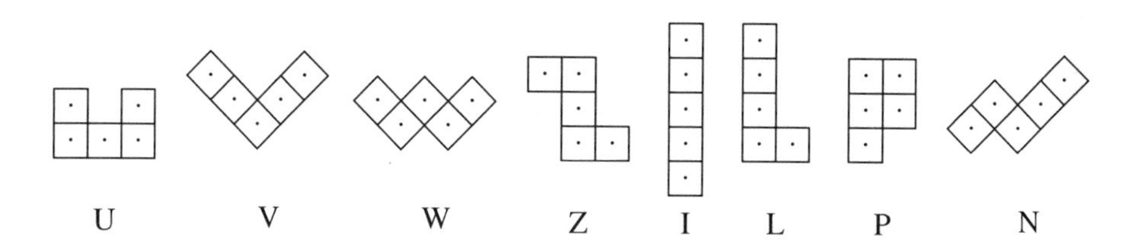

P164. Beware of frost

George owns a garden plot that has the shape of an equilateral triangle 20 yards on a side. He decides to cover the plot for the winter with five tarpaulins, and just to be different, he wants the five tarps to be equilateral triangles, each of the same size. He orders five sufficiently large tarps, but there is a mixup in the delivery and only four of the five arrive. He has a daughter, however, who is an excellent problem-solver, and she believes that if the lot could have been covered with five such tarps, it could also be covered with only four. Is this true? (H)

P165. Tower of pizza

Selma cut out a number of congruent equilateral triangles from cardboard, and numbered their corners $1, 2, 3$. Then she stacked them as shown. Can Selma stack a few of these and arrange them (by turning them over and/or rotating them) so that the sums of the numbers down the stack at each vertex is equal to 99? Can the number sums be equal to 98? (Q)

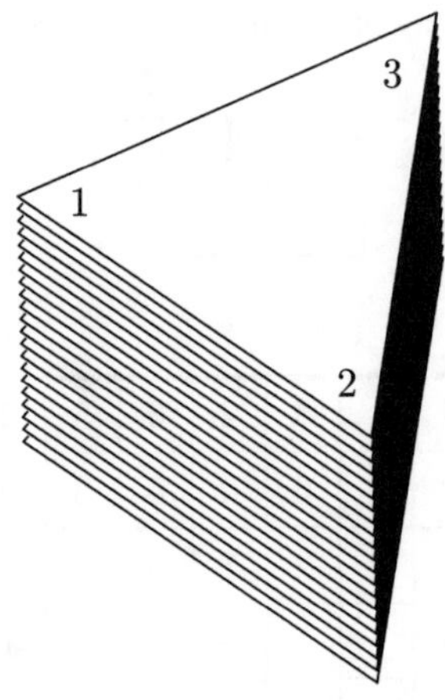

P166. Strange bedfellows

Several people gathered for a conference. Two rooms are available. The delegates are not all on good terms with each other, but no delegate has more than three others present who are difficult to tolerate. Is it possible to assign the delegates to the two conference rooms so that no person will be in the same room with more than one person with whom he/she cannot get along with? (Assume that the property of "not getting along with" is two-sided; if A doesn't get along with B then neither does B get along with A.) (H)

P167. Squaring the rectangle

The rectangle $ABCD$ in the figure is tiled by nine squares (the proportions are distorted). If the area of the smallest square is four square inches, is it possible to determine the area of rectangle $ABCD$? (H)

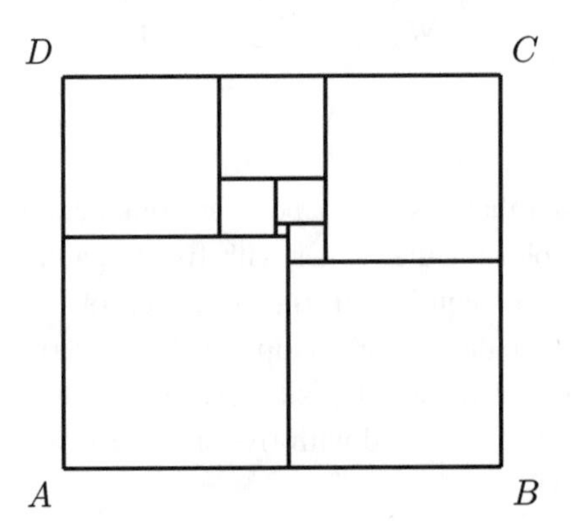

P168. Some good ones

If you add up the 99 decimal numbers $9, 99, 999, \ldots, \underbrace{99\cdots9}_{99 \text{ digits}}$, will the answer contain 99 ones?

P169. Fair and square

A number of people take part in a round-robin tennis tournament (each person plays against each other person once, and there are no ties). For each k, let W_k and L_k denote the number of wins and losses by player k. Clearly, the number of wins among the players is equal to the number of losses; symbolically, $W_1 + W_2 + \cdots = L_1 + L_2 + \cdots$. Must $W_1^2 + W_2^2 + \cdots = L_1^2 + L_2^2 + \cdots$ also always be true? (H)

P170. Try your luck

Can the numbers $1, 2, 3, \ldots, 13$ be placed in the circles so that the sums of the numbers in the six circles that surround each of the four small triangles are all the same? (H)

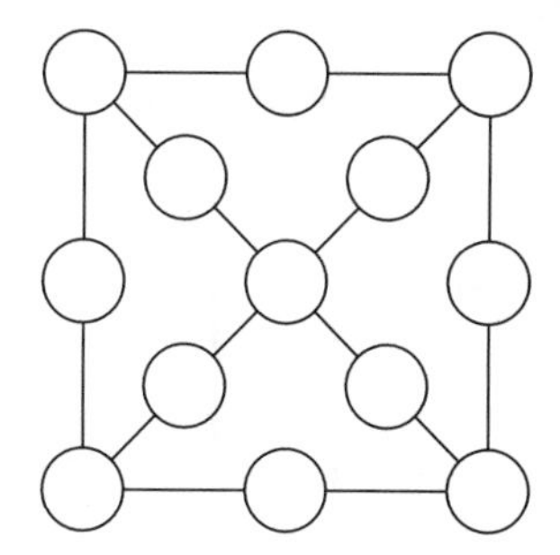

P171. Visitors from outer space

One of the differences between us and the outer space visitors from Tralfadom is that they have only one hand apiece; however, they have more than five fingers on their hand. As their delegation stood by, someone counted a total of 195 fingers among their group. At a press conference they stated that their planet consisted of four continents: one in the north, and one in the south, the east, and the west. More than 16 percent of their delegation came from the north continent, more than 27 percent came from the east, and more than 40 percent came from the west. Were there any representatives from the southern continent? (H)

P172. Find the dominoes

There are 28 dominoes in a full set. Each domino is divided into two square halves with 0 to 6 dots on each half. We will denote, for example, the domino with 2 dots on one half and 5 dots on the other by the pair of numbers $(2, 5)$ (or, equivalently, by the pair $(5, 2)$). Each possible combination of number pairs from $(0, 0)$ to $(6, 6)$ occur exactly once in a complete set of dominoes. The 56 squares of the 7×8 rectangular grid shown below are covered by these 28 dominoes. The numbers within the small squares indicate the number of dots on the domino half that covers that respective square. Can you reconstruct the actual domino placement? (H)

4	0	2	2	6	2	3	4
2	1	6	0	0	5	5	3
1	1	0	5	4	6	2	3
4	0	4	4	3	0	5	1
5	1	5	6	1	6	3	1
4	2	6	3	6	5	0	5
0	2	6	4	1	3	2	3

P173. Find more dominoes

Here are two more domino reconstruction problems (see the preceding problem for an explanation). (H)

1	1	6	4	1	5	1	4
6	3	6	3	4	6	0	5
4	4	1	3	0	0	1	5
6	3	0	3	5	5	2	3
4	1	2	6	1	6	2	5
0	2	4	2	2	5	3	5
0	3	2	6	0	2	0	4

(a)

2	0	0	1	1	5	0	4
2	3	6	5	5	2	6	2
1	3	6	4	2	0	4	2
4	4	0	3	3	2	4	1
2	3	0	6	5	3	6	5
0	0	4	6	6	1	3	5
1	3	4	6	5	5	1	1

(b)

P174. Banquet seating

Sixty-five distinguished guests are sitting around a banquet table.

(a) Must there be either two men or two women sitting next to each other?

(b) Must there be either two men or two women sitting with exactly one person between them?

(c) Must there be either two men or two women sitting with exactly two people between them?

(d) Is there some number k smaller than 33 for which we cannot say for sure that either two men or two women are sitting with exactly k people between them? (H)

P175. A planned neighborhood

Is it possible to place ten coins on each of the boards (at most one coin per square) according to the following rules:

(i) No coin is to be placed on a square with a number in it.

(ii) A number in a square indicates the number of coins contained on the neighboring squares. (In this problem, two squares are neighbors if they share a common edge *or* a common corner.) (H)

		2				
				1		1
		3				
1				4		1
		1				
3		2			1	
		1				

(a)

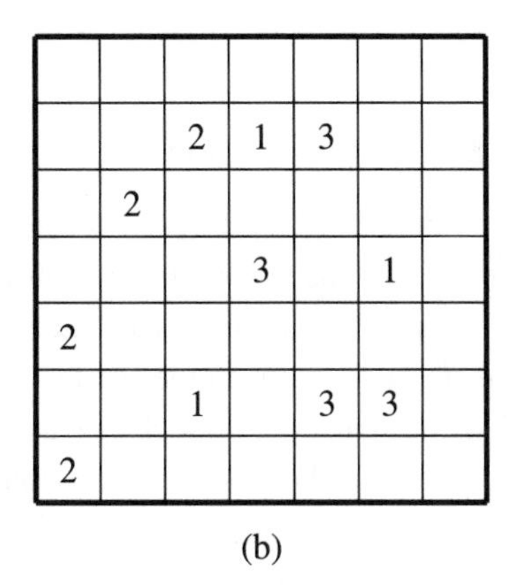

(b)

P176. A temperamental elevator

The house where Toby lives is equipped with an elevator. It has four floors, including the basement and the attic. The elevator is very unusual: it has only two buttons, "A" (toward the attic) and "B" (toward the basement) and goes only one floor up or down at a time. It takes 30 seconds between the time you press the button until the elevator reaches the corresponding floor above or below (at which time you can press a button again). Furthermore, the elevator doesn't work if you press two consecutive identical strings of two or more buttons. For example, suppose you start in the basement and proceed with the sequence AAABBAB. Then the elevator will stall, for if you press A, you have BA two times in a row, and if you press B, you have ABB two times in a row. If he starts in the basement, can Toby ride the elevator for more than 8 minutes before he is stymied by a stalled elevator? What if he starts on the first floor? (H, Q)

P177. Patternizing

Consider a 4×4 grid whose squares are colored black or white in some arbitrary pattern. In a single operation you are allowed to choose a row or column and change the color of all the squares in that row or column (from white to black and from black to white). You may repeat this operation as many times as you wish, and your goal is to obtain a coloring of the board with as few black squares as possible. The smallest number of attainable black squares will depend on the original coloring of the board. For example, the following board can be reduced to 0 black squares in four operations (choose rows

1 and 4, and columns 2 and 3). Is it always possible to get to less than 5 black squares, regardless of the initial coloring? (H, Q)

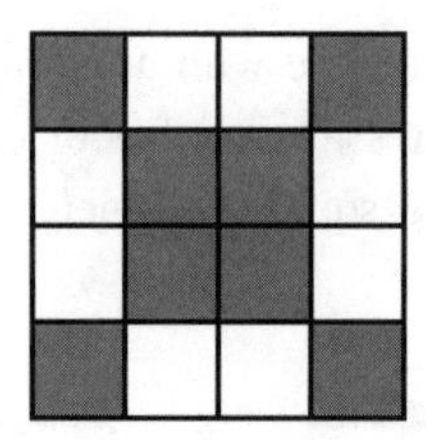

P178. Minimal extraction

The numbers from 1 to 2001 are written on a sheet of paper. Choose two of these numbers, say a and b, $a \geq b$, remove them from the list, and replace them with $a^2 - b^2$. Repeat this procedure over and over again. Each time, you take away two numbers from the list, square them, and adjoin to the list the difference (it might be 0) of their squares. After a number (how many?) of such operations, you will have only one number left on the paper. Can the numbers be chosen so that this last remaining number is zero? (H, Q)

P179. Tom and Jerry

When Tom and Jerry and their two friends took a table in the restaurant, Tom observed there were 23 men and 32 women guests in the room (including themselves). Because he sat by the door, Tom could observe everyone who came into the restaurant, or left. He noticed that for every two guests who left a new guest arrived, and furthermore, if the two who left were of the same gender, the new arrival was a woman, otherwise, the new arrival was a man. One of their friends left the restaurant (with another person) but came back later, remembering that the bill had not yet been paid. It was then that Jerry observed that his party of four (Tom, the two friends, and himself) were the only guests left in the restaurant. Were Tom's and Jerry's friends of the same gender? (H)

P180. Notched blocks

The figure below is a $1 \times 1 \times 3$ block with a $1 \times 1 \times 1/2$ notch cut out of the middle of one face.

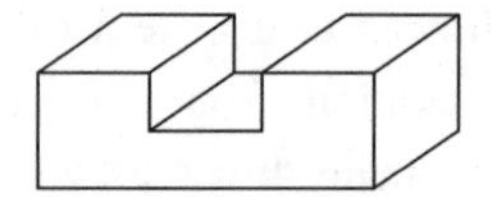

Each block has a volume of $2\frac{1}{2}$ cubic units, so it might be possible to pack 18 such blocks into a $3 \times 3 \times 5$ box. Do each of the figures opposite correspond to realizable packings?

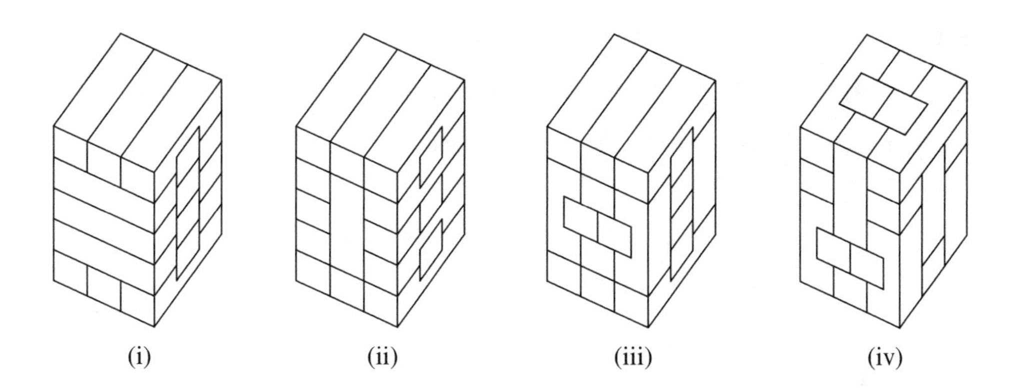

P181. Tiling with quadrilaterals

Let F be an arbitrary quadrilateral in which the angle at each vertex is less than 180 degrees. Is it always possible to tile the plane with copies of F? (H, Q)

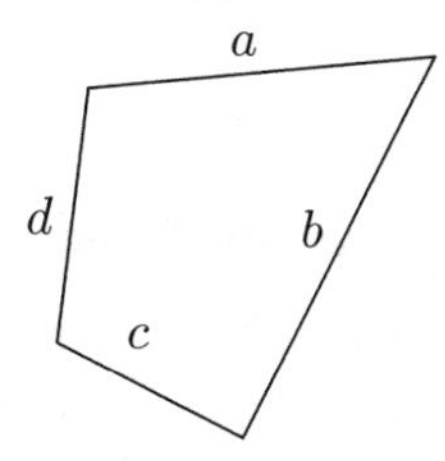

P182. Sum of digits of square

Is there a number A such that the decimal digits in A^2 add up to 44? (H)

P183. Number of the beast

Can the digits $1, 2, \ldots, 9$ be used to make several numbers which together add up to 666 (each digit is to be used exactly once)? What about 555?

P184. Divisible digits

(a) Given any three digits, no two the same, is it possible to take one, two or three of them and make a number that is divisible by 3 (no digit is to be used more than once)?

(b) Given any five digits, no two the same, is it possible to take one or more of them, and make a number that is divisible by 9 (no digit is to be used more than once)?

P185. Pretty products

The numbers

$$202020207070702525252474747475 = 45454545454545 \times 44444445555555$$

and

$$1212121333333366666668787879 = 36363636363637 \times 33333336666667$$

are examples of numbers which can be factored into two numbers with the same digits, but in a different order. Can the number

$$2020203434342525252383838384$$

be factored into two such numbers? (H)

P186. Meltdown

Let A_1 denote the product of all the numbers $1, 2, 3, \ldots, 1000$ (that is, $A_1 = 1000!$). Let A_2 be the sum of the digits of A_1, let A_3 be the sum of the digits of A_2, let A_4 be the sum of the digits of A_3, and so forth. The numbers in this sequence $A_1, A_2, A_3, A_4, \ldots$ steadily decrease until they eventually reach a single digit. From that point on, the terms in the sequence are the same. Will this digit be the same if you started with the product of $1, 2, 3, \ldots, 1001$? (H)

P187. A list of jousts (Barry Cipra and John Conway)

Eight (chess) knights are placed on a 3×4 chessboard, four along the top row and four along the bottom. Can you exchange the positions of A and a, B and b, C and c, and D and d by a judicious choice of no more than 40 knight moves? (H, Q)

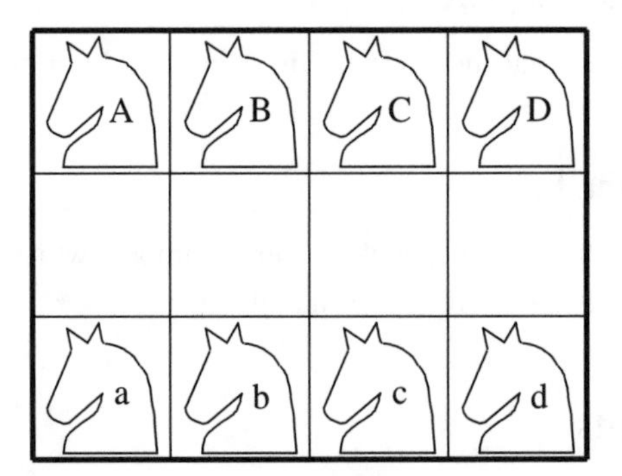

P188. Curious fences

Can a continuous path be drawn through the points in each of the square lattices shown on the opposite page, so that the following conditions hold?

(i) The path consists of horizontal and vertical segments connecting neighboring points of the lattice.

(ii) The path starts and ends at the same point and does not visit any point more than once (it need not visit every point of the lattice).

(iii) Each number indicates how many segments of the path form edges of the small square that surround the number. (H)

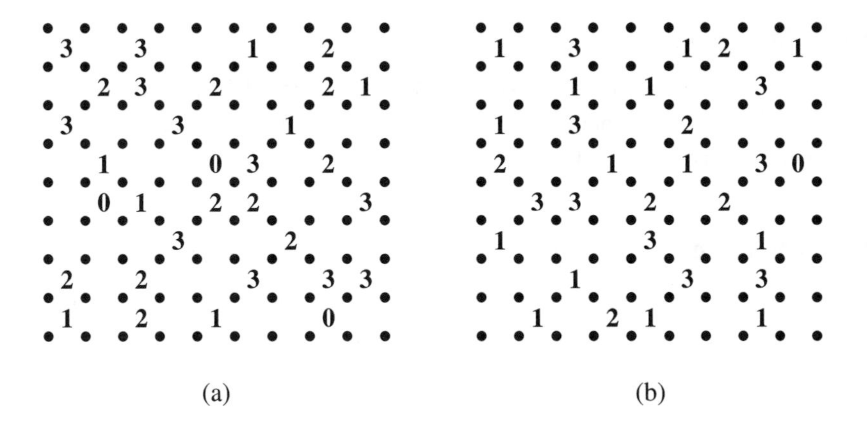

(a) (b)

P189. Cars or carts

A lottery sold 10000 tickets numbered from 0000 to 9999. Those holding a ticket in which the digits sum to 18 win a cart. Those holding a ticket in which the sum of the first two digits equals the sum of the last two digits win a car. If the ticket satisfies both conditions, the holder doesn't get anything. Will they give out more cars than carts? (H)

P190. Chain reaction

The owner of an apartment building allows families to exchange apartments, but only by making simple swaps. (A chain of moves involving more than two families, for example 1 to 2, 2 to 3, and 3 to 1, is prohibited.) A family is required to live in an apartment at least one week before becoming eligible for another exchange. Suppose fourteen families wish to move and their wishes form a chain 1 to 2, 2 to 3, $\ldots$, 13 to 14, and 14 to 1. Can this be accomplished within a month? Does the answer change if the chain were to involve only thirteen familites? (H)

P191. Two cryptarithms

Here are two more multiplication reconstruction problems.

(a) In the first one, different letters represent different digits. Can the letters be replaced by digits to make the computation correct?

(b) In the second problem, where multiplication is done from left to right so that the units product is below the tens product, D and E denote odd and even digits, respectively. Can the letters be replaced by odd and even digits to give a correct answer? (H, Q)

$$\begin{array}{r} A\ B\ C\ D\ E \\ \times \quad\quad 4 \\ \hline C\ D\ E\ B\ A \end{array} \qquad\qquad \begin{array}{r} D\ E\ E \\ \times\ E\ E \\ \hline E\ D\ E\ \\ E\ D\ E\ E \\ \hline D\ D\ E\ E \end{array}$$

P192. Odd neighbors

Can you place coins on a 6×6 chessboard, at most one coin per square, so that for each square, whether occupied or not, the number of coins on adjacent squares (sharing a common edge) is odd? (H, Q)

P193. The powers that be

In each case are there positive integers A and B which satisfy the given equations? (H)

(a) $A^2 - B^2 = 631$ (b) $A^3 - B^3 = 631$ (c) $A^4 - B^4 = 631$

P194. The towers that be

In each of the following five expressions, is the last decimal digit 4, 5, 6, 7, 8, respectively?

(a) $4^{4^{4^{4}}}$ (b) $5^{5^{5^{5}}}$ (c) $6^{6^{6^{6}}}$ (d) $7^{7^{7^{7}}}$ (e) $8^{8^{8^{8}}}$

P195. Blacksmithery

It takes a blacksmith five minutes to put on a horseshoe. Can eight blacksmiths shoe ten horses (that is, put on all forty horseshoes) in less than 30 minutes? (A horse cannot stand on two legs!) (H)

P196. Crossnumber puzzle

Can each square in the crossnumber puzzle be filled with a nonzero digit, $1, 2, \ldots, 9$, thereby making number-words (in place of ordinary words) so that the digits in each number-word will *sum* to the number indicated at the heading of the word? The digits in each number-word are all to be different. For example, a 2-digit word headed by the number 4 will be filled with 1 3 or 3 1, whichever fits the crossing words. (H)

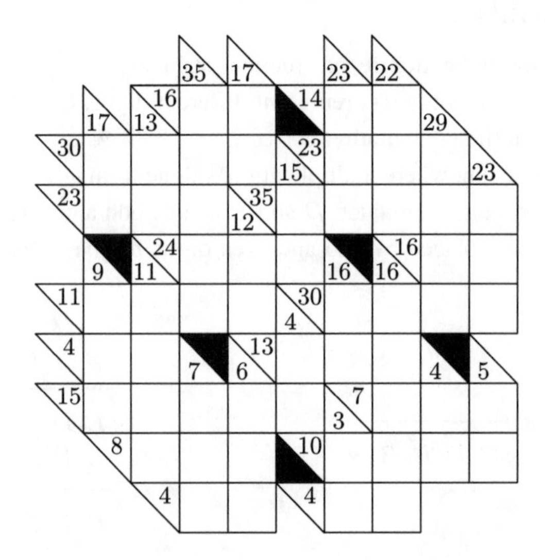

P197. Happy campers

(a) Not all of the 198 children at a summer camp had gotten acquainted with each other after one week. But is it true that at least two children had gotten to know the same number of other children at the camp?

(b) If each child had gotten acquainted with at least 132 others, is it true that at least four children had gotten to know the same number of other campers? (H)

P198. A regular icosagon

Place twenty points around the circumference of a circle so that successive points are equidistant (that is, the points are vertices of a regular icosagon). From among these points, choose any nine. Will three of these points form an isosceles triangle (a triangle with two sides of equal length)? (H, Q)

P199. Parquet flooring

It is not difficult to cover a 16×16 grid with 1×4 tiles in such a way that 32 of them are horizontal and 32 vertical.

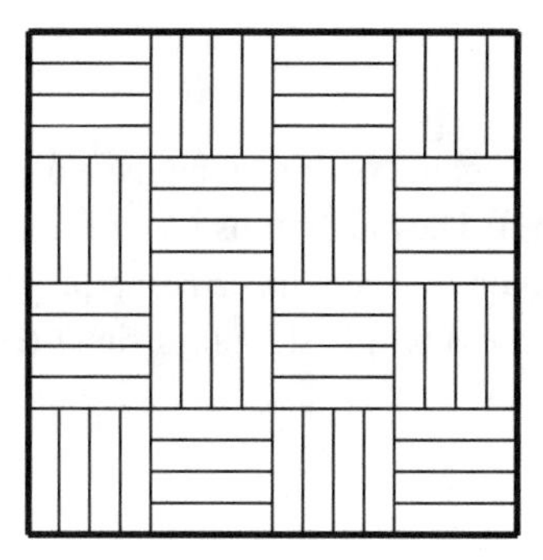

Can the tiles be rearranged so that the difference in the number of horizontal and vertical tiles is 4?

P200. Magic line (Barry Cipra)

From a strip of 25 squares, Barry cuts out the second and fourth fifths, leaving squares 1 through 5, 11 through 15 and 21 through 25. From each of these three shorter strips he then cuts the second and fourth squares, leaving squares 1, 3, 5, 11, 13, 15, 21, 23, 25. Can you write the numbers $1, 2, \ldots, 9$ on these squares so that any three equally spaced numbers add up to the same total? (Q)

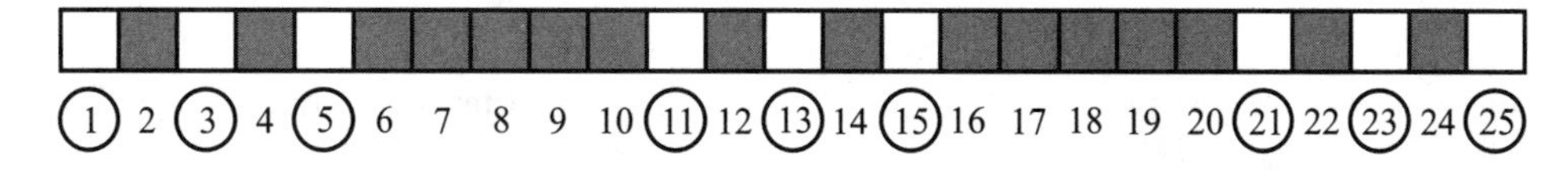

P201. Magic fifteen

Take nine slips of paper and number them from 1 to 9, then lay them face up on a table. Two players take turns choosing one of the numbers (that is, one of the slips of paper). The game ends when either one of the players (the winner) has some numbers that add to 15. Can the first player force a win? (H)

P202. Tic-Tac-Toe

A well-known game (Tic-Tac-Toe) is played by two players who take turns placing an X or an O, respectively, on the squares of a 3×3 grid. The first player to get three marks in a row (horizontal, vertical, or diagonal) is the winner of the game. It is well-known that there is no winning strategy for either player. Is there a winning strategy for one of the players if they play on the grid of squares shown below? (Q)

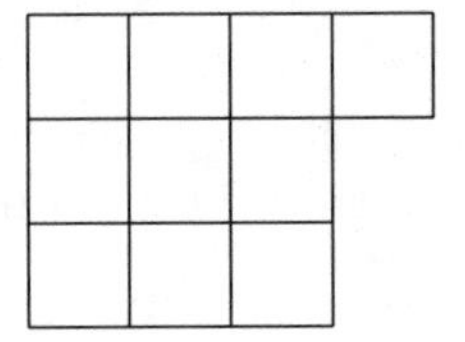

P203. All x-ed out

Nicole and Pryor are playing a game on a rectangular board with r rows and c columns, where r and c are larger than 6. They take turns placing an $\times$ in 1, 2, 3, 4, 5 or 6 adjacent squares of a single row or column. The winner is the player who puts an $\times$ into the last empty square. Can Nicole force a win if she plays first? (H, Q)

P204. Head on collision

In this game, the gameboard consists of a row of 222 squares. Ferdinand has a piece at the left end, on square 1, which he moves to the right, and Isabella has a piece on the right end, on square 222, which she moves to the left. Ferdinand starts and they take turns to move their pieces 1, 2, 3, 4, 5 or 6 squares at a time. A piece cannot jump over the other, or occupy the same square, so the game ends when they are on adjacent squares, and whoever's turn it is to move cannot do so and loses. Can Ferdinand force a win? (H)

P205. Staking claims

A rectangular board with r rows and c columns has a chess rook in the lower left square of the board. On a single turn, a player moves the rook horizontally or vertically to another square, but not to a square that has been previously visited. For example, if the first move is up to the sixth row, the second player could move to any other square in the sixth row or to any other square in the first column except the bottom square. Two players take turns moving the rook; the loser is the first player who cannot make a move. Can the first player force a win? (H, Q)

P206. One or both

(a) In this game, Nicole and Pryor alternate taking either one or two beans from a single heap. The game ends when the last bean is taken and the winner is the player who has taken an even number of beans. Can Nicole force a win if she has the first move and there are 19 beans in the heap?

(b) Here, there are two heaps of beans, one with 10 beans and one with 12. Nicole and Pryor alternately take one bean from one of the heaps or one bean from each of the heaps. The winner is the player who takes the last bean. Can Nicole force a win if she has the first move? Does the answer change if the heaps contain 10 and 11 beans?

P207. Help the king home

Nicole and Pryor play the following game on an 11×13 gameboard. The game begins with the chess king in the upper left corner of the board. They take turns moving the king one space, either to the right, or down, or diagonally down and to the right. The winner is the player who manages to get the king to the castle located in the lower right corner of the board. Can Nicole force a win if she plays first? (H, Q)

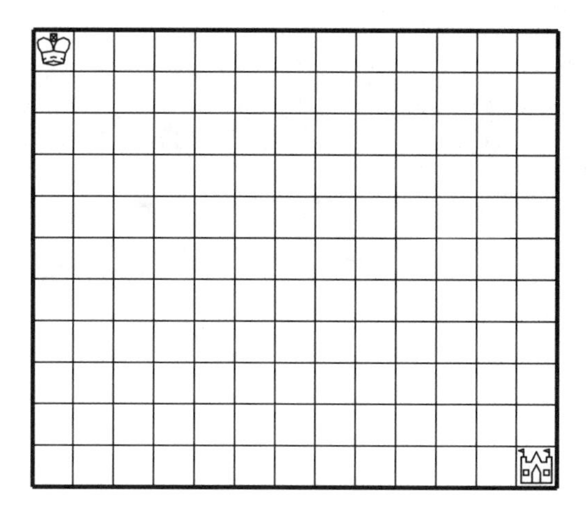

P208. Casting in the nines

(a) Adam and Eve take turns writing digits until they have twenty digits in a row. Adam starts, and they are allowed to use only the digits $1, 2, 3, 4, 5$. Eve wants the final 20-digit number to be divisible by 9. Can Adam prevent it?

(b) The same question as in part (a), except this time they continue until they write a 30-digit number.

P209. Operands and operants

Clemens and Sonja play the following game. On the blackboard, there is an expression with parentheses, large dots, and question marks, as shown overleaf. Sonja goes first and replaces the left-most question mark with a digit from 0 to 9. On the next move Clemens replaces the next dot by an addition or multiplication sign ($+$ or $\times$). They continue to move from left to right, Sonja replacing ? by a digit that hasn't already been used, and Clemens replacing a dot by $+$ or $\times$. Sonja wins if the nine operations on the ten different digits yield an odd number. Can she win regardless of Clemens's responses?

$$(((((((((? \bullet ?) \bullet ?) \bullet ?) \bullet ?) \bullet ?) \bullet ?) \bullet ?) \bullet ?) \bullet ?)$$

P210. Toad Road

A hexagonal pool has lilypads distributed in a hexagonal pattern, except for the center, as in the figure. A toad is sitting on the pad labeled A, and it makes knight-like jumps equal to the distance between pads A and B (two pads in one direction and one pad to the right or left at 120°). Can the toad make a succession of jumps (each of this same distance) that will take it to each of the lilypads exactly once and then return to its original position? (H, Q)

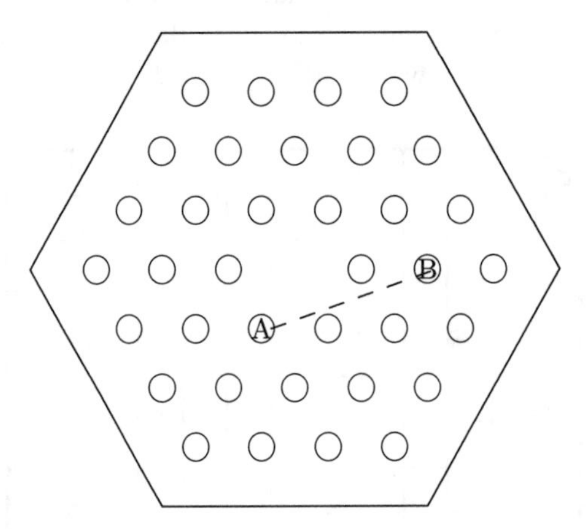

P211. Room for one more

Six tennis balls used to come in a box of size $1 \times 2 \times 3$, where the unit is the diameter of the ball. With this packing, it's easy to see that $2n$ tennis balls will fit into a $1 \times 2 \times n$ box. Is it possible to pack $2n+1$ balls into a $1 \times 2 \times n$ box if n is large enough? (Thanks to Eugene Luks for passing along this problem.) (H, Q)

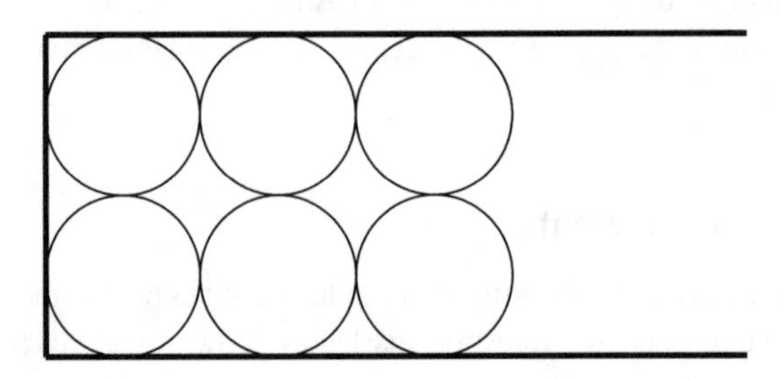

P212. Tiling with trapdoor

We have an 8×8 board, twenty-one 1×3 tiles and one 1×1 tile. The squares in the tiles are the same size as those on the board. Can the board be tiled with these pieces regardless of where the 1×1 tile is placed? (Q)

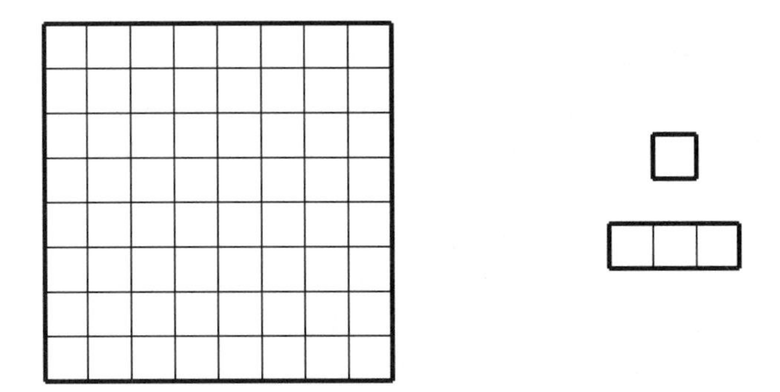

P213. Lucky sevens

Is the number 600! (the product of all numbers from 1 to 600) divisible by 7^{99}? (H, Q)

P214. Factorial digits

Are there any three-digit numbers A such that $A = a! + b! + c!$, where a, b, c are the three digits? (H)

P215. Two sums the same

Can the letters be replaced by digits to make the two addition problems encrypted below simultaneously correct? Different letters represent different digits.

$$\begin{array}{rcccc} & & a & a & a \\ & & b & b & b \\ + & & c & c & c \\ \hline & f & g & h & i \end{array} \qquad \begin{array}{rcccc} & & a & a & a \\ & & d & d & d \\ + & & e & e & e \\ \hline & f & g & h & i \end{array}$$

P216. The first shall be last

(a) If A is a five-digit number divisible by 41, does it follow that B, obtained by moving the first digit of A to the last position, is also divisible by 41?

(b) If A is a six-digit divisible by 37, does it follow that B, obtained by moving the first digit of A to the last position, is also divisible by 37? (H)

P217. Nonattacking queens

There are several ways of placing eight queens on an 8×8 chessboard so that no two of them are on the same row, column, or diagonal. Is it possible to do it so that an odd number of them are on the white squares? (H, Q)

P218. Magic triangle (Barry Cipra)

Can the letters be replaced with integers $1, 2, 3, \ldots, 9$ (no number more than once) so that the vertices of the equilateral triangles abc, dfg, ehi, ade, bfh, cgi, afi add up to the same number? (Q)

$$\begin{array}{ccccccc} & & & a & & & \\ & & b & & c & & \\ & d & & & & e & \\ f & & g & & h & & i \end{array}$$

P219. Far flung points

Twenty-one points on the circumference of a circle form the vertices of a regular 21-gon. Are there five points so that all the distances between them are different? (H, Q)

P220. By air or sea

There is a direct connexion, either by boat or plane, between each pair of islands in an archipelago. All connexions are two-way connexions (with the same means of travel).

(a) Can you go between any two islands either by one or two boat trips or by one or two plane trips?

(b) Is there a single means of transportation, either by boat or by plane, that will permit you to travel between any two islands with at most two stops (that is, three trips)? (H)

P221. Can you wing it?

The island of Mainstay has direct flights to and from seven of the islands in the archipelago, whereas the island of Wayside has direct flights to and from only three of the islands. All the other islands have either six or eight direct flights to and from the other islands in the group. Can you fly from Mainstay to Wayside (it may not be a direct connexion)? (H)

P222. Outward bound

There are 36 towns on the island of Mainstay besides the capital town of Lofton (37 towns in all). The governor, who wants to restrict driving on the island by making it more inconvenient, has declared that all roads shall henceforth be only one-way. Furthermore, each town, other than Lofton, is permitted to have just seven incoming roads and six outgoing roads. None of the roads on the island intersect each other (except within the towns). No two towns are directly connected by more than one road (there may be no direct connexion at all; accessibility from one town to another may require passing through other towns). One day the governor decides to travel around the island and then return home. Is this possible? (H)

P223. Build more overpasses

(Continuation of the preceding problem) Because of the experience in the preceding problem, the governor decides to build more roads. As a consequence of this construction, each pair of towns (including the capital) is directly connected by a single one-way road. (This is an expensive idea; it may require constructing up to $\binom{37}{4} = 66045$ bridges!) The befuddled Secretary of Transportation arbitrarily assigns directions to these roads. Regardless of how this assignment is made, will it nevertheless be true that there are two towns (one of them might be the capital), say A and B, for which one can travel from A to B and on the way visit every town on the island exactly once? (H)

P224. Point of no return

As in the preceding problem, each pair of towns is connected by a single one-way road. Might it happen that the road directions are assigned so that, regardless of your town of origin, should you leave the town you could not return?

P225. A better plan

(a) As in the preceding problem, there is a one-way road between each pair of the 37 towns on the island, and there are no road crossings except within the towns. The directions of the roads are assigned, this time, so that each town has the same number of incoming roads as outgoing roads (18 of each). Is it possible to travel from any one town to any other town on this road system by passing through at most one other town?

(b) The Secretary of Transportation decides to close some of the roads so that the number of incoming roads and the number of outgoing roads for each town is no more that 14. Is it possible to travel from any one town to any other town on this road system by passing through at most two other towns? (H)

P226. Business card exchange

Nine people besides myself were present at the last meeting of the club. A few of us exchanged business cards, and I observed that for every three people A, B, C, whenever A and B, and A and C, exchanged business cards, then B and C did not. Could there have been as many as 50 business cards exchanged at the club under these conditions? (H)

P227. Consecutive sums

(a) Is it possible to find four positive integers such that their pairwise sums give six consecutive numbers?
(b) Is it possible to find five positive integers such that their pairwise sums give ten consecutive numbers? (H)

P228. Incompetent advisors

The governor suspected that her top 68 advisors were deficient in their mathematical knowledge. She had them stand in a circle, with the top advisor in the first position, the second advisor in position 2, the third advisor in position 3, and so forth. Then she asked each of them a mathematical question, one at a time. She began the questioning with advisor 2, who answered the question incorrectly; this advisor was promptly taken away to the remediation center. The governor continued to ask each one of the remaining advisors, but did so systematically by asking every other person, in order: 2, 4, 6, 8, and so forth. Each of them answered incorrectly, and each of them in turn were taken away for remediation. After number 68 was taken away she continued around the tightened circle, asking every other person as before, so that the next questioned advisor was number 3, and then number 7. And so it continued until Bob was the only remaining advisor and he answered the question correctly! Immediately he was promoted to the highest position. Can you determine Bob's original position in the hierarchy before he was promoted? (H)

P229. A number scale

Can you decode the message 98 56 12 70 384 40 32 98, knowing that

$$\text{MI} \times \text{SI} = \text{SOL}$$

and

$$\text{LA} \times \text{DO} = \text{RE} \times \text{FA}?$$

(No number begins with 0, and different letters stand for different digits.) (H)

P230. Give and take

Each of the 20 mathematicians who had been invited to speak at a conference had gone to exactly nine of the talks given by these invited speakers. Can we say for certain that two invited speakers had heard each other's talks? (Q)

P231. Chess tournament scores

In a chess tournament each pair of players meet exactly one time. A player gets 1 point for a win, $\frac{1}{2}$ point for a draw, and 0 points for a loss. The final standing is determined by the total number of points accumulated in the tournament. Players with the same score are ranked within their group by a random drawing. The scores for the top four finishers were $4\frac{1}{2}, 3\frac{1}{2}, 3$ and $1\frac{1}{2}$. Is it possible to find the lowest score? (H)

P232. Equal and unequal shares

(a) Eight bowls are placed around a table and there are several coins in each bowl. At each turn, you are to add a coin to each of five consecutive bowls. Can you, by repeating this operation a number of times, get an equal number of coins in each bowl?

(b) Here you have seven empty bowls which are placed around a table. At each step you either add a coin to each of two bowls that are separated by exactly one other bowl, or you take out one coin from each of two bowls that are separated by a single bowl (provided neither of them is empty). Can you obtain a position where the bowls in clockwise order contain 1, 2, 3, 4, 5, 6, 7 coins respectively? (H, Q)

P233. Two more cryptarithms

(a) Can the nine stars be replaced by digits from 1 to 9 (each digit occurs once) so that the addition is correct?

(b) Can the stars be replaced by digits other than 1 or 3 (they may be repetitions) so that the multiplication problem (done from left to right) is correct? (H)

(a)

```
   * * *
   * * *
 + * * *
 -------
 2 2 9 5
```

(b)

```
        3 * *
      × * * *
    ---------
  * * * 3
    1 * * *
      * * *
  -----------
  3 * * * * *
```

P234. Peter makes his point

Peter says he has placed a point P inside a square $ABCD$ so that the perimeters of triangles PDA, PAB, and PBC are respectively 4.7 cm, 5.7 cm and 4.3 cm. Furthermore he claims that he can determine the perimeter of triangle PCD from this information. Is Peter correct about these claims? (H, Q)

P235. Covering a disk

Seven green circular disks, each of radius 1 cm, completely cover a black circular disk. Can the black disk have radius larger than 2 cm? (H)

P236. Sentries in the square

Three stationary sentries are guarding an important public square which is, in fact, square, with each side measuring 8 rods. If any of the sentries see trouble brewing at any location on the square, the sentry closest to the trouble spot will immediately cease to be stationary and dispatch to that location. And like all good sentries, these three are continually looking in all directions for trouble to occur. Will there always be, regardless of the position of the sentries, a potential trouble spot that is more than 4 rods from any sentry?. (This problem is adapted from one composed by David Savitt & Russell Mann for the ninth annual Konhauser Contest held at Carleton College in March 2001.) (H, Q)

P237. Forward thinking

(a) The numbers from 1 to 15 are written on a piece of paper. You may choose any two numbers from the list, erase them, and adjoin their sum to the list. After fourteen such operations only one number will be left on the list. Can the numbers be chosen so that the remaining number is 105?

(b) Again, the numbers from 1 to 15 are written on a piece of paper. You may choose any two from the list, say A and B, erase them, and adjoin the number $A+B+AB$ to the list. Can the numbers be chosen so that after fourteen steps the remaining number is less than one billion? (H)

P238. Tetromino tilings

Using four congruent squares, one can glue together five essentially different tetrominoes.

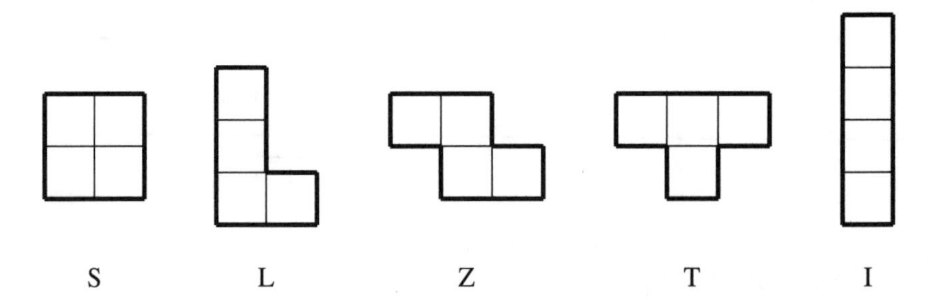

For each of these types, is it possible to tile a 10×10 grid with 25 copies of the tile? (H)

P239. Divisibility by 11

(a) A number A has an odd number of digits. Write this number down next to itself (two copies of A), and call this new number B. Is B necessarily divisible by 11?

(b) Write down a number A and let B be the number obtained by following the digits of A with the digits in A in reverse order. Is B necessarily divisible by 11?

P240. Painting by numbers

In each of the grids shown opposite, there are one or more numbers to the right of each row and under each column. The numbers in each row or column indicate the lengths of the blocks of black squares in that row or column. For example, in the fourth row of figure (a), the numbers 2 1 2 mean that there are three blocks, one of two consecutive black squares, one of a single black square, and another of two black squares, separated by, and perhaps preceded or followed by one or more white squares. For example, squares 1 2 5 8 9, or 2 3 6 9 10 might be black, with the others white. Is the information contained in the numbers sufficient to describe the color of each square? (H)

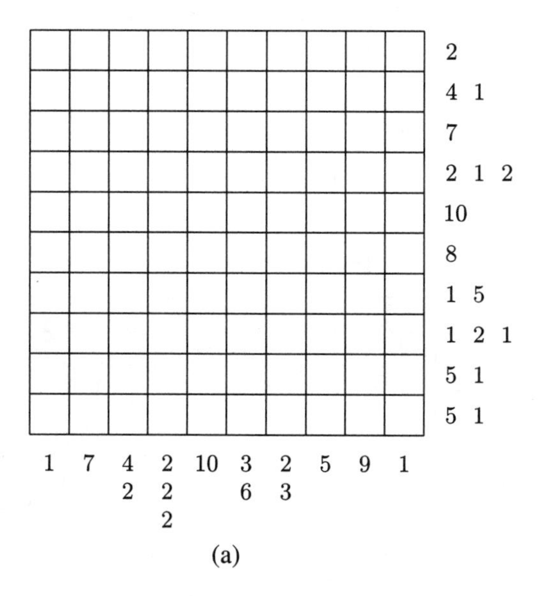

(a)

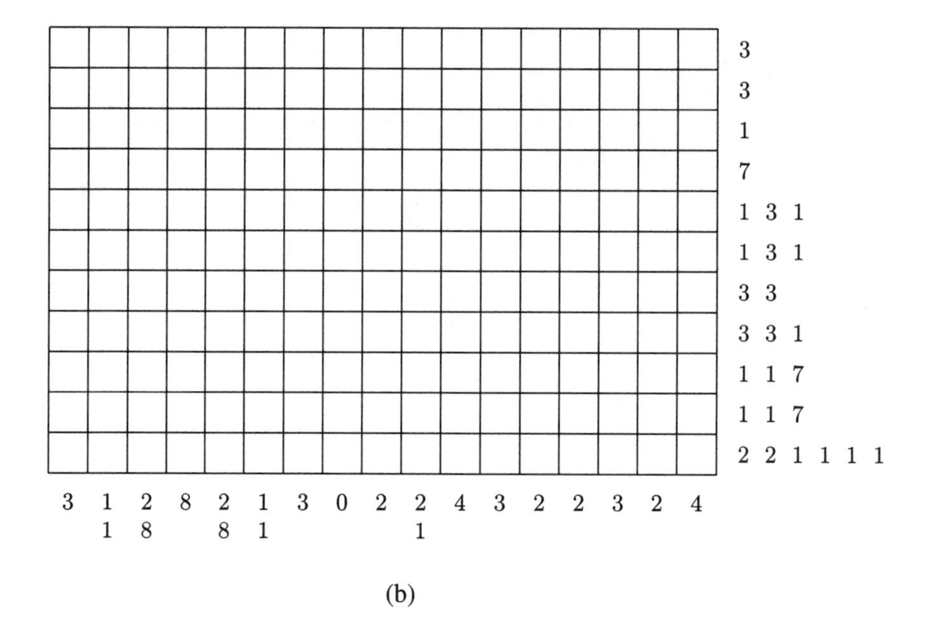

(b)

P241. Gasp! Asp!

Here's another problem of the same type as the preceding problem. Can you reconstruct the picture? (H)

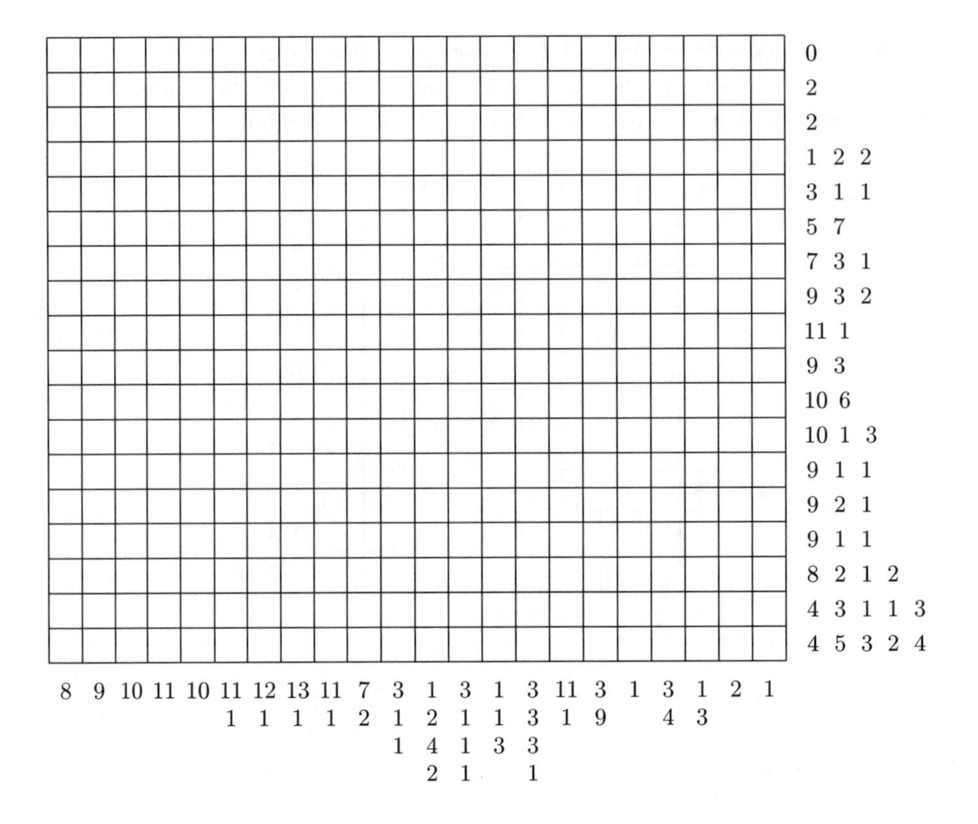

P242. Unlucky thirteen

The numbers from 1 to 500 are written on a piece of paper. In a single move you choose 2, 3, 4 or 5 numbers from the list, erase them, and adjoin to the list the remainder of the sum of the chosen numbers when it is divided by 13. After a number of moves there are only two numbers on the page. One of them is 102. Is it possible to find the other? (H)

P243. Animal sociology

In order to transfer eight animals to the zoo, the keepers constructed four cages, each large enough to accommodate two animals. It is known that certain animal pairs are incompatible and cannot be in the same cage. No animal is incompatible with more than three others (in the group). Is it possible to transport the animals in these four cages so that no two incompatible animals are in the same cage? (H, Q)

P244. Musical chairs

A round table is set for 16 people and each place is identified with a placecard. The guests sit down in a random manner, paying no attention to the placecards. Is it possible for each of the guests to move the same number of places to the right (or to the left) so that at least two persons are in their proper place? Is the same result true if 16 is replaced with 15? (H, Q)

P245. Taking the cake

A table is set for 25 people. On each plate (numbered clockwise from 0 to 24) there is a piece of cake, each of a different kind. Bruce gets to the table first and decides to take a taste of the cake on the zeroth plate. Then he moves one place clockwise to the first plate, and tastes the cake on that plate. Thereafter, he proceeds to taste other cakes at the table by moving (always clockwise) 2, 3, 4, 5, and so on, places (to the plates numbered 3, 6, 10, 15 and so forth).

(a) Will Bruce taste more than half of the different cakes?

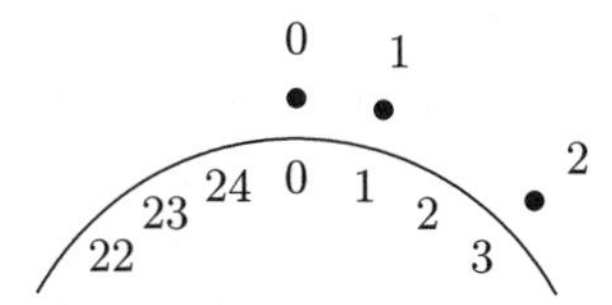

(b) Will Bruce taste more than half of the different cakes if he starts at the zeroth cake and then proceeds by successively moving 1, 3, 5, 7, and so forth, places in the clockwise direction?

P246. Domino tiling

(a) Are there more than 500 different ways of tiling the 2×14 rectangle with fourteen 1×2 tiles? (Each domino covers exactly two squares of the configuration.)

(b) Can you find a configuration of twelve 1×1 squares that can be tiled with six 1×2 dominoes in exactly seven different ways? (H, Q)

P247. Divisors of repunits

(a) Is there a number A which, in decimal notation, is written as a number of 1's followed by a number of 0's (that is, $A = 11\ldots100\ldots0$) which is divisible by 2010?

(b) Is there a number B which, in decimal notation, is written with only 1's (that is, $B = 11\ldots1$) which is divisible by 2011? (H)

P248. Modular triangles (Marc Roth)

In each of the figures

$t = 1$	$t = 3$	$t = 6$
		0
	0	2 4
0	1 2	5 3 1

the numbers $0, 1, 2, \ldots, t-1$, for a triangular number t, are arranged so that the sum of two adjacent numbers in a row is equal to the number in the row above, modulo t. For example, in the last figure, $5 + 3 = 2$ modulo 6. Is there such an arrangement for $t = 10$? (Q)

P249. Digital sums

(a) Are there exactly 10 numbers between 1 and 1000 whose digits sum to 3?

(b) Is there any other number k such that exactly 10 numbers between 1 and 1000 have digits which sum to k? (H)

P250. More gold medalists

In a round-robin ping-pong tournament (where each person plays each other person once), gold medals are given out according to the following principle: player A receives a gold medal if for each other player B, either A defeated B or for some other player C, A defeated C and C defeated B. If we use the notation $\mathrm{X} \rightarrow \mathrm{Y}$ to mean that X defeated Y, then A receives a gold medal if for every other player B, either $\mathrm{A} \rightarrow \mathrm{B}$ or $\mathrm{A} \rightarrow \mathrm{C} \rightarrow \mathrm{B}$ for some other player C.

(a) Will at least one gold medal will be awarded?

(b) Can it happen that exactly two gold medals will be awarded? (H, Q)

P251. House numbers in Straightsville

In Straightsville there are K streets which all run in different directions, so that every pair of streets intersects, and there's a house at every intersection, $K-1$ houses on each street. Each house on any street has a different number, of course, but only the numbers $1, 2, \ldots, K-1$ are used. Could Straightsville have only eight streets? (H)

P252. No two in a line

The figure has ten lines, including the outer edges, in each of three directions, meeting in 66 points. Can coins be placed on eight of these points so that no two coins lie on the same line? (H, Q)

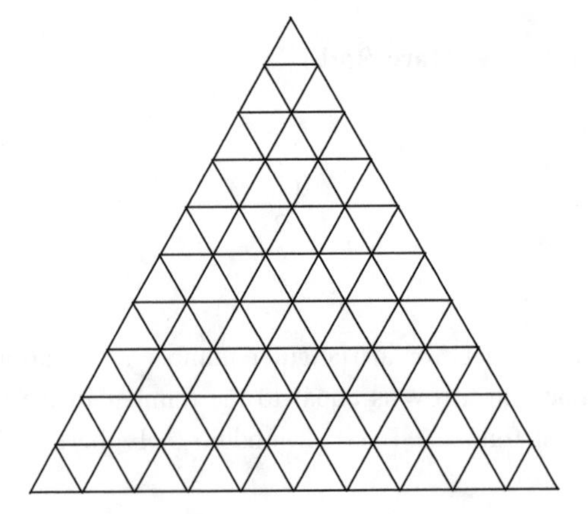

P253. Lie detection

I am thinking of a number from 1 to 16. You wish to determine this number by asking "Yes-No" questions. I am allowed to give one incorrect answer. Can you determine the number with at most seven questions? (H)

P254. Think of five numbers

I have chosen five positive integers, call them a_1, a_2, a_3, a_4, a_5, and you wish to determine what they are. You are permitted to give me five arbitrary numbers b_1, b_2, b_3, b_4, b_5 and ask me to give you back the value of the sum $a_1b_1 + a_2b_2 + a_3b_3 + a_4b_4 + a_5b_5$. Can you do it in less than five questions? (H)

P255. Dan's crazy computer

Dan built his own computer. It cannot do much; it can only perform one operation. If you enter a number a into the machine and then another number b (which is not 0; however, the numbers a and b can be negative), the machine prints out the value of $1-(a/b)$. Dan claims that his computer can multiply, divide, subtract, and add any two numbers. For example, to divide a by b, he can compute $(a*b)*1$, where $*$ means the operation which his computer can do. Here, one enters a into the computer, and then b; the computer prints the value of $1-(a/b)$. Then, one enters this number, and then the number 1. The result is the value $1-(1-(a/b))$, which is equal to a/b, as claimed. Is Dan right in his claim that his computer can multiply, subtract and add any two numbers?

P256. The 13-coin problem **(Herbert Wright)**

After we had compiled these problems and solutions, Cedric Smith brought to our attention a remarkable problem that he had received from Herbert Wright around 1960 and he graciously consented to let us use it as a suitable climax for our collection. Unfortunately the original statement of the problem has been mislaid but what follows is essentially the same as Herbert Wright's formulation.

The Rev. Hal rashly decided to show how clever he was by forging a £1 coin and proudly showing it to his son. But his bishop, Hog, got word of this, summoned him, declaring:

> "A vicar forging coins is rather shocking.
> The usual penalty is a defrocking.
> Empty your pockets, Hal."

Hal produced 13 £1 coins. "Which is the fake?" asked Hog. Hal replied "I don't know. They all look the same. But I know that the fake coin differs slightly in weight from the others."

"Then," said Bishop Hog, "I know what to do." He called for his assistant and asked him to bring a pair of scales, a box of weights, and a piece of chalk. He wrote a different letter on each of the coins, to identify them. Then he told the assistant, "Here are directions for three weighings; in each case I've indicated which coins to put on the left-hand pan of the scales and which to put on the right-hand pan, and you are to add weights to make the scales balance. When you're done with all three weighings, come to me and tell me what weights you added to restore the balance."

Then the bishop asked, "Which of these weights (restoring the balance) refers to the first weighing, which to the second, which to the third?" The assistant replied, "But you didn't ask me to tell you that." "Fool!" said the bishop, "But tell me, in which weighings did you put the balancing weights on the left-hand pan, and in which on the right-hand pan?" "But you didn't ask me to tell you that." "Idiot!" said the bishop, "But, all the same, I now have enough information to tell which is the fake, how much it weighs, and how much a genuine coin weighs."

Then the bishop looked at Hal, a little more kindly, and said, "I'm told you are very clever. Well, if you can tell me how I find which is the fake coin, I will forgive you."

After some thought, Hal correctly discovered a procedure. Can you do the same? (H)

2
Hints

H19. Two of the figures can be traced with a single pencil path. One of them can be done with a path that begins and ends at the same place. In the other one, the starting and ending points are different. Can you find a rule which tells you whether or not a figure can be traced by a single path without lifting the pencil?

H26. By suitably rearranging the pieces after each cut, it is possible to do it in six cuts. Is this the best you can do?

H32. Yes, it is possible and one can start from any square on the chessboard.

H37. Are the first two numbers in the row both odd, both even, or one of each?

H39. It is obvious that we can obtain the desired row in eight moves: in each move, replace a single penny with a dime. Can it be done in fewer than eight moves? Consider the number, S, of pairs of adjacent coins that are different from each other. In the beginning S is equal to 15 (every coin except the first and last is included in two different pairs).

H41. In order to keep better track of the transactions, let (a, b) mean that the 7-liter bucket contains a liters of water and the 11-liter bucket contains b liters. We begin at $(0, 0)$ and after the first step we have either $(7, 0)$ or $(0, 11)$.

H42. First consider the case where all four darts hit the target. Then answer the question assuming one or more darts miss the target.

H43. How many cells were there on the evening of May 16? May 15? May 14? and so forth.

H44. What is the lowest possible score for the second place finisher?

H45. Consider the first multiplication. The last digit in the multiplier must be either 1 or 6 because the units product ends with a 4. It cannot be 1 because the units product is a four-digit number. We conclude that the last digits in the other two products are 2 and

8. The 1 in the second place of the final product tells us that the digit just above it is 0 or one of 8 or 9 (the carry from the neighboring column could be 2 or 1). If the digit is 8 or 9 then the hundreds product begins with 18 or 19. This product is twice the multiplicand which means that the multiplicand begins with 9. But then the tens product, which is 3 times the multiplicand, would begin with 2 rather than 1 (as given). Thus the hundreds product begins with the digits 10, and the multiplicand begins with 5.

$$\begin{array}{cccccc}
 & & & 5 & \ast & 4 \\
 & & \times & 2 & 3 & 6 \\
\hline
 & & \ast & \ast & 2 & 4 \\
 & 1 & \ast & \ast & 2 & \\
1 & 0 & \ast & 8 & & \\
\hline
1 & 1 & \ast & \ast & 4 & 4
\end{array}$$

Now consider the second multiplication. It is clear that the last two digits in the units product must be 25. We also see that the second digit in the multiplier must be even. It cannot be larger than 4 for that would mean that the hundreds product would have seven digits. Because this product ends in 70, the second digit of the multiplier can only be 2. This gives us 4 for the leading digit of the hundreds product and 6 or 7 for the leading digit of the thousands product. It follows that the multiplier begins with 3 and therefore the thousands product ends with 55.

$$\begin{array}{ccccccccc}
 & & & 2 & 3 & \ast & \ast & 8 & 5 \\
 & & & \times & & 3 & 2 & \ast & 5 \\
\hline
 & & \ast & \ast & \ast & \ast & \ast & 2 & 5 \\
 & & \ast & \ast & \ast & 7 & \ast & \ast & \\
 & 4 & \ast & 9 & 5 & 7 & 0 & & \\
\ast & \ast & 4 & \ast & 5 & 5 & & & \\
\hline
7 & \ast & \ast & \ast & \ast & \ast & \ast & \ast & 5
\end{array}$$

H50. Color the squares alternately black and white as on a checkerboard. Suppose that the corner squares, which have the same color, are black. How many squares are black and how many are white? What is the color of the middle, forbidden, square?

H56. Can it happen that B is a liar? (If you have solved the previous problem you may use the conclusions.)

H57. If E is a liar, then E's statement is false. Therefore E's hair is blue and C and D have different hair colors. But, if E's hair is blue then E is truthful. This is a contradiction. Therefore E is truthful.

H60. Add together the numbers in the middle row, the middle column, and the two diagonals.

H62. Let the digits of A, from left to right, be denoted by a, b and c. Then $A = 100a + 10b + c$ and a must be larger than zero. Suppose we leave off the last digit. Then $100a + 10b + c = 7(10a + b)$, or equivalently, $30a + 3b + c = 0$. This equation has no solutions because $a > 0$, so there is no A in this case.

H64. Because there were nine games in the first set, the player who served first served in five games and the other player served in four games.

H66. For a product of numbers to be odd, each of the factors must be odd.

H67. Let K denote the number of people in the tournament. Each contestant took part in $K - 1$ matches. If Jack had m wins and no draws among his matches, then he has $K - 1 - m$ losses, for a total of $m - (K - 1 - m) = 9$ points. From this equation, $2m - K = 8$, we see that K must be an even number.

H69. Let $d = s - t$ be the updated difference after each game, where s is the number of games John has won at that point and t is the number of games that Jane has yet to win.

H71. Let $A = 10a + b$ with $0 \leq b \leq 9$ and a any positive integer.

H73. Observe that \$17 cannot be made from \$4 and \$7 coins, but that \$18, \$19, \$20 and \$21 can. It follows that any amount greater than or equal to \$18 can be made up in at least one way using only \$4 and \$7 coins, and no amount less than \$28 can be made in more than one way from \$4 and \$7 coins. Also, note that seven \$4's is equal to four \$7's.

H74. Yes, it is possible, even if it is a rather drawn-out procedure. The idea is to show that you can get from floor k to floors $k + 1$ and $k - 1$.

H77. How many of the row and column products are -1?

H81. It is possible.

H82. One can make a triangle from three wire pieces if and only if the length of each piece is less than the sum of the lengths of the other two pieces.

H84. The smallest three-digit number is 100 and $100 \times 100 = 10000$, so the product of two three-digit numbers has at least five digits. If all digits are to be different, one would need $3 + 3 + 5 = 11$ different digits, which of course is not possible. This shows that at least one of the factors in the product must have two digits.

Suppose the other factor has four digits. The smallest two-digit number is 10 and the smallest four-digit number is 1000, so the product has at least five digits. Again, if all digits are to be different, one would need $2 + 4 + 5 = 11$ different digits, which is impossible.

Suppose both factors have two digits. Their product can have no more than four digits because their product is less than $100 \times 100 = 10000$. In this case, there are at most $2 + 2 + 4 = 8$ different digits. We conclude that if there are examples for either part, one of the factors is a two-digit number and the other is a three-digit number.

H85. The Venn diagram represents the 100 responses, where the circular regions labeled A, B, C, represent the people who answered "yes" to questions A, B, C, respectively. The labels a, b, c, denote the number of people who answered "yes" only to questions A, B, C, respectively, x, y, z, are the number who answered "yes" only to questions A and B, B and C, C and A, respectively, and w is the number who answered "yes" to all three questions. Three people (represented by 3 in the lower-left corner) answered "no" to all three questions.

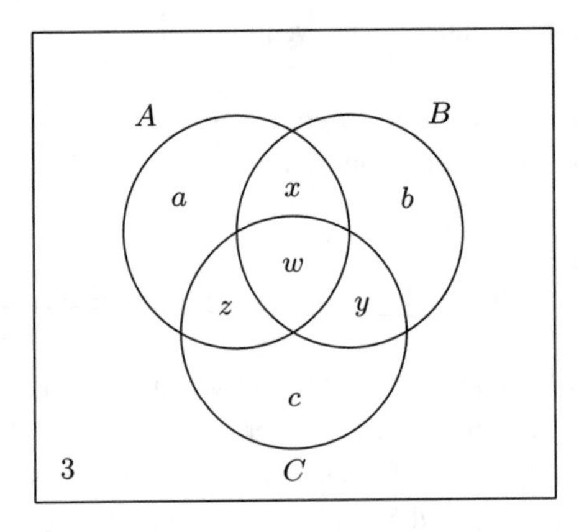

H86. Label the squares as shown below.

b	11	10	m
c	a	h	l
d	7	i	k
e	f	g	j

The twelfth segment must be either on square a or square b. But not on a, since that would violate the *cul-de-sac principle* in that the cul-de-sac $b-c-d-e-f-g$ cannot be occupied by either $16-15-14-13-$ or $1-2-3-4-5-6$ (b is the end of a cul-de-sac which means it must be either 16 or 1). In fact we need only look at the former, since the latter violates the *checkerboard principle* that even numbers go on white squares and odd numbers on black squares (when chessboards are placed in playing position, black is always the color of the lower-left corner).

H87. Label the figures as shown.

m	l	k	j	i
n	15	g	h	6
o	p	d	e	f
r	q	2	b	c
s	t	u	24	a

t	s	h	e	b	a
u	26	r	f	c	10
v	x	q	g	d	17
w	y	p	k	j	i
*	z	o	n	m	l
*	*	*	*	*	*

(a) From the cul-de-sac principle (see the previous hint), $a = 25$. Also $i = 7$, not 5.

(b) Segments 11 and 9 occupy a and c, but not in that order, since that would require $b = 12$, $e = 13$, $f = 14$, $g = 15$, $d = 16$ and we have cut the 1-end off the worm. So $a = 9$, $c = 11$, $b = 8$, $e = 7$. Then $f = 12$ (not $d = 12$, else 17 will be in a cul-de-sac), $h = 6$. Now d and i are 16 and 18 in some order. Which?

H88. The circuit exists and is unique. Begin by putting in the corners that are forced.

H89. Show that the number of horizontal segments in the circuit must be equal to the number of vertical segments.

H91. Let P denote the sum of the points given out each week, and w the number of weeks since beginning the agreement. Then, $P \times w = 40$ (the children have accumulated $9+9+22 = 40$ points). The three awards must be at least 1, 2, and 3 points, respectively, which means that P must be at least 6. This leaves three possibilities for P: 8, 10, or 20 ($P = 40$ cannot happen for that would imply that $w = 1$, whereas we know that w is at least 2).

H92. If b and g denote the number of boys and girls, respectively, who went to the dance, then $22 < b + g < 44$. Furthermore, every boy danced with four girls, and this implies that $4b$ couples danced during the evening.

H93. Label the circles a, b, c, d, e, f as shown.

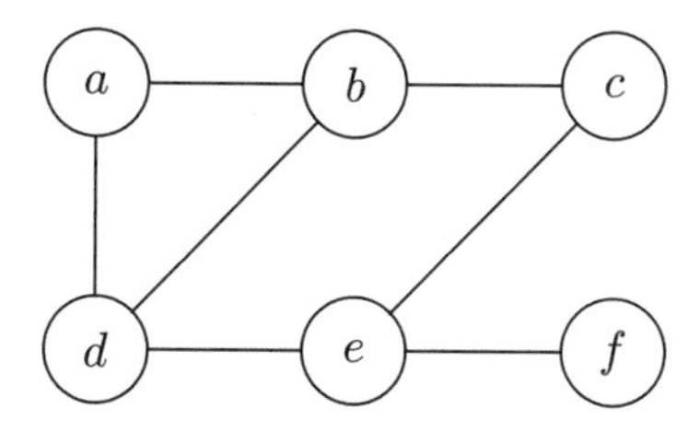

The sum of the numbers connected to the circle containing 0 is required to equal 4. Since, with these numbers, 4 can only be written as $1 + 3$, or 4, 0 must go into circle a or c (if 4 is the result of $1 + 3$) or f (if 4 is just 4). If 0 is put into a or c then 1 must go into circle b, d or e. But we know that $1 \to 12$, so 12 is the sum of the numbers in the circles connected to 1. One of these circles contains 0 and there are only two other connecting circles. But 12 cannot be written as the sum of two different numbers the largest of which is 5. Therefore 0 must be placed in circle f.

H94. These problems can be worked heuristically as in the last problem, or more algebraically. For the latter approach, label the circles as shown.

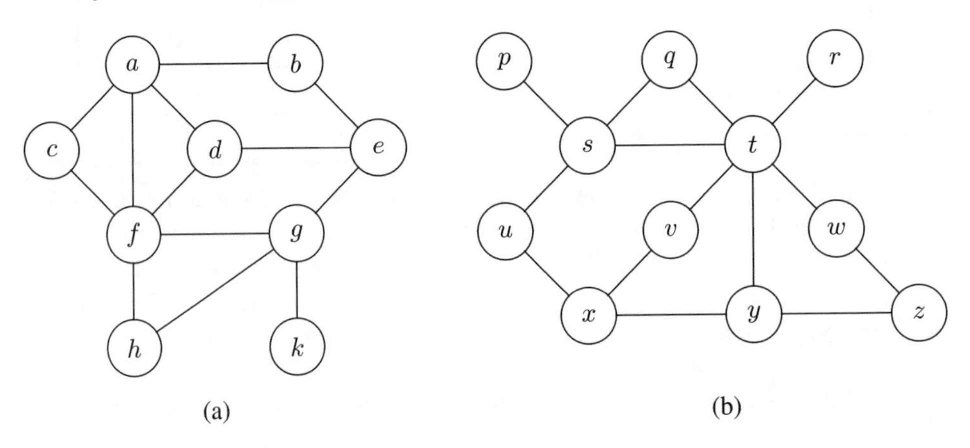

(a) (b)

In (a), $a \to b+c+d+f$, $b \to a+e$, $c \to a+f$, $d \to a+e+f$, $e \to b+d+g$, $f \to a+c+d+g+h$, $g \to e+f+h+k$, $h \to f+g$, and $k \to g$. Also,

$$a+b+c+d+e+f+g+h+k=45, \tag{1}$$

$$4a+2b+2c+3d+3e+5f+4g+2h+k=140. \tag{2}$$

In (b), 5 is at least 5-valent (why?), so $t=5$; $r \to t$, so $r=0$; $p \to s$ so $p=2$, 4, 6, 8, or 9, so $s=7$, 8, 10, 9, or 6 (from $s \to q+p+t+u$), and $q+p+t+u=23$, 9, 12, 6, or 10. Thus, $q+u$ is 16, 0, 1, -7, or -4, but only the first of these makes sense, because $q+u \geq 1+2=3$. Therefore, $p=2$, $s=7$, and $q+u=16$, and it follows that $q=10$ and $u=6$.

H95. In a single second, Romeo walks $1/36$-th of the ring, and Juliet $1/45$-th. Together they separate from each other $1/36+1/45=1/20$-th of the ring each second, which means that they first meet each other after 20 seconds (and then take time to do their waltz).

H96. Let $S=A+B+C=C+D+E=E+F+G=G+H+J$. Then $4S=(A+B+C+D+E+F+G+H+J)+(C+E+G)$. The first summand is equal to $1+2+\cdots+9=45$ and the second is at most $7+8+9$. Thus $4S$ is at most $45+24=69$. But 69 is not divisible by 4, so $4 \times S$ can be at most equal to 68.

H97. Let P denote the common product. If any of the numbers is 0 then P is equal to 0. But 0 can occur in at most two of the products. Thus, none of the letters can equal 0. Can any of the letters equal 5 or 7?

H98. Note that if you find a solution, then it's still a solution if you interchange A and B, so you may has well put A in the top-left corner. If you reflect about the main diagonal (from top-left to bottom-right), this swaps the row words with the column words and keeps the diagonal the same. So the other diagonal must be the same and is therefore a palindrome (reads the same backwards and forwards). It can't be ABBA, else we would have six words beginning with A (five words begin with A and five with B), so it's BAAB.

H100. Notice that the three corner triangles, together with the middle number, m, account for all the numbers 0 to 9, so that if s is the sum of a triangle, $3s + m = 45$. Then m must be a multiple of 3, so $m = 0$, 3, 6, or 9, and $s = 15$, 14, 13, or 12, respectively. In addition, if you find a solution, there's another one with each number x replaced by $9 - x$, so we need only consider $s = 15$ or 14, and $m = 0$ or 3.

H102. Why aren't there as many as six countries represented?

H103. Label the squares as shown.

16		d	
f	e	11	8
g		c	
a	15	14	b

The magic sum (the common sum of the rows, columns, and diagonals) is $(1+\cdots+16)/4 = 34$. The last row gives $a + b = 5$, so there are four possibilities for a and b. The third column and the main diagonal give $c + d = 9$ and $c + e = 18 - b$, which determine c, d and e.

H104. Color the squares in the usual checkerboard fashion (white square on the lower right) and consider the difference between the sum of numbers in the black squares and the sum of numbers in the white squares. What happens to this difference after a single operation?

H106. On Monday, Skip passed Harry for the first time after his father had run one-third of a lap. How much faster is Skip than Harry?

H107. The sum S is the same regardless of how the minus-signs are distributed. Prove this and determine the value of S.

H108. The remaining numbers are those that can be written with the digits 0, 1, 2, 4, 5, 7, and 8. To find how many numbers less than 1000 are left, insert leading zeros to make them all into 3-digit numbers. Then it's easy to see that there are $7 \times 7 \times 7 = 343$ of them.

H109. Disregard the streets and think of the friends as living on the same street. Then their locations are distributed as shown.

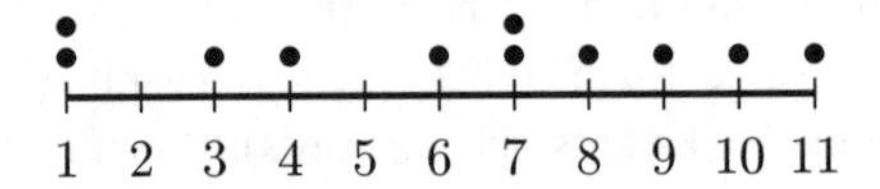

On which avenue should the friends meet to minimize their total walking distance? Now disregard the avenues and think of the friends as living on the same avenue.

H110. Originally there were 100 cells in the containers. The number of cells increased during the month (31 twenty-four hour periods)) by $653 - 100 = 553$. If we divide 553 by 31 we get $553 = 17 \times 31 + 26$. If there were 17 containers originally then the three new containers contained 26 cells on the evening of June 1. If the new containers were in the incubator for $k, k-1$ and $k-2$ days, they would contain $k+1, k$ and $k-1$ cells, respectively, on the evening of June 1. The total number of cells in these three containers would be $(k+1)+k+(k-1) = 3k$. We therefore have the equation $26 = 3k$, which does not have an integer solution. How can the argument be modified?

H111. The first arrangement can be made with six circles and the second with twelve.

To see how Euler might have arrived at this observation, suppose there are c circles required for the first arrangement. Then there are $V = 4c/2 = 2c$ points of contact (vertices of the planar graph) and $E = 4c$ edges (each circle is cut into four arcs). Suppose there are $F = c+t$ regions (that is, c circles and t other regions, t yet to be determined). Then **Euler's formula** (see the Treasury), $V - E + F = 2$, gives $2c - 4c + (c+t) = 2$, or $t = 2 + c$. The c circles each have 4 edges, and the t regions each have at least 3 edges, so $4c + 3t \leq 2E$ (each edge is counted twice in the count on the left) = $8c$. Therefore, $4c \geq 3t = 3c + 6$ (from above), so $c \geq 6$.

For the second arrangement, $V = 5c/2$, $E = 5c$, so if $F = c + t$, Euler's formula gives $5c/2 - 5c + (c+t) = 2$, so that $t = (3c+4)/2$. The c circles each have 5 edges and the t regions each have at least 3 edges, so $5c + 3t \leq 2E = 10c$. Therefore, $5c \geq 3t$, or from above, $5c \geq (9c+12)/2$, so $c \geq 12$.

H112. It is possible to construct a closed polygon whose six edges each intersect with exactly one other edge, and it is possible to construct a closed polygon whose seven edges intersect each other exactly twice. Polygons with six edges and two intersections or seven edges and one intersection cannot be constructed. Think about why this should be so.

H116. Each digit has to appear five times in the grid. It is therefore best to begin by placing the digits which already appear most often. In the first grid, 2 is such a digit, and in the second grid, 1 is such a digit.

H117. Show that such a number cannot have more than one digit.

H120. Let t_1, t_2, t_3, t_4, denote the four lines through one of the given intersection points, and s_1, s_2, s_3, s_4, s_5, be the five lines through the other special intersection point. Consider two cases: (a) one of the lines in the first set is also a line in the second set, and (b) no lines in the first set are in the second set.

H122. Let C, B, and K be, respectively, the number of coins, baseball cards, and pieces of candy in Elizabeth's hiding place (these values will change with time). How do the sums $C+B$, $C+K$, and $B+K$ change each time Dan visits her hiding place?

H123. Think of the chips as lying on a row of squares whose colors alternate between black and white.

H127. Yes, it can be done.

H129. Color every other space red. What happens to the number of pieces on the red squares after each move?

H131. When one undertakes to solve a problem of this sort, it's best to begin with a consideration of the digit which already appears most often in the original grid. For the first grid the digit 5 appears seven times. It is an easy matter to determine where the other two 5's must be placed (second row, seventh column, and third row, sixth column). Now one can continue to reason with the digits $3, 6$ and 8, each of which appears five times.

For the second grid, 5 again appears most frequently, followed by 1, 4, 6 and 8.

H132. An answer to the first question is given by the number 1213121 (this is the smallest number that satisfies the requirements of the problem). We can easily check that the number remains balanced when we add any of the digits 1, 2, 3 to the end of the number: 12131211, 12131212, 1213 1213. The answer to the second question, however, is another matter.

H133. (a) Denote the digits in the grid by a, b, c, d, as shown. The digits in the rows and columns can be read as 2-digit numbers, A, B, C, D, where $A = 10a + b$, $B = 10c + d$, $C = 10a + c$, and $D = 10b + d$.

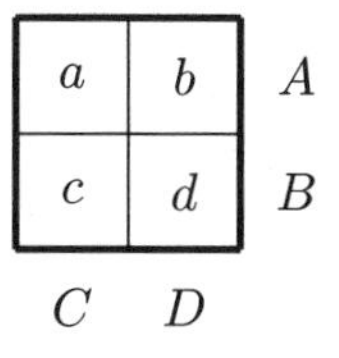

H134. If each Neptune has k heads then each Siren has $2k$ heads. The total number of heads must therefore be a number divisible by k. But 23 is a prime number, and since k cannot be 23, it must be 1. Therefore, a Neptune has one head and a Siren has two.

H136. Suppose the common row and column product, the *magic product,* is of the form p^n, where p is a prime. Then the numbers in the squares must each have the form p^i for some integer i, and if the numbers in the grid are different, one number would have to be at least p^8, which makes the overall common product much larger than 150.

Suppose the the magic product has the form $p^n q$ where p and q are distinct primes. The numbers in the grid have to be divisors of $p^n q$, so they must be among the numbers $1, p, p^2, \ldots p^n$, $q, pq, p^2q, \ldots, p^n q$. Therefore, the product of the 9 numbers in the grid must be at least as large as $1 \cdot 2 \cdot 4 \cdot 8 \cdot 16 \cdot 3 \cdot 6 \cdot 12 \cdot 24 = 5308416$. However, the cube root of this number, the magic product, is larger than 150.

The smallest number of the form $p^n q^m$ with p and q distinct primes and $m, n \geq 2$ is $2^2 3^2 = 216$. So, if such there is such an example, the magic product must have at least three prime divisors.

H139. Label the dimensions of the fields on the right and underneath as shown. Using this notation, $av = 12$, $aw = 8$, $bu = 20$, $bt = 25$, $cu = 14$, $cw = 15$, and $dt = 21$. The area of the acreage on which the farmhouse is built is dv, which we will also denote by x.

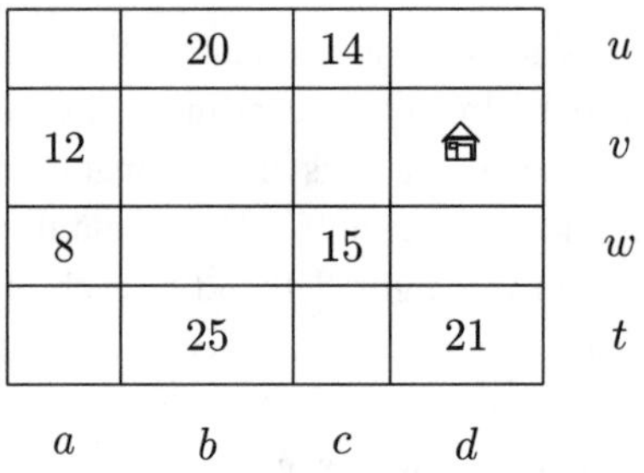

	20	14		u
12			🏠	v
8		15		w
	25		21	t
a	b	c	d	

H141. Because no number is divisible by 0, it cannot be one of the digits. Show that 5 cannot be one of the digits either. Can 9 be left out of the seven-digit number?

H142. (a) The first empty square in the last row must be filled with 1, 2, or 3 (otherwise the sum would be more than 16). It can't be 1 because then the equation in the second column would not be true. If it is filled with 2, then the equation in the second column would have to be $(12 + 6)/9 = 2$, and in the second row, $(9 + 6)/3 = 5$. Consequently, the last column would be $(12 \times 5)/x = 16$, which doesn't have an integer solution. Therefore, the last row must be $(4 \times 3) + 4 = 16$. Also, the equation in the first row implies that its two missing numbers must be equal.

(b) The sum of the two first numbers in the first row is at most 18 (each number is at most 9). After division by the number in the third empty square, we must get 7, and therefore, the third empty square must be filled with 1 or 2. If we fill it with 2, then the equation in the third column is $(2 + 9)/x = y$. Because 11 is a prime number, either x or y must be 11, which is not permissible. Consequently, the third empty square in the first row is 1 and the numbers in the first two squares must sum to 7.

In a similar manner, we can argue that the third number in the first column must be 1 or 2. If we place a 2 there, the sum of the first two numbers in that column must be 16, which is impossible since it implies that the top left number is 7, 8, or 9, whereas we know it to be at most 6. So the equation in the third row reads $1 \times 3 - 2 = 1$.

H143. Consider the players in two successive games. It should be clear that each boy played at least once in these two games.

H145. Denote the numbers by a, b, c, d, e, f, g, and suppose that they are written in increasing order; that is, $0 < a < b < c < d < e < f < g$. If the sum of the three largest numbers is at least 51, we're done. Can the sum be less than 51?

H146. Will there always be three rows with at least six dimes on them?

H147. First, for balance, try to assign a 0, 1, and 2 to the three edges adjacent to each vertex. The result, if it can be done, will require six 0's, six 1's, and six 2's. Once this is done, the numbers $1, 2, \ldots, 18$, numbers can be arbitrarily assigned in a manner consistent with their congruence class modulo 3 (for example, place 4 on any edge having value 1, 5 on any edge with value 2, 6 on any edge having value 0, and so forth).

H150. Note that each sheet has two (consecutive) page numbers. For (c), consider the remainder when one divides the sum of the torn-out sheets' numbers by four.

H151. We will first determine the common sum, S, of the numbers about each face. The sum of all the numbers is $1+2+\cdots+12 = 78$. But each edge belongs to two faces, so the sum of the six numbers about the faces, $6S$, is equal to $2 \times 78 = 156$. Therefore, the common sum, S, is 26.

H154. The data can be organized in a table. Let a, b, c, d, e, f denote the number of left and right shoes in the respective colors. The sum of the numbers in each column is 80 and in each row is 120. What are the possibilities?

	Color 1	Color 2	Color 3
Left	a	b	c
Right	d	e	f

H156. No, it is not possible. First, find a way of coloring the board so that each 2×2 square covers exactly one black square, whereas each 1×4 piece covers either 0 or 2 black squares.

H157. We need to know if the number of committee members is divisible by four. First show that the number of pairs of people of different gender who sit next to each other is an even number.

H158. Yes, it can be done.

H159. It can be done. Label the coins A, B, C, D, E. In the first weighing, compare coins A and B. Then, compare coins C and D.

H161. A total of $(8 \times 7)/2 = 28$ games took place in the tournament. Some team must have won four games (if not, then the number of games would be at most $8 \times 3 = 24$). Let A be a team with the most wins (there may be more than one team with the highest score).

H162. You should first determine the sum of the numbers each pentomino should cover.

H164. Draw an equilateral triangle and place six points along the perimeter as shown: three of the points are at the vertices of the triangle and three at the midpoints of the sides. If this triangle represents George's garden plot, then the distance between any two of these six points is at least ten yards. How large must the five tarps be in order to cover the entire garden plot?

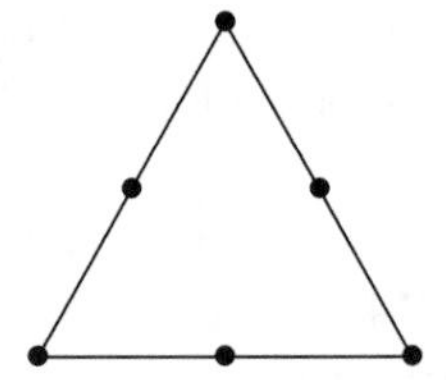

H166. Begin by distributing the delegates into the two conference rooms in an arbitrary manner. Let S denote the total number of bad relations in the two rooms. Can you decrease the value of S by moving people between rooms?

H167. Label the squares a, b, c, ... as shown. Let s be the side length of square (b). Find the side length of square (g) in two different ways.

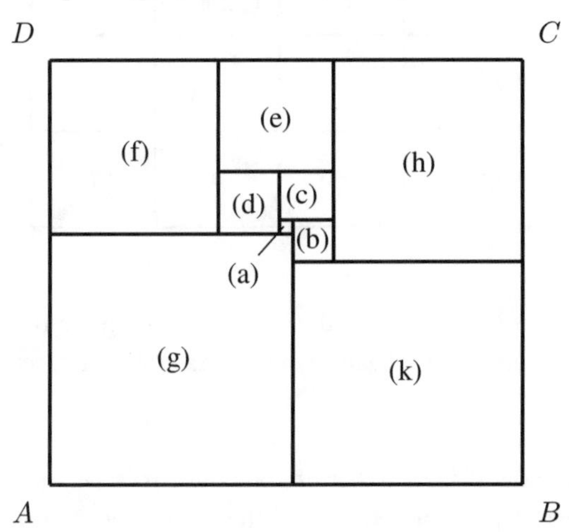

H169. Use the fact that if $n + 1$ players participate in the tournament then each plays n matches and the total number of matches is $n(n + 1)/2$.

H170. Let the numbers at the **a**pices (plural of apex), **b**orders, **c**enter, and **d**iagonals, be as indicated, and let S be the sum of the six numbers around each of the four triangles.

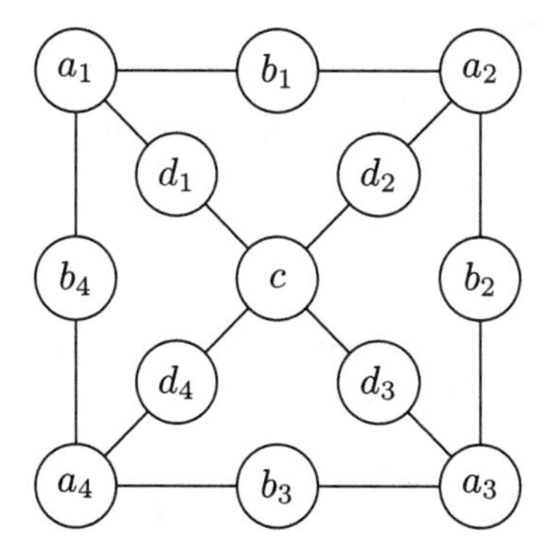

If we add all the numbers around each of the four triangles, we will have counted the apices and the diagonals each twice, the borders once, and the center four times. So $4S = 2(A + D) + B + 4c$, where we've written A for $a_1 + a_2 + a_3 + a_4$, and so forth. Of course, $A + B + c + D = 1 + 2 + \cdots + 13 = 91$, so that $4S = 182 - B + 2c$.

It is helpful to find the range of possible values for S. We'll make S as large as possible by making c as large as possible and B as small as possble. By taking $c = 13$ and $B = 1 + 2 + 3 + 4$, we get $4S = 198$, which doesn't have an integer solution for S. However, we can take $c = 13$ and $B = 1 + 2 + 4 + 5$, or $c = 12$ and $B = 1 + 2 + 3 + 4$, and this gives $4S = 196$, or $S = 49$.

For the smallest possible value for S, we note that every solution yields another one by replacing each number x by $14 - x$. This replaces S by $(6 \times 14) - S$, so that the smallest possible sum is $84 - 49 = 35$.

Can you find a solution with the maximum possible sum? the minimum possible sum? something in between?

H171. What are the possible sizes of the delegation?

H172. You can make the work easier by making a list of the full set of dominoes (note that (3,4) and (4,3) denote the same domino, so we use the convention that the first digit will not be larger than the second).

$(0,0)$	$(0,1)$	$(0,2)$	$(0,3)$	$(0,4)$	$(0,5)$	$(0,6)$
	$(1,1)$	$(1,2)$	$(1,3)$	$(1,4)$	$(1,5)$	$(1,6)$
		$(2,2)$	$(2,3)$	$(2,4)$	$(2,5)$	$(2,6)$
			$(3,3)$	$(3,4)$	$(3,5)$	$(3,6)$
				$(4,4)$	$(4,5)$	$(4,6)$
					$(5,5)$	$(5,6)$
						$(6,6)$

The best way to begin this problem is to find the dominoes that occur only one time. For example, which of the doubles (0,0), (1,1), ..., (6,6) can your find for sure? Not (1,1) or (2,2). But when you've found the others, you'll know where (3,4), (2,6), and (2,2) are.

H173. (a) Can you locate the positions for dominoes (1,1), (4,4), and (6,6)? And then, what covers the 6 below (4,4)? When you have fixed a domino, say (X,Y), then it helps to separate other occurrences of X,Y pairs in the rectangle, say by drawing a thick line between them.

(b) Can you locate the positions for dominoes (0,5) and (1,6)? Then place (2,4), (2,6), and (2,5).

H174. (d) Let k denote such a number and let $K = k + 1$. Pick one of the men at the table and label him with an M. Now go around the table, say clockwise, and place an M or W on every Kth person, depending on whether that person is a man or a woman, respectively. Assume that there do not exist two men or two women with exactly k people between them. Then the labels will be assigned alternately: $M, W, M, W, \ldots$.

H175. Label the rows and columns as shown.

	a	b	c	d	e	f	g
G			2				
F					1		1
E			3				
D	1				4		1
C			1				
B	3		2			1	
A			1				

(a)

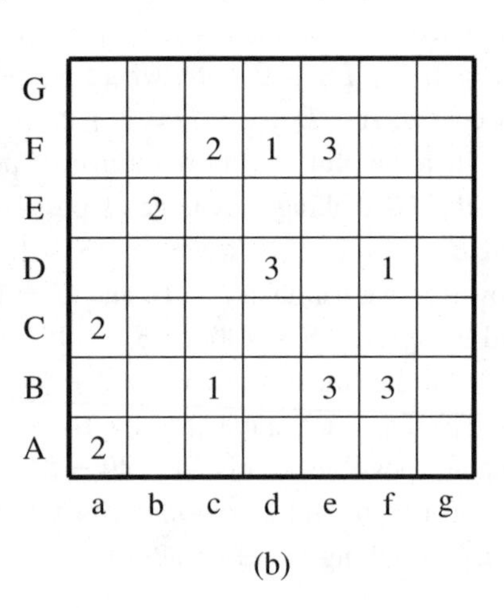

(b)

Write Ab = 0 or 1 according to whether Ab (the square in the bottom row, second column) is empty, or contains a coin. The number 2 on square Bc implies that Ab + Bb + Cb + Ad + Bd + Cd = 2, and we'll abbreviate the left side of this expression by (Bc). Thus, (Ac) = 1 is an abbreviation for Ab + Bb + Ad + Bd = 1, and (Bc) − (Ac) = Cb + Cd = 1. Put this in (Cc) = 1 (that is, Bb + Cb + Db + Dc + Bd + Cd + Dd = 1), and we see that Bb = Db = Dc = Bd = Dd = 0.

For (b), the equation (Bc) = 1 (that is, Ab + Bb + Cb + Ac + Cc + Ad + Bd + Cd = 1) implies that Ab + Bb is at most 1, whereas (Aa) = 2 (that is, Ba + Ab + Bb = 2) implies that Ab + Bb is at least 1. Therefore, Ba = 1 and Cb = Ac = Cc = Ad = Bd = Cd = 0.

H176. Keep track of the floor numbers to eliminate impossible sequences; for example, AABAAA is an impossible sequence. Thus, for every initial string of symbols, there should always be more A's than B's, but not more than three more. One way to keep track of this is to draw a picture of the position of the elevator after each step. For example, the diagram shows an elevator ride that lasts $5\frac{1}{2}$ minutes before stalling.

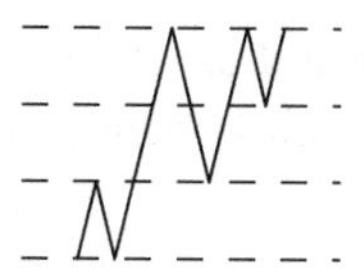

Elevator sequence: ABAAABBAABA

H177. Take any 2×2 square on the board and suppose that either one or three of its squares are black. It is not difficult to see that the number of black squares on this 2×2 square will always be an odd number.

The grids shown below contain one, two, three, and four black squares, respectively. By the preceding observation, the number of black squares on these boards cannot be reduced by our operation, so that $1, 2, 3, 4$, are minimal numbers for certain colored boards. Can 5 or more black squares be minimal?

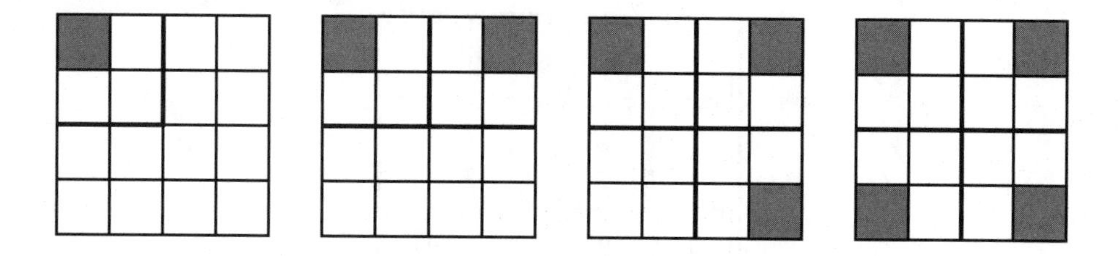

H178. The key idea is to think about how the number of odd numbers changes after each operation.

H179. The method of solution is the same as in the preceding problem.

H181. Make a hexagon from two copies of the quadrilateral, one rotated through $180°$ about the midpoint of an edge.

H182. Make use of the rule for divisibility by 3. Write A in the form $A = 3a + r$, where r, the remainder, is 0, 1, or 2. What remainders are possible when A^2 is divided by 3?

H185. If there's one more 3 in the pattern (near the middle) it can be done,

$$202020343434\underline{3}25252523838384 = 18181818181818 \times 11111118888888$$

but this isn't our number.

H186. Think about the rule for divisibility by 9.

H187. Label the squares of the board as shown on the left, and draw a graph of 14 edges, on the right, representing the 28 possible knight moves. The original problem is seen to be equivalent to a sliding block puzzle in which the pieces can slide one square to the left or right in the top and bottom rows or up and down in each of the four columns.

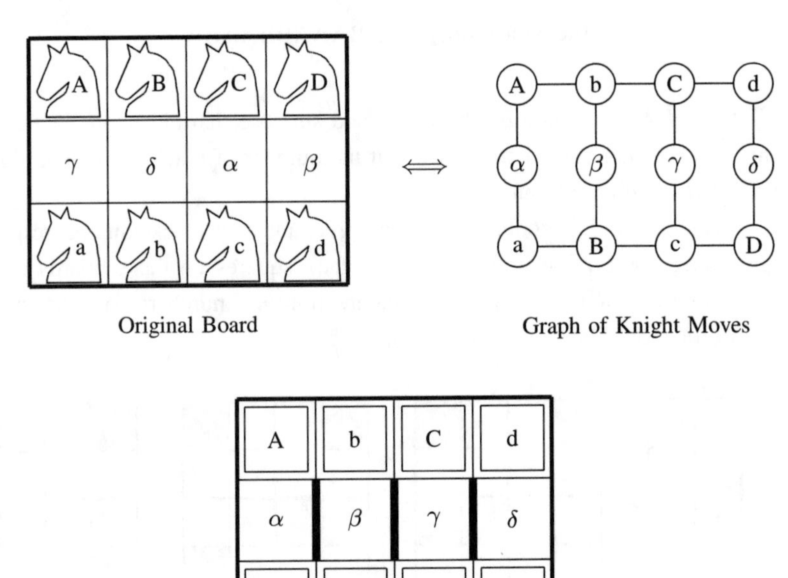

Original Board

Graph of Knight Moves

Sliding Block Puzzle

On either board the moves are simply described as: Move between squares with corresponding roman and greek letters or between squares with neighboring roman letters of differing case. The object on each board is to swap the pieces on the roman letter squares with the pieces on the roman letter squares of the other (upper or lower) case.

H188. Instead of trying to draw a continuous path from the start, it is better to draw those segments that are strictly determined, and when that is done, one can attempt the complete solution. It is clear, for example, how the segments should be drawn around a small square containing the number 3 if the square next to it contains the number 0, as in (i). It is also clear that if two neighboring squares both contain the number 3, then there must be a segment between these two squares and an additional two segments, one on opposite sides of the two squares, see (ii). One can discover several other rules for example (iii) and (iv); here's just one more. If a small square contains the number 2 and one of the corners of that square already has two segments attached to it as in (v), then the segments that belong to that square are strictly determined.

0 3 ⟹ 0 3   3 3 ⟹ 3 3 or 3 3

(i)   (ii)

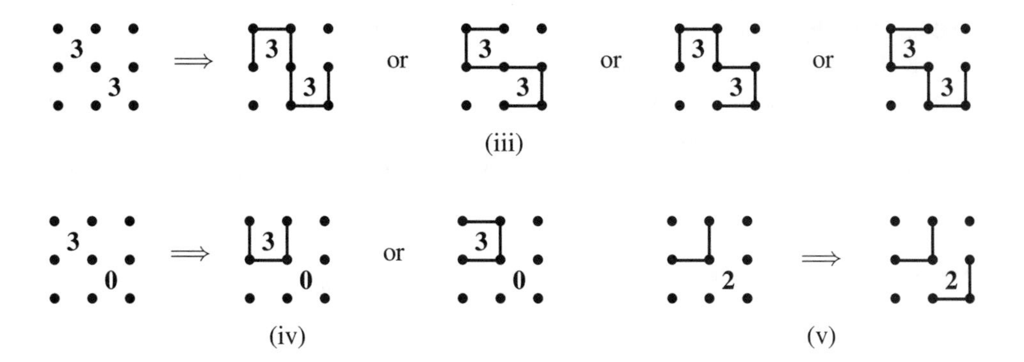

H189. Can you find out which of the two numbers is larger without actually counting them?

H190. The apartments and the families can be numbered from 1 to 14. The numbers within the circles represent the families, and the numbers beside the circles represent the apartment numbers.

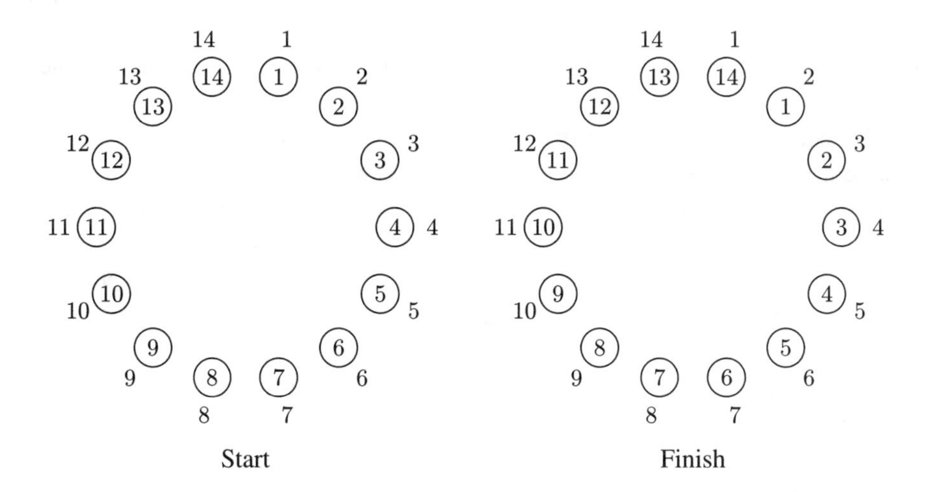

It is possible to organize the exchanges in pairs so that no resident will have to move more than twice.

H191. (a) Because the product is a five-digit number, A must be 1 or 2. The A occurring in the answer must be even, so $A = 2$. It follows that E is 3 or 8, and therefore B must be odd (the carry from the product $E \times 4$ is odd), and less than 5 (otherwise the product would be a six-digit number). Furthermore, C is 8 or 9.

(b) Because the tens product is a three-digit number and the multiplier is at least 22, the first digit of the multiplicand is 1 or 3. It cannot be 1, because the units product begins with at least a 2. Thus, the multiplicand begins with 3, and the multiplier with 2. It follows that the units product begins with 2, the tens product with 6, and the final product with 9. Furthermore, the units digit of the multiplier is 6 or 8.

H192. Start by placing coins in some arbitrary way on the bottom row. For example, you might place coins on the first, second, fourth, and fifth squares (from the left) in the bottom row. Now work your way up row by row, placing coins as necessary to remain consistent with the stipulations of the problem. Either you will reach an inconsistency, in which you can backtrack to the last step where you had a choice, or you will reach a solution.

H194. Do you remember the formulas $x^2 - y^2 = (x - y)(x + y)$ and $x^3 - y^3 = (x - y)(x^2 + xy + y^2)$? (See **Algebra** in the Treasury.)

H195. It takes 20 minutes to shoe one horse. In each five minute interval the eight blacksmiths can work on eight different horses, while two horses are unattended. Is it possible to arrange the work assignments so that a single horse will be unattended for at most one five-minute interval?

H196. It is best to begin by considering "words" where there are not so many alternatives, for example, the second word in the last row. The number 4 can be written as $3 + 1$ or $1 + 3$. The first alternative can be rejected because the vertical word labeled by the number 3 must have the sum $1 + 2$ or $2 + 1$. Therefore, we are left with $4 = 1 + 3$, and the last vertical word in the sixth column is 21. We can now continue with the last words in the two last columns.

We could also begin in the first row. There, the first word must be 9 7 or 7 9. But because the fourth column must be 8 9 or 9 8, the choice is easy. It follows that the first column begins with 9 8 (not 8 9, else there would be two 8's in 30 across).

H197. (b) Divide the campers into two groups: let group A be those children who had gotten to know an odd number of other children, and let group B be the other children. Suppose that each child makes a mark on a piece of paper for each camper he/she has gotten to know. The children in group A will therefore have an odd number of marks on their papers, whereas those in group B will have an even number of marks on theirs. Now collect the papers in group A, and separately, the papers in group B. Let A_1 and B_1 denote the number of marks on the papers in A and B, respectively. Consider the number $A_1 + B_1$. This must be an even number because each "acquaintance" among the campers contributes two marks (one for each of the two children who are acquainted with each other). Furthermore, B_1 is an even number (a sum of even numbers), and, because $A_1 = (A_1 + B_1) - B_1$, A_1 is also an even number. But A_1 is a sum of odd numbers, so there must be an even number of terms, that is, an even number of children in group A. We have therefore shown that the number of children with an odd number of acquaintances at the camp is an even number.

H198. Label the vertices A, B, C, D, A, B, C, D, ... around the icosagon. How many of the nine points will have the same label?

H201. In how many ways can you choose numbers from 1 to 9 which add to 15?

H203. The answer depends upon whether either r or c is odd.

H204. Would you like to have a multiple of seven empty squares in front of you?

H205. The answer depends upon whether either r or c is even.

H207. We can analyze the problem by working backwards. Beginning at the lower right square, mark it with a P because the Previous-player has just won. Then mark each neighboring square with an N because the Next-player can win, and so on.

H210. Think of the hexagonal grid as made up of three sets of parallel lines, each set at 60° with each other; the toads jumps are similar to the jumps of the knight on a chessboard: two in one direction and one in another. For example, from square A the toad can jump to the pads as shown.

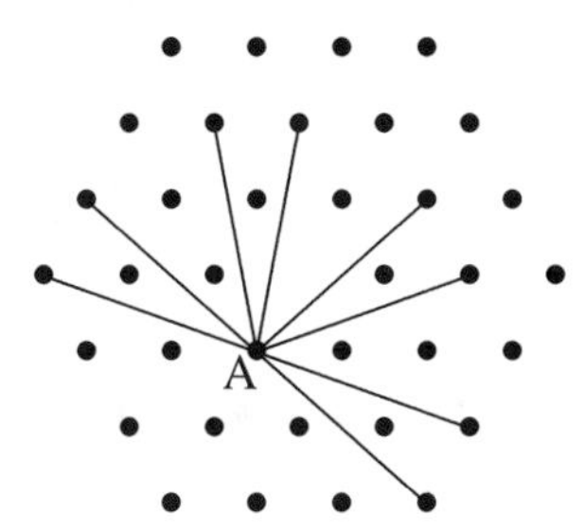

H211. Astonishingly, there is such an n. As a first step, start with the natural packing of two identical rows of n balls, and then, disregarding the boundary on the right, nudge the top row over as shown.

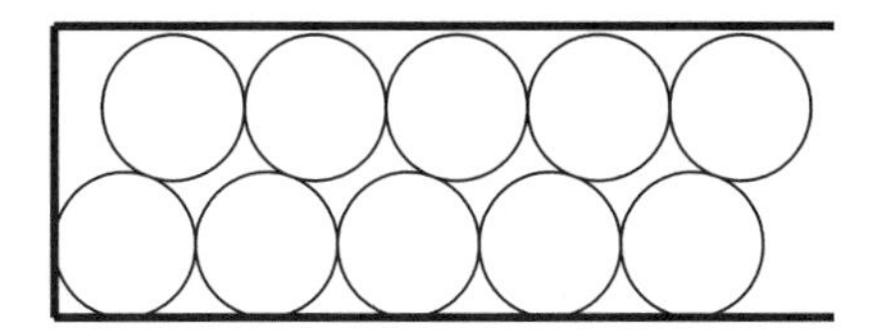

H213. Because $600 = 85 \times 7 + 5$, there are 85 numbers between 1 and 600 divisible by 7, which means that 600! is divisible by 7^{85}.

H214. Because $7! = 7 \times 6 \times 5 \times 4 \times 3 \times 2 \times 1 = 5040$ is a four-digit number, the digits of A must be smaller than 7. If one of the digits is 6, then $a! + b! + c!$ must be larger than $6! = 720$, but the largest possible value for A is 666. Therefore, no digit can be more than 5.

H216. (a) Let a denote the first digit of A. Show that $B = 10A - 99999a$.

H217. Number the columns of the chessboard from left to right and the rows from bottom to top with the numbers $1, 2, \ldots, 8$. Each square on the board can be identified by an ordered pair (a, b) of numbers, where a is the column number and b is the row number. It is easy to check that the squares with odd coordinate sums, $a+b$, are white (recall, by convention, the lower-left corner of the chessboard is always black). What happens if there is an odd number of queens on the white squares?

H219. Number the points from 0 to 20 as in the figure. The distances between these points have only ten different values. The figure shows these ten different values, by considering the possible distances achievable from the point labeled 0. The distance between point 0 and point 14, for example, is the same as the distance between 0 and 7.

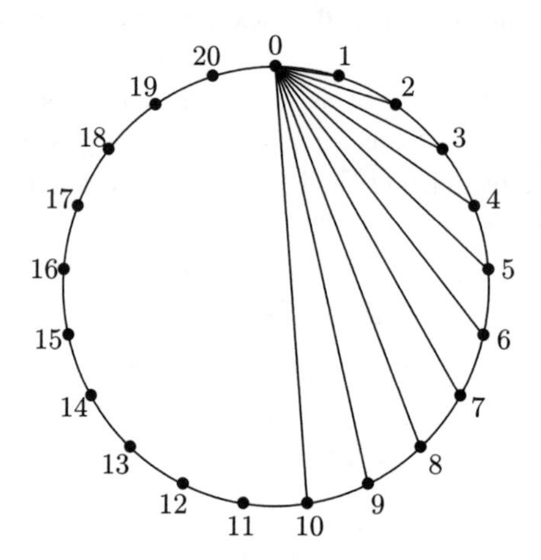

There are three distances between the vertices of a triangle; there are six distances between vertices of a quadrilateral, and there are ten distances between the vertices of a pentagon. The figure shows that we have ten different distances between the vertices of the 21-gon. Is it possible to find five points so that all ten distances between them are different?

H220. (a) If there is an island B that you can't boat to from A with at most one stop, then you can fly from A to B. Now what?

(b) Suppose it were not possible. Then there are two pairs of islands, (A,B) and (C,D) such that we cannot get from A to B by boat with at most two stops, and we cannot get from C to D by plane with at most two stops. Show that this leads to a contradiction.

H221. Consider Mainstay and all the islands to which you can fly to from Mainstay, either directly or indirectly. Let S be the total number of outgoing flights from the islands in this group. Is S odd or even?

H222. There is no information given concerning the number of roads leading into or out of Lofton.

H223. For two towns X and Y, let X $\to$ Y mean that the road is one-way from X to Y. Take an arbitrary town, and give it the number 1. Now choose another town, say X. Either 1 $\to$ X or X $\to$ 1. In the first case, give X the number 2; in the second case, let X be numbered 1, and re-number the first town 2.

Now choose a third town Y. There are three cases: Y $\to$ 1 or 2 $\to$ Y or 1 $\to$ Y $\to$ 2 (the first two cases are not mutually exclusive, since it might happen that both Y $\to$ 1 and 2 $\to$ Y).

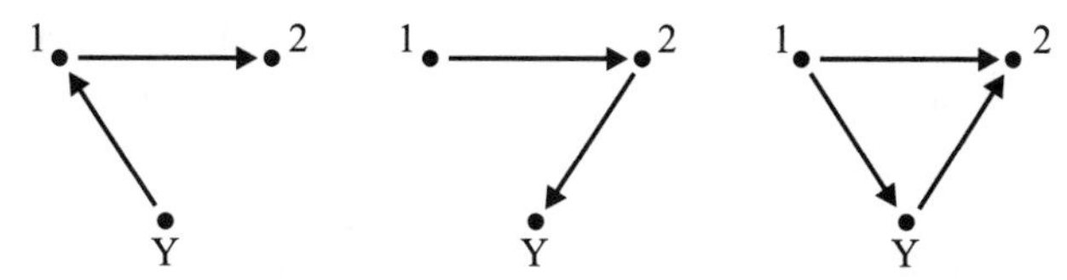

In the first case (or even if both first and second cases hold), give Y the number 1 and increase the numbers of the other two towns by 1. In the second case, give Y the number 3. In the third case, let Y be number 2 and increase the number of the town which had number 2 by 1. Now the three towns are numbered so that 1 $\to$ 2 $\to$ 3.

H225. Let A and B be two towns on the island, and let $\mathcal{A}$ and $\mathcal{B}$ be the sets of all towns to which one can *get to* directly from A, and *from which* one can get directly to B, respectively.

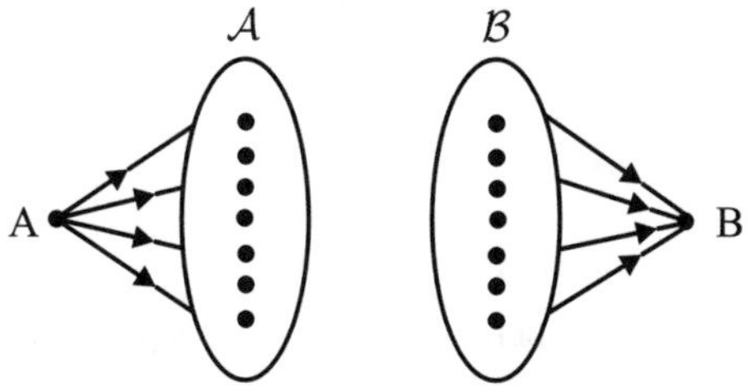

(a) Because 36 roads meet in each town, 18 of them are incoming and 18 are outgoing. It follows that both $\mathcal{A}$ and $\mathcal{B}$ contain 18 towns.

(b) In this case, $\mathcal{A}$ and $\mathcal{B}$ each contain 14 towns. Show that if there is no town in both $\mathcal{A}$ and $\mathcal{B}$ then there must be a town C on the island, for which, using the notation of **P223**, E $\to$ C $\to$ F for some towns E in $\mathcal{A}$ and F in $\mathcal{B}$.

H226. We will first establish that at least one person exchanged business cards with at most five others at the club. To show this, we will assume otherwise, and suppose that A exchanged cards with six people B, C, D, E, F, G, and possibly others. According to the conditions of the problem, no two of B, C, D, E, F, G exchanged cards with each other. Because we are assuming that no one exchanged cards with fewer than six others, there are five others (besides A) who exchanged cards with B. But with these five persons, the number of people present at the club would have to be twelve, which is contrary to our assumption that only ten club members were present. Consequently, there must be someone who did not hand out more than five cards.

H227. (b) If we add the same thing to each of the numbers, then we add twice that something to the sum of each pair, so we may as well assume that one of the numbers is 0.

H228. After going around the circle one time, 34 advisors had been taken away, and 34 remained, specifically, those advisors numbered $1, 3, 5, \ldots, 67$. The next round of questioning went just as before, starting at advisor 3 and again questioning every other remaining advisor. To simplify matters, re-number the advisors, by adding 1 to their present number and then dividing by 2. Their new numbers (those who remain) are $\frac{1+1}{2}, \frac{3+1}{2}, \frac{5+1}{2}, \ldots, \frac{67+1}{2}$, that is to say, $1, 2, 3, \ldots, 34$. Now the questioning proceeds as before, starting with advisor 2. In the second round, just as before, those with numbers $2, 4, 6, \ldots$ answered incorrectly and were taken away for remediation. Those who remain have numbers (their latest numbers) $1, 3, 5, \ldots 33$. We can go through this same procedure one more time. However, the next time around is slightly different.

H229. $\mathrm{MI} = \dfrac{\mathrm{SOL}}{\mathrm{SI}} \le \dfrac{100\mathrm{S} + 98}{10\mathrm{S} + 1} \le 10 + \dfrac{88}{11} = 18$. So $\mathrm{M} = 1$. Now find I and S, knowing that they are different and greater than 1, from the multiplication:

$$\begin{array}{r} \mathrm{S\ I} \\ \times\ 1\ \mathrm{I} \\ \hline \mathrm{S\ I}\ \ \\ ?\ \mathrm{L} \\ \hline \mathrm{S\ O\ L} \end{array}$$

H231. The top four players have accumulated $4\frac{1}{2} + 3\frac{1}{2} + 3 + 1\frac{1}{2} = 12\frac{1}{2}$ points. If there are k players in the tournament, there will be $\binom{k}{2} = k(k-1)/2$ games for a total of $k(k-1)/2$ points. For $k(k-1)/2$ to be at least $12\frac{1}{2}$, k must be at least 6.

H232. (a) We do not know how many coins are in each bowl at the start. Nevertheless, we wish to get an equal number of coins in each bowl. Consequently, we need to describe a way to add coins so that the number of coins increases the most in the bowls which contain the fewest coins. It suffices, therefore, to show how we can repeat the operation a number of times so that the number of coins in one specified bowl increases by, say, $k+1$ coins, while the number of coins in the other six bowls increases by only k coins.

H233. (a) Because $1+2+3 > 5$ and $7+8+9 < 25$, the sum of the units column is 15. What are the sums of the tens and hundreds columns, and the sum of all three columns?

(b) Because the units product has only three digits, and a is not 1 or 3, $a = 2$.

$$\begin{array}{rrrrr} & & 3 & * & e \\ & \times & d & * & a \\ \hline b & c & * & 3 & \\ & 1 & * & * & * \\ & & * & * & * \\ \hline 3 & * & * & * & * \end{array}$$

Because b cannot be 3, it must be 2. At most 2 can be carried from the thousands column, so $c \ge 7$. Then $d \ge 7$ and $de \equiv 3$ modulo 10, so d and e are 7 and 9 in some order.

H234. Notice that Peter has made two claims.

H235. Let points A, B, C, D, E, F divide the circumference of a disk into six equal arcs.

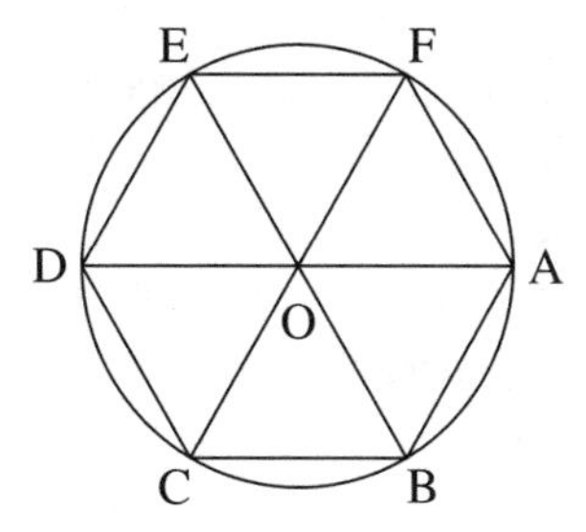

Each of the small triangles (for example, triangle AOB) is equilateral.

H236. Because there are four corners of the square and only three sentries, at least two of the corners must be guarded by the same (closest) sentry.

H237. The final answer does not depend on the order in which the pairs of numbers are chosen.

H238. It is clear that the 10×10 grid can be tiled with 25 copies of the square tetromino. It is, however, not possible for any of the other four types. Why?

H240. We need to find a way to attack this sort of problem. We can begin, for example, by looking for blocks that are completely determined, for example, the fifth column and the fifth row in the first problem, and columns three and five in the second problem, where $2 + 8 = 10$ and at least one white square between them accounts for all 11 squares in those columns). One can also fill in parts of "large" blocks. For example, in the second column of the first problem, the middle four squares must be in the same block of seven. Similar arguments work elsewhere. For example, the fourth and fifth squares of row ten in the first problem must belong to the same block. One can also sometimes fill in squares between two already colored squares. For example, in the first row of the second problem, if we had colored columns three and five, then we should color column 4, since there is only one block in this row. Now we can complete column four.

H241. It is easiest to start with the last row. There are 18 black squares in this row, divided into five blocks, which means there are four white squares in this row, exactly the four needed to separate the five blocks from each other.

H242. Here it's best to work modulo 13 (see **Modular arithmetic** in the Treasury).

H243. Yes, it is possible. First, suppose we disregard the restrictions on the number of animals that can be put into each cage. Now place animals in the following way. Place them one at a time, and suppose we now need to place animal D. Because at most three animals are incompatible with D, and these can be in at most three different cages, there must be a cage in which D can be placed.

Once all the animals have been placed in cages in a compatible way, we can think about how to move them about so that there will be exactly two animals in each cage.

H244. Suppose the result is not true. Then if all the guests simultaneously move either $0, 1, 2, \ldots, 15$ places to the right (say), *at most* one person will be in their proper place. Because each of the sixteen guests come to their proper place exactly once in this procedure, it must be the case that after each of these 16 moves, *exactly one* person will be in their proper seat. If we then determine, for each person, the "distance" from their proper seat (all measured from the beginning position, in the direction to their right), we get each of the numbers $0, 1, 2, \ldots, 15$ one time.

Imagine, now, that one of the guests goes (to their right) to the proper seat. Then imagine that the person in *that* place goes (always to their right) to the proper place, and continuing, each person going to their proper place after getting "bumped" by another, until eventually one of the persons in the chain arrives at the empty chair vacated by the first guest. Each person in this chain moves their own "distance" (defined above), and the total of the distances in this chain of moves will be a multiple of 16 (because the chain ends at the same place in which it started).

H246. First compute the number for a smaller rectangle, for example, a rectangle of dimensions 2×2, 2×3, or 2×4. Can you find a relationship between the number of ways of tiling the 2×4 rectangle and the number of ways of tiling the 2×2 and 2×3 rectangles?

H247. (a) Consider the remainders of the numbers 1, 11, 111, 1111, 11111, and so forth, when they are divided by 2010. The remainders are 1, 11, 111, 1111, 1061, 561, 1591, and so forth. At some point, two of these remainders will have to be the same.

(b) Use the same reasoning as in (a) and observe that 2011 is a prime number.

H249. (b) The sum of the digits, of course, cannot be larger than 27. We will consider three cases, depending on which remainder we get when n is divided by 3: $n = 3k + r$, where r is 1, 2, or 0.

Suppose that $n = 3k + 1$. If $k = 0$, then $n = 1$ and there are only four numbers whose digits add to 1: 1, 10, 100, and 1000. If k is between 1 and 7, there are six different numbers whose digits are $k - 1$, k, and $k + 2$ (if a, b, c are three different digits, there are six ways to arrange them, and each has the same digit sum). In addition, there are three different numbers with digits $k - 1$, $k + 1$, and $k + 1$, and three others with digits k, k, and $k + 1$. Already we have twelve different numbers, and there are others (for example, $k - 2$, $k + 1$, and $k + 2$ if k is larger than 2), all of which have the same digit sum. Because we are only interested in those n with exactly ten different numbers which satisfy the stated condition, we must consider other alternatives. If $k = 8$, then $n = 25$. But in this case there are only six different numbers whose digits sum to 25: namely, 889, 898, 988, 799, 979, and 997. Thus there are no numbers of the form $3k + 1$ which satisfy the conditions of the problem.

Now consider the cases $n = 3k + 2$ and $n = 3k$.

H250. (a) Choose a player A who wins the most games in the tournament. Show that if A loses to another player B then there is some other player C such that A $\to$ C $\to$ B.

(b) Suppose there are exactly two gold medal winners A and B, and that A $\to$ B. Then there is another player C so that B $\to$ C $\to$ A (because B received a gold medal). Consider the set $\mathcal{V}$ of all those players V who defeated A.

H251. First, suppose there are three streets that intersect at some place. Then they each intersect only $K-3$ other streets, so they couldn't have $K-1$ houses on them. Therefore, each house is on exactly two streets. There's a house numbered 1 on each street, and as we've seen, each of these houses (those numbered 1) can be uniquely paired with two streets. This means that K, the number of streets, must be an even number. Now show that K can be any even number.

H252. Number the lines as shown in the figure. Each intersection point can be identified by a triple of numbers: the first number indicates which down diagonal it lies on, the second number which up diagonal it lies on, and the third which horizontal it lies on. For example, the triple $(6, 1, 3)$ identifies the position of the marked point in the figure. Notice that the three numbers, which are called *trilinear coordinates*, always add up to 10.

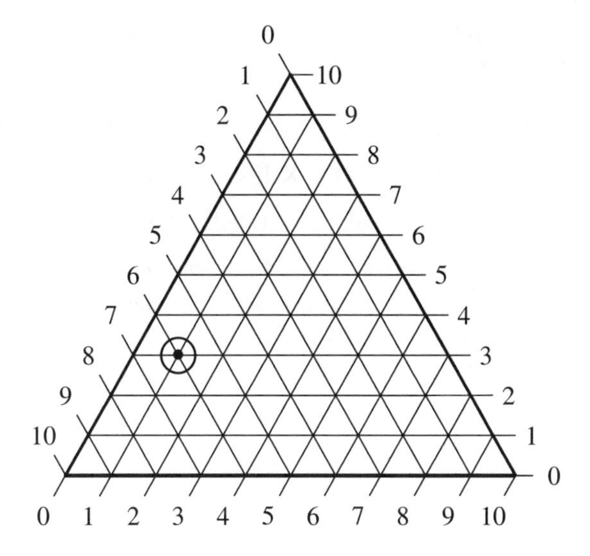

Suppose that we have placed k coins according to the rules, and that their positions are given by the triples $(a_1, b_1, c_1), (a_2, b_2, c_2), \ldots, (a_k, b_k, c_k)$. The first coordinates, $a_1, a_2, \ldots, a_k$ must all be different, because the coins are on different lines, and so they must be at least as large as $0, 1, 2, \ldots, k-1$, and their sum, $a_1 + a_2 + \cdots + a_k$, must be at least $0 + 1 + 2 + \cdots + (k-1) = k(k-1)/2$.

H253. One possible strategy for finding the number is as follows: Begin by writing down the numbers from 1 to 16 in a row.

Question 1. "Is your chosen number between 9 and 16?" If my answer is "No," draw a line under each of the numbers $9, 10, \ldots, 16$. If my answer is "Yes," put a line under the numbers $1, 2, \ldots, 8$. Because of symmetry, let us consider only the case where my answer is "Yes." You then have the following diagram.

$$\underline{1}\ \underline{2}\ \underline{3}\ \underline{4}\ \underline{5}\ \underline{6}\ \underline{7}\ \underline{8}\ 9\ 10\ 11\ 12\ 13\ 14\ 15\ 16$$

Question 2. "Is your chosen number between 5 and 12?" If the answer is "No," then draw a line under the numbers $5, 6, \ldots, 12$, and if the answer is "Yes," draw a line under numbers $1, 2, 3, 4$, and $13, 14, 15, 16$. Again, because of symmetry, suppose my answer is "Yes." You then have the following row.

$$\underline{\underline{1}}\ \underline{\underline{2}}\ \underline{\underline{3}}\ \underline{\underline{4}}\ \underline{5}\ \underline{6}\ \underline{7}\ \underline{8}\ 9\ 10\ 11\ 12\ \underline{13}\ \underline{14}\ \underline{15}\ \underline{16}$$

Because I have twice indicated that my number is not $1, 2, 3$ or 4, we can suppose that it is true. Therefore, you can eliminate these numbers from the list, and consider only the numbers in the following list.

$$\underline{5}\ \underline{6}\ \underline{7}\ \underline{8}\ 9\ 10\ 11\ 12\ \underline{13}\ \underline{14}\ \underline{15}\ \underline{16}$$

H254. It is clear that five questions will suffice. If, for example, you give me the numbers $b_1 = 1$, $b_2 = b_3 = b_4 = b_5 = 0$, then I will return the value of a_1. With four other such questions you can find the four other numbers. But there is a method for finding my five numbers with fewer questions.

H256. Hal noticed that Bishop Hog had written on top of the 13 coins the letters

D E F R O C K I N G H A L

He also noticed that he had asked his assistant to do the following weighings

	Coins put on the	
	Left-hand pan	Right-hand pan
First weighing	C O I N	F A K E D
Second weighing	F O R	A C H I L D
Third weighing	——	F I N D E R H O G

3
Solutions

S1. When Sarah steps onto the scale with Dan, the indicator increases by Sarah's actual weight, and therefore Sarah weighs $22.5 - 10 = 12.5$ kilograms. Thus, the scale is set $14 - 12.5 = 1.5$ kilograms too high.

S2. A formula is not necessary. Place the bricks as shown in the figure below. With the ruler, measure the distance between points A and B, which is the desired length.

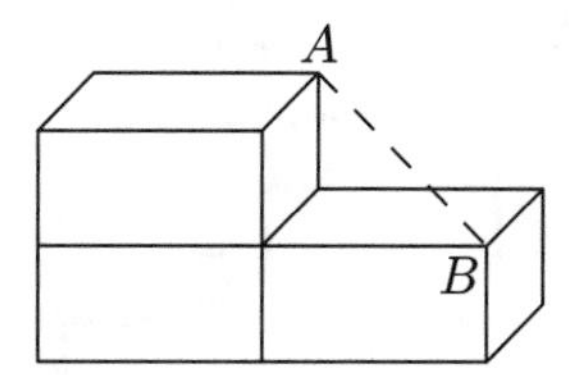

S3. Because Sven finished in the middle, there must have been an odd number of runners (Sven, plus those ahead of him, and an equal number behind him). Because Lars finished sixteenth, there must have been at least seventeen runners. There couldn't have been more than seventeen runners because Sven placed ahead of Dan, who finished tenth. Therefore, Sven placed ninth, and there were seventeen runners.

S4. The dominoes in each of the following arrays are forced in the order indicated on the dominoes. That is to say, those dominoes labeled 1 are forced at the first stage of consideration, and these in turn force those labeled 2, which force those labeled 3, and so forth.

(i)

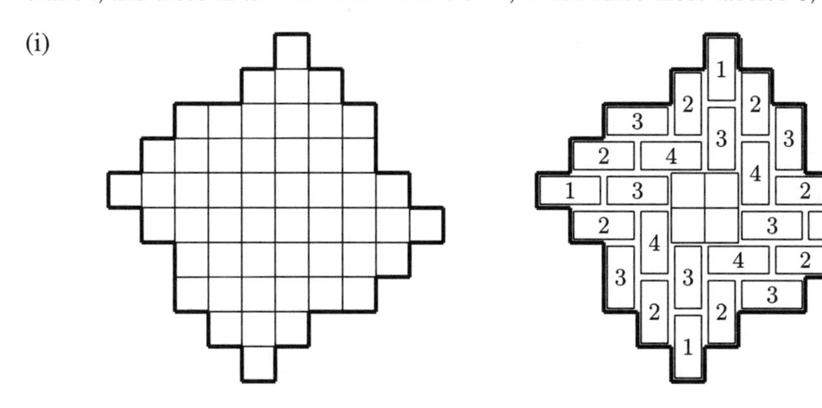

The remaining four squares can be covered in two different ways by two dominoes.

(ii)

In this case there is a unique covering.

(iii)

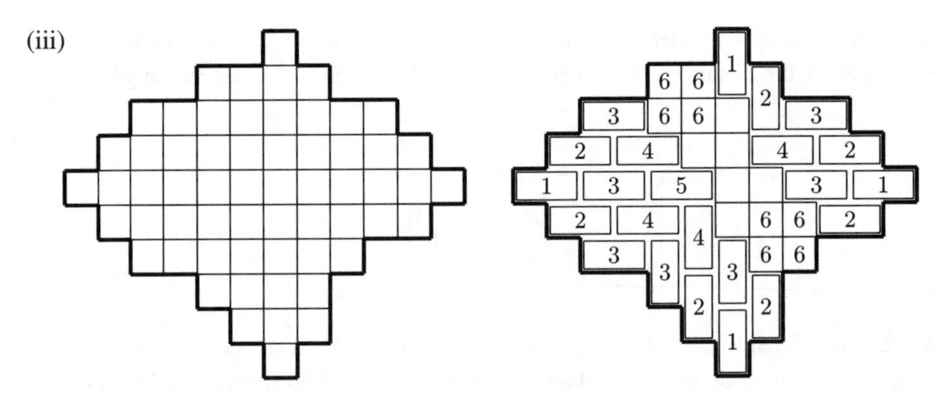

If there is a covering, each of the two sets of four squares labeled 6 must be tiled with two horizontal or two vertical dominoes. It is then easy to check that the remaining six squares cannot be covered.

S5. One cannot turn A, or B, at all! The overall system of gears cannot work. The number of gears in the ring at the right is nine, an odd number. For such a ring to function, there must be an even number of gears, because every other gear rotates clockwise, and those in between rotate counterclockwise.

S6. Yes, it is possible to measure a thirty-minute time interval with these two sand-glasses. Here's one way: begin by setting the 13-minute-glass. When it runs out, set the two glasses simultaneously. When the 9-minute-glass runs out (after $13+9$ minutes), turn it over and start it again. After four more minutes, the 13-glass runs out (after $13+9+4$ minutes), and now, turn the 9-glass over again. There are four minutes of sand on the bottom, and when it runs out we will have $13+9+4+4=30$ minutes.

The following diagram describes the steps more succinctly. Here, the pair (a,b) means that there are a minutes of sand *remaining* (in the top chamber) in the 9-glass, and b minutes *remaining* in the 13-glass. At time 0, we are in state $(0,0)$.

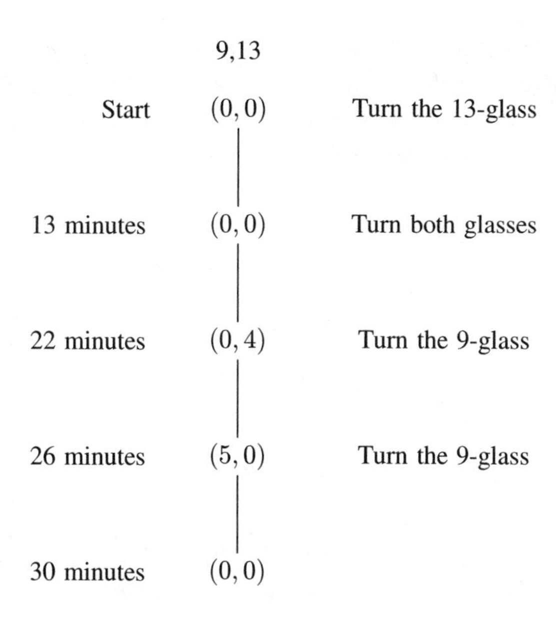

Q6. What lengths of time can be measured under these stipulations?

S7. Let n denote the number of school children who play both soccer and tennis. Then seven times as many, $7n$, play soccer, and nine times as many, $9n$, play tennis. It follows that $6n$ play soccer only, and $8n$ play tennis only; so there are $n + 6n + 8n = 15n$ children in total. The proportion of tennis players to children is $9n$ to $15n$, or $3/5$ (more than half).

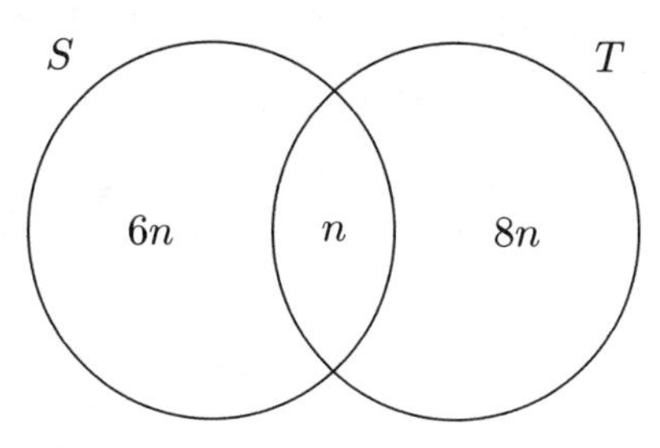

S8. We see that each day (from the second), the parrot repeated the same word as the day before followed by a similar word with "O" and "K" interchanged. Consequently, on the fifth and sixth days the words were:

Fifth Day: "OKKOKOOKKOOKOKKO," and

Sixth Day: "OKKOKOOKKOOKOKKOKOOKOKKOOKKOKOOK."

On the sixteenth day the parrot's word contained $2^{15} = 32768$ letters. There are

$$60\ \frac{\text{seconds}}{\text{minute}} \times 60\ \frac{\text{minutes}}{\text{hour}} \times 24\ \frac{\text{hours}}{\text{day}} = 86400\ \frac{\text{seconds}}{\text{day}},$$

so, yes, the parrot finished squawking on the sixteenth day—but, knowing what to expect on the next day, we returned the parrot to the shop.

Q8. What was the 2020th letter of the parrot's sixteenth word?

S9. Three edges meet at each corner of the cube, so there must be at least one wire end at each corner. There are eight corners, so there are at least eight ends and four pieces of wire. But four pieces suffice, as shown in the following figure. Each piece is at least one inch long, so the longest piece is at most 9 inches. This can be achieved, as shown in the following figure.

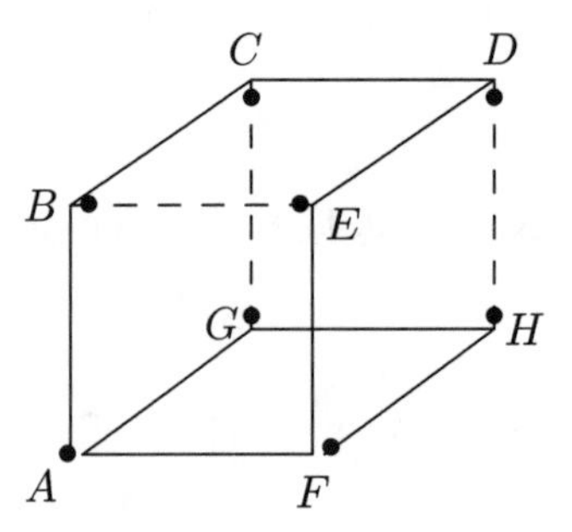

The wire of length 9 begins at A and continues $ABCDEFAGHF$.

Q9. How else might you divide the wire into four pieces to make the frame? List all of them.

S10. Each of the cube's corners border three squares and each square borders exactly one of the cube's corners. Consequently, we can divide the squares into eight groups of three squares where the squares in each group share a common corner. The three squares in the same group must be painted differently, which means that the number of squares with the same color cannot be more than eight.

Eight squares of the same color is possible, in many different ways. Here's one:

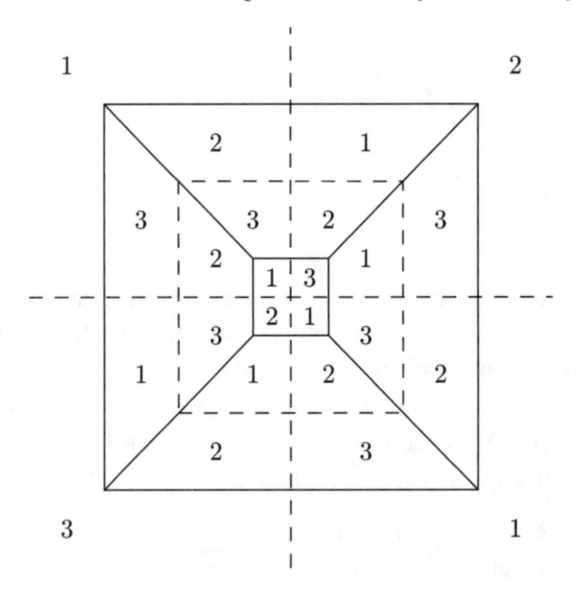

S11. A ruler is not necessary. You need to cut off a piece which is $\frac{2}{3} - \frac{1}{2} = \frac{1}{6}$ meter long. Therefore, to find where to make the cut, fold the original piece in half twice: the first yields a piece one-third meter long, and the second yields a piece one-sixth meter long.

Q11. What other lengths may be cut off without using a ruler?

S12. Yes, it is possible. The following figure shows one possible solution.

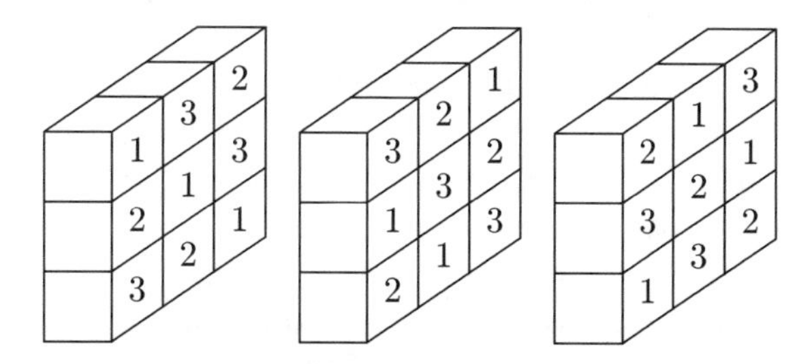

The pattern of colors in this solution demonstrates how the problem can be generalized to building an $n \times n \times n$ cube from n sets of n^2 $1 \times 1 \times 1$ blocks, so that each row of n blocks will have one block from each of the sets.

S13. To separate the numbers 1 to n into two sets, neither of which contains both a number and its double, put the odd numbers in the first set, their doubles in the second set, four times the odd numbers in the first set, eight times the odd numbers into the second set, and so on. That is, if $m = 2^k d$, with d odd, then m goes into the first set or second set according as k is even or odd. Alternatively, write m in binary notation, and then m goes into the first set or second set according as the number of final zeros is even or odd.

If n is even and it is further required that the two sets should be equal in size, then the above process allocates $n/2 + n/8 + n/32 + \cdots \approx 2n/3$ to the first set and $n/4 + n/16 + \cdots \approx n/3$ to the second. Transfer the largest $n/6$ odd numbers from the first set to the second. These have sizes $\geq 2n/3$ and their doubles are $\geq 4n/3 > n$, and do not occur in either set.

Another method is to put m into the first or second set according as its binary representation has an even or odd number of bits. For example,

$$1;\ 4 \text{ to } 7;\ 16 \text{ to } 31;\ 64 \text{ to } 127;\ \ldots \quad \text{go to the first set}$$

$$2 \text{ to } 3;\ 8 \text{ to } 15;\ 32 \text{ to } 63;\ \ldots \quad \text{go to the second set}$$

Again, to equalize the size of the two sets, move enough of the largest odd numbers from the larger set to the smaller, which can't contain any of their halves.

A third method is to put m into the first or second set according as the number of zeros in its binary representation is even or odd. For example

$$1, 3, 4, 7, 9, 10, 12, 15, 16, \ldots \quad \text{go to the first set}$$

$$2, 5, 6, 8, 11, 13, 14, 17, 18, \ldots \quad \text{go to the second set}$$

Here, the sets will already be of almost equal size.

S14. The counterfeit coin cannot be found among 30 coins in three weighings. For in the first weighing either at least 10 coins are on each pan or at least 10 coins are left off the weighing, but either way there is a set of 10 coins that could contain the counterfeit. In the second weighing, either at least 4 coins are on each pan or at least 4 coins are left off. The counterfeit could be in this set of 4. In the third weighing, either at least 2 coins are on each pan or 2 coins are left off. The counterfeit could be in this set of 2, so that after three weighings, we haven't succeeded.

The counterfeit can be found in the case of 27 (or fewer) coins. To see how, first divide the coins into three groups of nine coins. In the first weighing, compare two of the groups. If it balances, the counterfeit coin is in the third group, otherwise it is in the lighter group. Divide the group with the counterfeit coin into three groups of three coins and proceed with the second weighing in the same way as the first. In this way you will find the group which contains the counterfeit coin. Now you can take two of the three coins and compare them. If they balance the counterfeit is the third coin, otherwise it is the lighter coin.

For references to the literature, see **Counterfeit coins** in the Treasury.

S15. Suppose that C beat A in the Lund-Uppsala match. If every team that beat C also beat A, then they would have unequal scores. Therefore, there must be a third team, B, that beat C but lost to A.

S16. Perhaps the simplest way to solve the problem is to count the number of faces of the cubes that have been glued together in the construction of each of the three models. One way to count the number of glued faces is to count the glued faces on every other cube. The numbers on individual cubes shown below give this information (the "3" in the back corner in models (i) and (iii) is on the back lower-left corner cube).

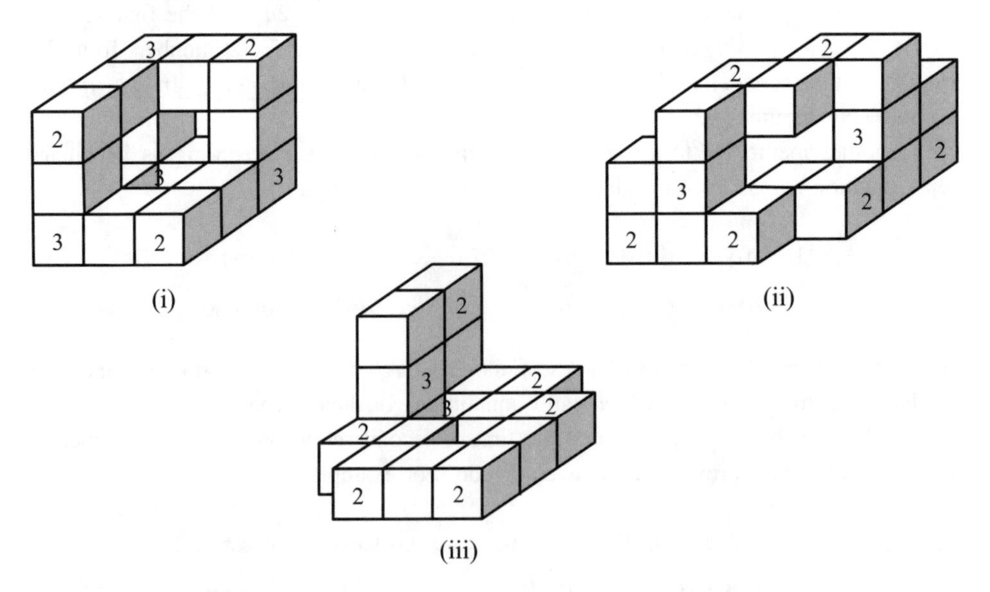

We find that the number of glued faces is 18 in each model, which implies that all three surface areas are the same: $16 \times 6 - 18 \times 2 = 60$ unit squares.

S17. It is helpful to think of the sum of n consecutive integers as equal to n times their average value. The average of an odd number of consecutive integers is the "middle" integer in the set (the other numbers are balanced on either side). The average value of an even number of consecutive integers is the "half-integer" between the two middle integers. So, if A denotes the average of a set of n consecutive integers, we need to investigate the equation $A \times n = 105$, where A is an integer or a half-integer. In this way we find there are only seven ways to write 105 as the sum of two or more consecutive positive numbers, namely:

$$\begin{aligned}
105 &= 52 + 53 \\
&= 34 + 35 + 36 \\
&= 19 + 20 + 21 + 22 + 23 \\
&= 15 + 16 + 17 + 18 + 19 + 20 \\
&= 12 + 13 + 14 + 15 + 16 + 17 + 18 \\
&= 6 + 7 + 8 + 9 + 10 + 11 + 12 + 13 + 14 + 15 \\
&= 1 + 2 + 3 + 4 + 5 + 6 + 7 + 8 + 9 + 10 + 11 + 12 + 13 + 14
\end{aligned}$$

Q17. How will the answer change if we don't restrict the consecutive integers to be positive?

S18. The dance cannot be properly done no matter how many Varmlanders are invited. The delegation from Venus has altogether an odd number of hands (the sum of seventeen odd numbers) and together with an even number of hands from the Varmlanders, the total number of hands in the combined group is an odd number. For the dance to be done properly there must be an even number of hands.

Q18. If Varmlanders have seven hands, what is the least number of Varmlanders needed?

S19. The Swiss mathematician Leonhard Euler was the first to study this sort of question, in about 1730 (in connexion with the well-known problem concerning the bridges of Königsberg). He showed that a figure can be traced by a single path (without lifting the pencil) if and only if the number of vertices of odd valence (those vertices with an odd number of edges connected to it) is either zero or two. If there are no vertices of odd valence, one can find a path that traces the entire figure and returns to the starting point. If there are exactly two vertices of odd valence (as, for example, in the first figure), there is a path that begins at one of these vertices and ends at the other. In no other case is it possible to find such a path. Thus, the last figure cannot be traced, because it has four vertices of odd valence.

A path for the first figure is *ABCDEB FGHIJKLMF LENCJHDA N*.

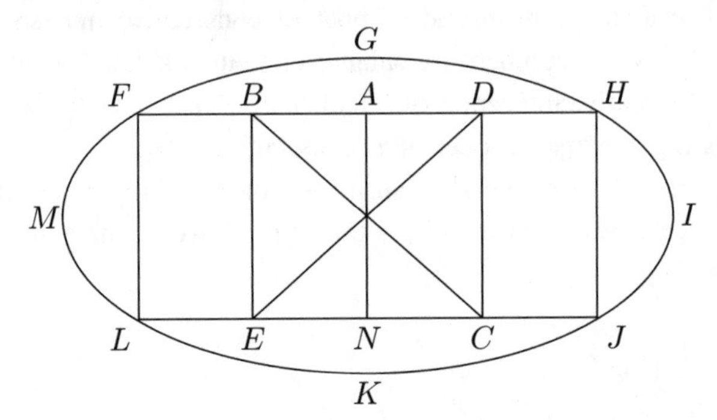

A path for the second figure is *ABCDEF GBH AGIDJ CIEJH FA*

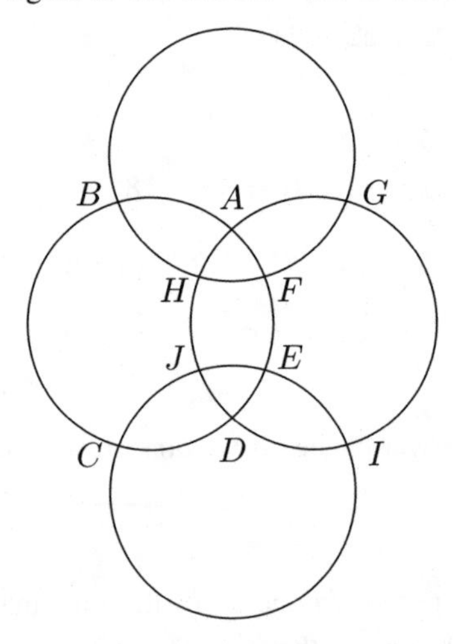

S20. Any figure in the plane with more than two vertices of odd valence will suffice. The following figure shows such an example.

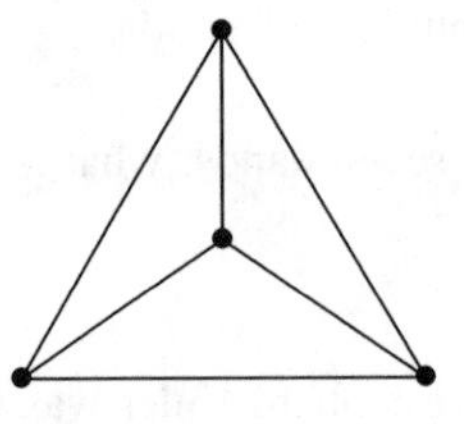

Tracing each edge twice doubles the valence at each vertex, so that all vertices are even. This is equivalent to tracing an Euler circuit in the following graph (see the preceding problem).

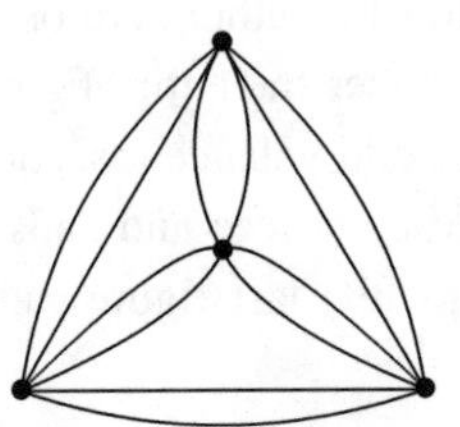

S21. The numbers can be so arranged. Because $1+2+3+4+5+6+7=28$, the sum of the numbers on each side of a line must be 14. It now remains to place the numbers in an appropriate way. One solution is shown below.

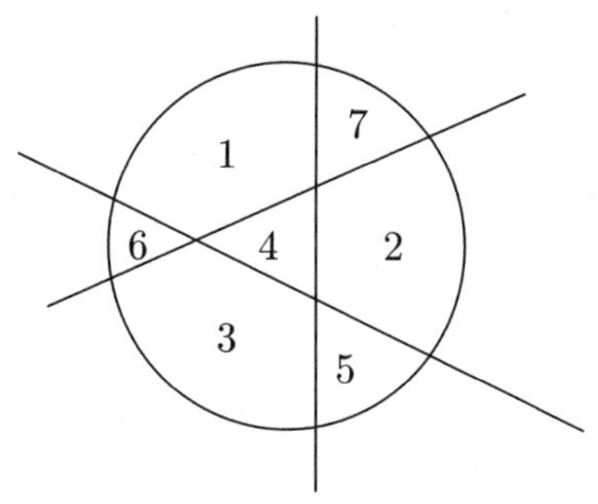

Q21. How many different solutions can you find?

S22. Suppose there are n employees. Then the number of women, which we will denote by w, is more than $\frac{60}{100}n$ and less than $\frac{65}{100}n$. That is to say,

$$\frac{60n}{100} < w < \frac{65n}{100},$$

or equivalently,

$$12\,n < 20\,w < 13\,n\,.$$

The smallest positive number n such that the interval from $12n$ to $13n$ includes a multiple of 20 is $n=8$, namely,

$$8 \times 12 = 96 \ < \ 20 \times 5 \ < \ 104 = 13 \times 8.$$

Therefore, the fewest number of employees required for such a claim is 8.

Q22. What is the next fewest number of employees that will make this a valid claim? And the next? And the next? ...

S23. (a) This can be answered most easily by first arranging the numbers in a table as shown below. The numbers with first digit larger than second are below the diagonal.

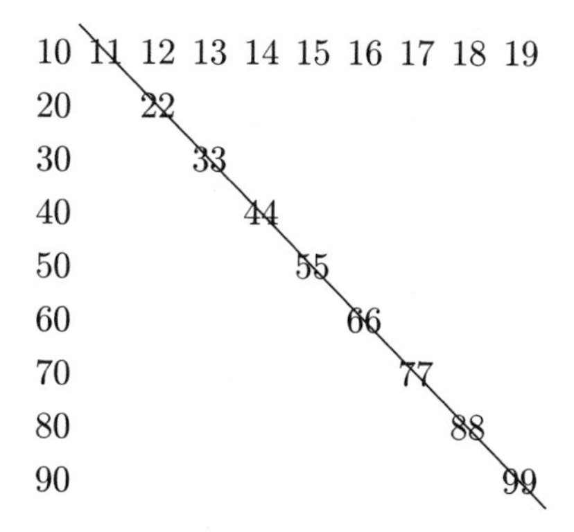

There are clearly more, in fact nine more, numbers below the diagonal than above.

(b) Every even number is followed by an odd number, except for 123456. Each even number differs by one in the sum of its digits from the next odd number. The sum of the digits of 123456 is 21, so there is one more number with an odd digit sum than there are with even digit sums.

S24. All but one of the pairs differ by more than 250. The one exception is 50123 and 49876, whose difference is 247.

S25. Every hop of neighboring frogs exchanges the positions of these two frogs. For these frogs to return to their original order they must exchange positions a second time. It follows that the number of hops which return each frog to its original order must be an even number. Thus the frogs returned to their original order on the eighteenth jump, or equivalently, after 90 seconds.

S26. One can accomplish it in six cuts. First divide the cube into eight $2 \times 2 \times 2$ cubes (three cuts), and then arrange these cubes into a $2 \times 2 \times 16$ stack. With two cuts (from top to bottom) we obtain 32 blocks with dimensions $1 \times 1 \times 2$. Arrange these blocks together in a stack with 1×2 base. A sixth cut gives the desired solution.

The six cuts can be made in other ways. For example, we can make two $2 \times 4 \times 4$ blocks with one cut, then four $1 \times 4 \times 4$ blocks with a second cut, then eight $1 \times 2 \times 4$ sixteen $1 \times 1 \times 4$, thirty-two $1 \times 1 \times 2$, and so forth.

One cannot do it in less than six cuts. To see this, consider the eight small cubes surrounding the center of the large cube. To cut this $2 \times 2 \times 2$ cube free requires at least six cuts, one for each of its faces. Alternatively, it is clear that each cut will at most double the number of pieces, so to produce 64 pieces will require at least six cuts ($64 = 2^6$).

Though one of us is a keen craftsman, we admit to never having carried this out.

S27. Let's say that a sequence of one or more, but not all, consecutively placed numbers is a *proper* sequence. The sum of the numbers in a proper sequence is more than 0 if and only if the remaining numbers, also a proper sequence, is less than or equal to 0. Therefore, each positive proper sequence corresponds uniquely to a nonpositive proper sequence (the other numbers), and vice-versa. Thus there is an equal number of positive proper sequences and nonpositive proper sequences. The sequence of *all* numbers is positive, so there is one more positive sequence than nonpositive.

S28. Yes, the number of people who took part in the election is determined. Suppose that $17n$ men and $15n$ women took part. Then

$$\frac{17n - 90}{15n - 80} = \frac{8}{7}$$

$$119n - 630 = 120n - 640$$

$$n = 10$$

It follows that 320 took part.

S29. (a) If 51 boxes are 6 cm wide then $150-51=99$ boxes are 3 cm wide. Altogether the total widths are $51\times 6+99\times 3=603$ cm, which is more than the available shelfspace. Therefore the shelving cannot be done.

(b) In this case, Soren needs $50\times 6+100\times 3=600$ of shelfspace. Suppose a single shelf accommodates A and B boxes of widths 6 cm and 3 cm respectively. The total width of these boxes is $6A+3B=3(2A+B)$, which is a multiple of 3. Therefore, the boxes on this shelf will not fill the shelf completely. Consequently the six shelves will not accommodate all the recordings.

(c) Here, the total width of the boxes is $49\times 6+101\times 3=597$ cm, but since at least one centimeter is unused on each shelf, Soren only has $6\times 99=594$ cm available. Therefore, the boxes will not fit.

(d) In this case it is possible to place all the recordings on the shelves. He can, for example, put 8 boxes of width 6 cm on every shelf together with 17 boxes of width 3 cm. These will use up $8\times 6+17\times 3=99$ cm on every shelf.

S30. (a) Yes. One can write, for example, $50, 100, 49, 99, \ldots, 3, 53, 2, 52, 1, 51$.

(b) No. In any arrangement of the numbers, the two numbers on either side of 50 would have to differ from it by at least 50. But there is only one such number available, namely 100.

Alternatively, draw a graph with $1, 2, 3, \ldots, 100$ as vertices. Join two vertices if they may be neighbors.

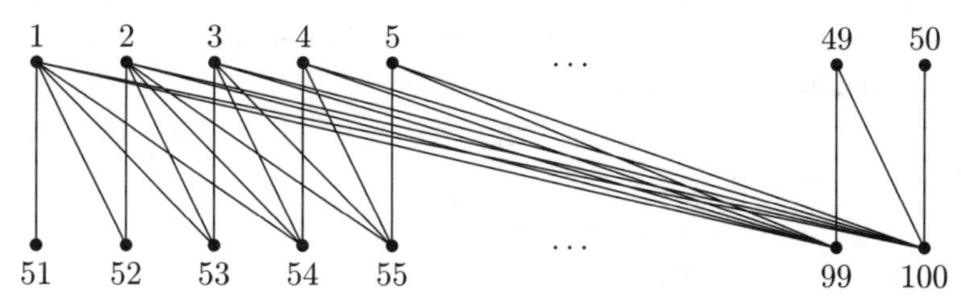

There is a Hamilton path that visits each vertex exactly once (for example, start at 50 and end at 51 as in (a) above. But there is no circuit because vertices 50 and 51 only have valence one.

S31. The four corners of the figure need only be visited once, whereas the eight corners of the central "cross" which lie on the outside perimeter must each be visited at least twice. If one starts at one of the "cross" corners and ends at another, then one must trace at least three segments (between these corners) two times. The path therefore must be at least $24+3=27$ inches long. The following figure shows that the grid can be traced by a path of length 27 inches.

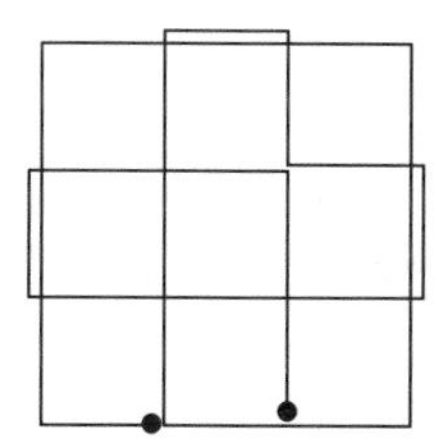

S32. One idea is to label the squares of the board as shown at the left, then draw a graph of all possible moves, as in the middle. On the middle graph a circuit is easy to find: from 00 to 02, clockwise around the middle portion, then from 32 to 30, and counterclockwise back to the starting point. Working back to the grid, the path in the middle gives a solution at the right.

03	13	23	33
02	12	22	32
01	11	21	31
00	10	20	30

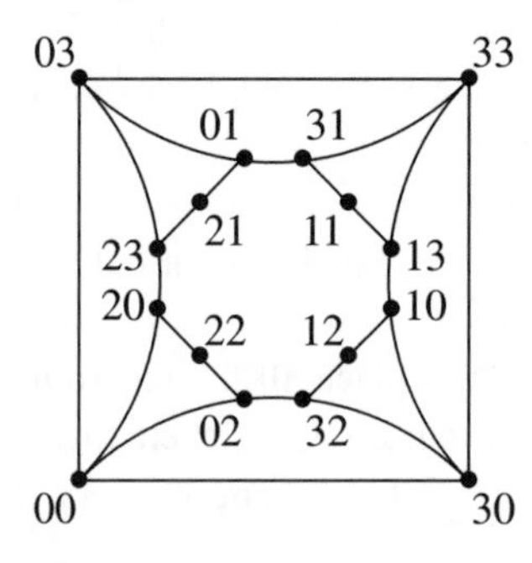

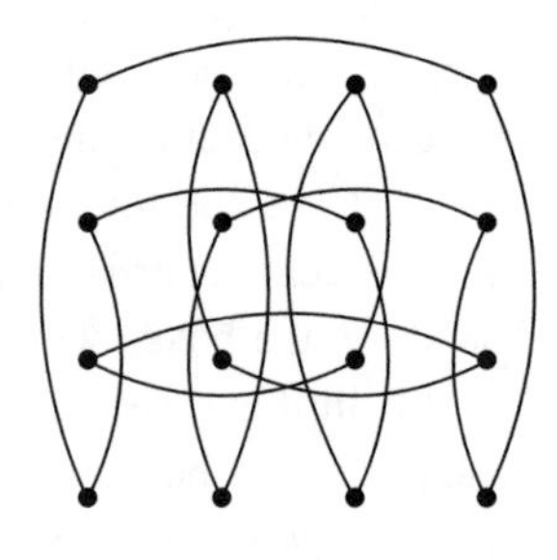

For further discussion of vertex touring, see **Hamilton circuit** in the Treasury.

Q32a. Can you find a circuit that is symmetric with respect to *both* the horizontal and vertical axes through the center of the board?

Q32b. Can you find a solution for a 5×5 board?

S33. The answer to the first three parts is "yes." You can do it in many different ways, for example as shown below (the number of stones in the respective piles is written within parentheses, and the arrows denote the moves).

(a) $(12, 123, 1234) \longrightarrow (0, 111, 1222) \longrightarrow (611, 111, 611) \longrightarrow (500, 0, 500) \longrightarrow (250, 250, 500) \longrightarrow (0, 0, 250)$.

(b) $(12, 123, 1234) \longrightarrow (0, 111, 1222) \longrightarrow (611, 111, 611) \longrightarrow (500, 0, 500) \longrightarrow (250, 250, 500) \longrightarrow (250, 500, 250) \longrightarrow (0, 250, 0)$.

(c) $(12, 123, 1234) \longrightarrow (0, 111, 1222) \longrightarrow (611, 111, 611) \longrightarrow (500, 0, 500) \longrightarrow (500, 250, 250) \longrightarrow (250, 0, 0)$.

(d) In this case, the answer is "no." Observe that when we remove stones from the piles we take away the same number from each of them, and therefore we take away $3n$ stones at a time (n a positive number). Thus, if it were possible to remove all the stones, the total number of stones would have to be a multiple of 3. But $12 + 123 + 1234 = 1369$ is not a multiple of 3.

S34. A key idea is to realize that jars with the same number of cookies can be treated as if they were just one jar: for if you've got two (or more) jars with k cookies, the cookie monster might as well do the same thing with them at each subsequent step—it won't help to deplete them at different rates. With this in mind, we can describe the moves by recording the *different* cookie numbers (in the respective jars) at each stage. Here is a four-move solution: the monster takes 8 cookies from every jar he can, then 4 from every jar he can, then 2, and finally 1.

$$\{15, 14, \ldots, 2, 1\} \rightarrow \{7, 6, \ldots, 2, 1\} \rightarrow \{3, 2, 1\} \rightarrow \{1\} \rightarrow \emptyset$$

It can't be done in fewer than 4 moves, because the best he can do (assuming the numbers are consecutive) is reduce the different numbers from c to $c/2$ if c is even or to $(c-1)/2$ if c is odd. So, even eight jars require four moves, for example

$$\{8, 7, \dots, 2, 1\} \to \{4, 3, 2, 1\} \to \{2, 1\} \to \{1\} \to \emptyset.$$

This "bisection method" works optimally when the cookie numbers are in an arithmetical progression, but what about other situations?

Q34. What is the fewest number of moves required to empty jars containing the following numbers of cookies?

(a) $\{9, 5, 4, 2\}$
(b) $\{20, 11, 10, 8, 7\}$
(c) $\{36, 33, 5, 1\}$
(d) $\{11, 5, 4, 2\}$

S35. We first determine how many points each contestant scored:

$$\frac{50 + 2\times 25 + 3\times 20 + 3\times 10 + 2\times 5 + 2\times 3 + 2\times 2 + 3\times 1}{3} = \frac{213}{3} = 71.$$

Bjorn made $71 - 22 = 49$ with his last few shots, which number at least four: for example, 25, 20, 3, 1, or 25, 20, 2, 2. But 22 needs at least two shots, 20, 2, and there were only two two's, so Bjorn scored 25, 20, 20, 3, 2, 1. Whoever didn't score the bull's-eye must have scored 25, 20 and at least two tens, but not more than two, and so 25, 20, 10, 10, 5, 1. The bull's-eye scorer made 50, 10, 5, 3, 2, 1, and so was Camilla.

Alternatively, the only way to partition the given numbers into three sets of six, each of which adds to 71, is $\{25, 20, 10, 10, 5, 1\}$, $\{25, 20, 20, 3, 2, 1\}$, and $\{50, 10, 5, 3, 2, 1\}$. The second sum must correspond to Bjorn's scores because it is the only one that allows for a score of 22 points among the first few shots. The last set of scores must be Camilla's (she scored 3 points on her first shot). Therefore, it was she who hit the bull's-eye.

S36. No. The price after two years was $49/100$ of the original, so the intermediate price was $7/10$ of the original; that is, the annual markdown was 30%. The price after three years was therefore $(49/100)(1-3/10) = 343/1000$ of the original price, which means it was 65.7% off the original price.

S37. There are four even numbers and six odd numbers in the list, and the parity of the terms in the sequence is completely determined by the parity of the first two entries. There are four parity choices for the first two entries: $0\,0$, $0\,1$, $1\,0$, and $1\,1$. Only the second of these leads to a sequence of ten numbers four of which are even and six odd. Thus, the parity sequence must be $0\,1\,1\,0\,1\,1\,0\,1\,1\,0$

There are two essentially different solutions to the problem: $4\,5\,1\,4\,3\,1\,2\,3\,5\,2$ and $4\,1\,5\,4\,1\,3\,2\,5\,3\,2$. Two other solutions can be found by writing these sequences in reverse order.

Q37. Find all solutions to the same problem for the case of $2n$ numbers $1, 1, 2, 2, \dots, n, n$, where n is an arbitrary integer greater than 1.

S38. (a) It is impossible to partition the gifts among Alice, Bertha and Carol, so that all receive the same total value. This is because the value of all the gifts is $1+2+\cdots+13 = 91$, and this is not divisible by 3.

(b) In this case, $1 + 2 + \cdots + 14 = 105$ and $105 = 3 \times 35$. The sum of each subset must therefore be 35. There are many ways to partition the gifts as desired; one way is $(1, 2, 3, 5, 7, 8, 9)$, $(4, 6, 11, 14)$, and $(10, 12, 13)$, or, more evenly, $(1, 5, 6, 11, 12)$, $(2, 3, 7, 10, 13)$, and $(4, 8, 9, 14)$.

> **Q38.** Suppose you have gifts of values $1, 2, \ldots, 15$ dollars, of which those worth $10, 11, \ldots, 15$ dollars are special presents. Can you divvy these up among Alice, Bertha and Carol, so that each gets the same number of presents, the same number of special presents, and the same total value?

S39. Following the hint, let S denote the number of pairs of adjacent coins that are different from each other. Let P be an arbitrary sequence of adjacent coins on the table. When we replace every penny in P with a dime and every dime with a penny, the number of pairs of adjacent coins in P that are different from each other remains unchanged. Thus, in a single move, the number S (for the entire row) will decrease by at most 2, depending on which coins are adjacent to P. At the beginning S is equal to 15, so we need at least eight moves to reduce S to 0. That is to say, the final desired sequence cannot be reached in fewer than eight moves.

There are many ways of doing the problem, but at each stage (with just one exception when you turn the end penny) you must be sure that you turn an interior set whose ends each differ from their unturned neighbors. One of the more oddball ways is to turn the middlemost dime into a penny, then 3 pennies into dimes, then 5 dimes into pennies, then 7 pennies into dimes, 9 dimes into pennies, 11 pennies into dimes, 13 dimes into pennies, and finally 15 pennies into dimes.

S40. (a) Pair the first and second, the third and fourth, and the fifth and sixth rows. For each of these, pair a child in one row with the neighboring child in the other row. Then at the signal these paired children can simply exchange places with each other.

Another idea is to line the children up along the path shown below. At the signal, each child can jump one space forward along this path. After 36 jumps, each child will return to the original position, having visited all the squares of the grid.

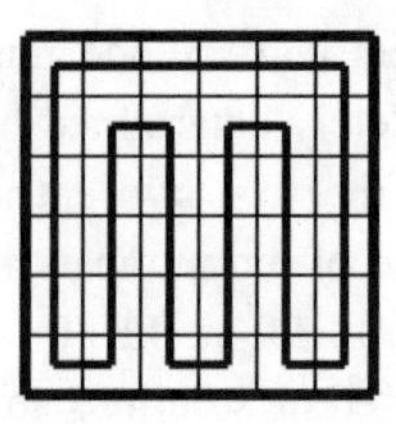

(b) If we color the squares of the 5×7 board in the usual checkerboard manner, then 18 squares will be one color (for example, black) and 17 will be the other (white). Each child will jump from a square of one color to a square of a different color. But it is impossible for the 18 children on the black squares to jump to different white squares because there are only 17 available white squares. However we organize the jumps, some square will end up with at least two children on it.

S41. The first problem can be solved in eight steps. Here's one way of doing it (where (a, b) means a and b liters of water in the 7-liter and 11-liter bucket, respectively):

$$(0,0) \longrightarrow (0,11) \longrightarrow (7,4) \longrightarrow (0,4) \longrightarrow (4,0) \longrightarrow$$
$$(4,11) \longrightarrow (7,8) \longrightarrow (0,8) \longrightarrow (7,1)$$

The second problem can be done in fourteen steps:

$$(0,0) \longrightarrow (7,0) \longrightarrow (0,7) \longrightarrow (7,7) \longrightarrow (3,11) \longrightarrow (3,0) \longrightarrow$$
$$(0,3) \longrightarrow (7,3) \longrightarrow (0,10) \longrightarrow (7,10) \longrightarrow$$
$$(6,11) \longrightarrow (6,0) \longrightarrow (0,6) \longrightarrow (7,6) \longrightarrow (2,11)$$

In the first solution, we repeatedly fill the 11-liter bucket and transfer it to the 7-liter bucket which we empty when it fills. Altogether, we fill the 11-liter bucket twice, and empty the 7-liter bucket three times.

In the second solution, we repeatedly fill the 7-liter bucket and transfer it to the 11-liter bucket which we empty when it fills. Altogether, we fill the 7-liter bucket five times, and empty the 11-liter bucket three times.

Q41. Reformulate the problem as a diophantine equation (see **Diophantine equation** in the Treasury). How many solutions can you find?

S42. To solve the problem we have to determine the number of different ways 62 can be written as a sum of at most four numbers from among the numbers on the target: 1, 9, 10, 11, 13, 18, 25, and 35. The numbers in the sum can be used more than once. There are only two even numbers so the scores contain 18, 18, or 10, 18, or 10, 10 or all odd numbers. This helps cut down the search and we find eight different ways to score 62 when all four darts hit the target:

$$\begin{aligned} 62 &= 35 + 25 + 1 + 1 \\ &= 35 + 13 + 13 + 1 \\ &= 35 + 9 + 9 + 9 \\ &= 25 + 25 + 11 + 1 \\ &= 25 + 18 + 18 + 1 \\ &= 25 + 18 + 10 + 9 \\ &= 25 + 13 + 13 + 11 \\ &= 18 + 18 + 13 + 13 \end{aligned}$$

There is one additional possiblility assuming not all darts hit the target, namely $\{35, 18, 9\}$. This shows that nine is the maximum number who could have played the game.

S43. Because there were a million cells in the incubator on the evening of May 17, there must have been 500000 cells there on the evening of May 16, and 250000 cells on the evening of May 15. Continuing by working backwards, there were 125000 cells on May 14, 62500 cells on May 13, 31250 cells on May 12, and 15,625 cells on May 11. But 15625 is an odd number, so a new cell must have been added on that date.

We could continue in this manner, but it is instructive to think about it as an application of binary notation. Write a million in base two and write 17 under the units digit, 16 under the twos digit, 15 under the fours digit, and so on:

$$\begin{array}{cccccccccccccccccccc}
1 & 1 & 1 & 1 & 0 & 1 & 0 & 0 & 0 & 0 & 1 & 0 & 0 & 1 & 0 & 0 & 0 & 0 & 0 & 0 \\
 & & & 1 & 2 & 3 & 4 & 5 & 6 & 7 & 8 & 9 & 10 & 11 & 12 & 13 & 14 & 15 & 16 & 17
\end{array}$$

Then the number of cells present on the evening of any date is gotten by truncating the binary number at that date. So we can see that an extra cell was added on the evenings of the 11th, 8th, and 3rd of May, and that on the 1st there were $1111_2 = 15$ cells in the incubator.

S44. No, it is not possible. The least combined score for the second and fourth players is 91 so that their average is at least $45\frac{1}{2}$. But if one scores 46 and other 45 there isn't a score available for the third place finisher. Therefore, the second place finisher had to have made at least 47 free-throws, so the lowest possible top score is 48. In this case, the fourth best score is 44, and the lowest third place score is 45. It follows that the largest lowest score is $200 - (48 + 47 + 45 + 44) = 16$, and the difference between the top and bottom scores is 32.

S45. We continue our analysis with the diagram obtained in the hint.

$$\begin{array}{cccccc}
 & & & 5 & * & 4 \\
 & & \times & 2 & 3 & 6 \\
\hline
 & & * & * & 2 & 4 \\
 & 1 & * & * & 2 & \\
1 & 0 & * & 8 & & \\
\hline
1 & 1 & * & * & 4 & 4
\end{array}$$

Because the units product in the first multiplication problem ends with 24, the middle digit of the multiplicand must be either 5 or 0. The digit 5 does not work because this forces the hundreds product to begin with 11 rather than 10. Therefore, the only possibility is for this digit to be 0 and this choice agrees in every digit: $504 \times 236 = 118944$.

In the second problem we begin by considering the hundreds digit in the multiplicand.

$$\begin{array}{ccccccccc}
 & & & 2 & 3 & * & * & 8 & 5 \\
 & & & \times & & 3 & 2 & * & 5 \\
\hline
 & & * & * & * & * & * & 2 & 5 \\
 & & * & * & * & 7 & * & * & \\
 & 4 & * & 9 & 5 & 7 & 0 & & \\
* & * & 4 & * & 5 & 5 & & & \\
\hline
7 & * & * & * & * & * & * & * & 5
\end{array}$$

It can only be 2 or 7 because the hundreds product ends in 570. If we take it to be 2 the product would not end in 9570. Therefore we can put 7 for the hundreds digit of the multiplicand, and at the same time place 3 in the hundreds position of the thousands product. Since the third digit in this product is 4, the only missing digit from the multiplicand must also be 4. We now wish to determine the only missing digit from the multiplier. It cannot be larger than 4 (the thousands product has only six digits) and it can be neither 2 nor 3 (the 7 in the tens product does not check with the hundreds digits in the last two products. Therefore the digit is 1 or 4. It is easy to check that 4 does not work, and this leaves us with digit 1. This answer checks in all positions:

$$234785 \times 3215 = 754833775.$$

This problem presents an opportunity to point out that multiplications (and divisions), contrary to additions and subtractions, are best done from left to right, rather than right to left as we are usually taught. Try it with your friends, classmates, pupils, and yourself, going as fast as you can, by doing examples in both ways. For example, try the product 47×8 in your head: it's easiest to figure $40 \times 8 = 320$ first and then add $7 \times 8 = 56$ to get 376.

In multiplying left to right, you're dealing with the most significant digits immediately, and indeed, for approximations, there may be no need to multiply the less significant digits.

Let's consider these two problems again, this time in the left-to-right format. We have replaced some of the asterisks by labels to make it easy to discuss the solution.

$$\begin{array}{cccccc}
 & & & * & * & 4 \\
 & & & 2 & 3 & a \\
\hline
1 & * & * & b & & \\
 & 1 & * & * & c & \\
 & & d & e & 2 & 4 \\
\hline
* & 1 & * & * & * & *
\end{array}$$

The digit a is either 1 or 6, but not 1 because the units product has four digits. Therefore, the multiplier is 236 and it follows that $b = 8$ and $c = 2$. If the multiplicand is x, then the hundreds and tens products imply that $2x \geq 1008$ and $3x \leq 1992$, or equivalently,

$$3024 \leq 6x \leq 3924.$$

It follows that $d = 3$ and $e = 0, 3, 6$, or 9 (the units product is a multiple of 3, so the sum of the digits of the number must be a multiple of 3). Dividing by x yields

$$x = 504 + 50t, \qquad t = 0, 1, 2 \text{ or } 3.$$

$$236x = 118944 + 11800t$$

The second digit in the answer is 1, and therefore $t = 0$ and we get the solution: $504 \times 236 = 118944$.

In the second problem we have the form

$$
\begin{array}{ccccccccc}
 & & & 2 & 3 & a & b & 8 & 5 \\
 & & & & & c & d & e & 5 \\
\hline
\ast & \ast & 4 & \ast & \ast & \ast & & & \\
 & \ast & \ast & 9 & 5 & 7 & 0 & & \\
 & & \ast & \ast & \ast & 7 & \ast & \ast & \\
 & & \ast & \ast & \ast & \ast & \ast & \ast & \ast \\
\hline
7 & \ast & \ast & \ast & \ast & \ast & \ast & \ast & \ast
\end{array}
$$

From the size of the final product, $c = 3$, and the final digits in the hundreds product show that $d = 2$, and it follows that $b = 7$ and a is 4 or 9, so the multiplicand is $234785 + 5000t$, $t = 0$ or 1. Then the 7 in the tens product implies that $e = 1$, so the multiplier is 3215. Then $3x = 704355 + 15000t$ and now, as the third digit in the thousands product is 4, $t = 0$. Thus, the solution is $234785 \times 3215 = 754833775$.

Q45. Reconstruct the two multiplications shown below when they are arranged left to right.

$$
\begin{array}{cccccc}
 & & & \ast & \ast & 7 \\
 & & \times & 2 & 5 & \ast \\
\hline
1 & \ast & \ast & \ast & & \\
 & 3 & \ast & \ast & \ast & \\
 & & \ast & \ast & 9 & 6 \\
\hline
\ast & 6 & \ast & \ast & \ast & \ast
\end{array}
\qquad
\begin{array}{ccccccccc}
 & & & 2 & 5 & \ast & \ast & 4 & 3 \\
 & & & \times & & \ast & \ast & \ast & 5 \\
\hline
\ast & \ast & 9 & \ast & \ast & \ast & & & \\
 & \ast & \ast & 6 & 2 & 8 & 6 & & \\
 & & \ast & \ast & \ast & 4 & \ast & \ast & \\
 & & \ast & \ast & \ast & \ast & \ast & \ast & \ast \\
\hline
8 & \ast & \ast & \ast & \ast & \ast & \ast & \ast & \ast
\end{array}
$$

S46. The following figure shows seven lines and eleven triangular regions.

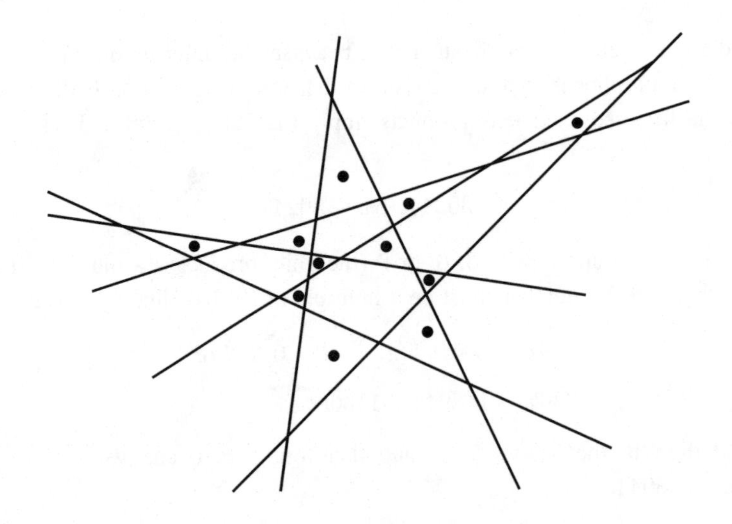

Q46. Is there an arrangement of six lines that make four triangular regions? five? six? seven?

S47. Label the digits as shown.

$$\begin{array}{cccc} & a & b & c \\ & d & e & f \\ \hline g & h & i & j \end{array}$$

Solutions come in sets of eight, got by swapping a and d, b and e, and c and f, so assume that $a < d$, $b < e$, $c < f$. If the summands are x and y, and the total z, then $x + y = z$, and, by 'casting out the nines' (see **Divisibility** in the Treasury), $x + y + z \equiv 0 + 1 + 2 + \cdots + 9 \equiv 0 \bmod 9$, so $2z \equiv 0 \bmod 9$ and z (and $x + y$) are multiples of 9 and its first digit is 1. One can find many solutions.

$$437 + 589 = 1026, \;\; 246 + 789 = 1035, \;\; 589 + 473 = 1062,$$

$$432 + 657 = 1089 = 324 + 765, \;\; 269 + 784 = 1053, \;\; 356 + 742 = 1098$$

Q47. Exactly how many different solutions are there?

S48. Label the houses A, B, C, and D as shown in the figure. Under the proposed plan, the gazebo should be built at the intersection P of the diagonals AC and BD. To see this, let Q be any point in the plane.

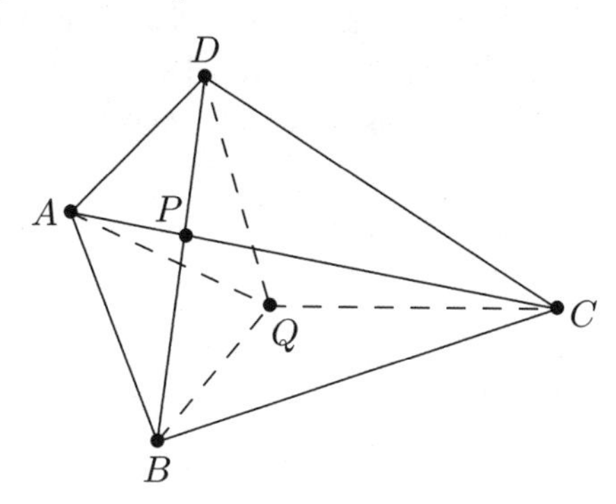

We know that the sum of the lengths of two sides of a triangle is at least as large as the length of the third side. Therefore, $d(QA) + d(QC) \geq d(AC)$, where d denotes length, with equality if and only if Q lies on the diagonal. In the same way, $d(QB) + d(QD) \geq d(BD)$, with equality if and only if Q lies on the diagonal. By adding these inequalities, we find that

$$\begin{aligned} d(QA) + d(QB) + d(QC) + d(QD) &\geq d(AC) + d(BD) \\ &= d(PA) + d(PB) + d(PC) + d(PD) \end{aligned}$$

with equality if and only if Q lies on both diagonals.

Q48a. Where should the gazebo be if one of the four houses is inside the triangle formed by the other three houses?

Q48b. What if there are only three houses?

This problem is closely related to the following problem which still may be unsolved in the general case. In a letter to Schumacher in March 1836, Gauss said that it seemed like a good problem.

Q48c. Suppose there is no gazebo at all, but you wish to construct a paved trail that will connect the four houses. What is the minimal total length of such a trail?

S49. No, it is not possible. The number 23 is the only number between 1 and 25 that is divisible by 23. Thus, only the subset containing 23 will have a product divisible by 23.

The problem is in fact impossible for any positive value of n. That is to say, the numbers from 1 to n cannot be partitioned into two or more subsets so that the products in each subset is the same. This is because of a famous theorem known as Bertrand's postulate that asserts that for $n \geq 2$ there exists a prime p such that $n/2 \leq p \leq n$. Since there can be only one integer from 1 to n divisible by this prime p, only the subset containing p will have a product divisible by p.

S50. If we color the rectangular grid as described in the hint, we find that there are 200 black squares and 199 white squares. The middle square is white. Regardless of where the path starts, it must pass through 200 black squares and 198 white squares. It is clear that the path will always pass from a black square to a white square, or vice-versa. Even if the path starts on a black square and ends on a black square and visits all 198 white squares, it will have only visited 199 black squares. Thus the path can never pass through all the black squares. Therefore, the answer to the problem in all cases is that it cannot be done.

S51.

$3A > 35$									12	$\cdots$	20	21	22	23	$\cdots$	48	49	50	$\cdots$
$7A > 43$				7	8	9	10	11	12	$\cdots$	20	21	22	23	$\cdots$	48	49	50	$\cdots$
$2A < 99$	$\cdots$	5	6	7	8	9	10	11	12	$\cdots$	20	21	22	23	$\cdots$	48	49		
$A > 21$													22	23	$\cdots$	48	49	50	$\cdots$
$5A > 51$								11	12	$\cdots$	20	21	22	23	$\cdots$	48	49	50	$\cdots$
								$\uparrow$											

Only one column has exactly three numbers in it; $A = 11$. Conditions 1 and 4 are false.

S52. No. Let N denote the number of pounds of chocolate sold in July. The income for July was $2.56 \times N$ dollars. Thus the income for August must have been $(2.56 \times N) \times \left(\frac{87.6}{100}\right)$ dollars, and the number of sales in August was $N \times \left(\frac{76.8}{100}\right)$. The price of a pound of chocolate in August can be found by dividing the first of these by the second, and this gives the answer: \$2.92.

More succinctly, 76.8% of the new price is 87.6% of \$2.56, so the new price is \$2.92.

S53. The supposition is that

$$100(a+d)+10(b-d)+(c-d)=A\times d.$$

Rearranging the terms gives $100d-10d-d=A\times d-(100a+10b+c)=A\times(d-1)$, or equivalently, $89d=(d-1)A$. Because

$$A=\frac{89d}{d-1}$$

is a whole number, and 89 is a prime number, and $d-1$ doesn't divide d for d larger than 2, it must be the case that $d=2$. We conclude that $A=178$.

S54. If the stars are all replaced with plus (+) signs the resulting sum is equal to 55. Now suppose that some of these numbers, whose sum we shall denote by a, are replaced by their negatives. Then the sum of the new expression is $55-2a$. Because we want the sum to be 29, we have $55-2a=29$. Solving for a yields $a=(55-29)/2=13$. There are exactly eight different subsets with sum equal to 13; each of these leads to a solution. Specifically, put minus signs in before before the numbers (i) 10 and 3, (ii) 9 and 4, (iii) 8 and 5, (iv) 8, 3, and 2, (v) 7 and 6, (vi) 7, 4, and 2, (vii) 6, 5, and 2, (viii) 6, 4, and 3. Substitute positive signs for all the other asterisks.

Q54. What nonnegative sums, other than 29, are possible by some choice of plus/minus signs?

S55. There are only two possibilities to consider: (i) the green-haired community are the truth-tellers (symbolized by GT) and the blue-haired community the liars (BL), or (ii) the reverse, GL and BT. In the first case the communities would each answer the question "Is your hair green?" with a "yes." In the second case, they would each answer "no."

Because these are the only cases that can happen, we conclude that if both answer "no" then the blue-haired people tell the truth; if they both answer "yes," the green-haired people tell the truth; the case where one answers "yes" and the other answers "no" cannot happen.

S56. If B is a truth-teller then A is a liar, and so also is C. But then the first statement from D is false, whereas the second is true, which is not consistent. We must conclude that B is a liar (and therefore has green hair), A and C are truth-tellers (and have blue hair), and D is a liar (with green hair).

A more mechanical approach to the problem is to use truth-tables. The four left columns, taken as a whole, consider the sixteen ways A, B, C, D can be configured as truth-tellers (T) or liars (L). The crosses in the subsequent columns mean that the corresponding statement is consistent with the circumstances in that row.

A	B	C	D	B to A You are a liar.	C to B You are a liar.	D to C A and B are liars.	D to C You are a liar.
T	T	T	T				
T	T	T	L			x	x
T	T	L	T		x		x
T	T	L	L		x	x	
T	L	T	T	x	x		
T	L	T	L	x	x	x	x
T	L	L	T	x			x
T	L	L	L	x		x	
L	T	T	T	x			
L	T	T	L	x		x	x
L	T	L	T	x	x		x
L	T	L	L	x	x	x	
L	L	T	T		x	x	
L	L	T	L		x		x
L	L	L	T			x	x
L	L	L	L				

There is only one case in which all four statements are consistent. That is the case in which A and C are truth-tellers (and therefore have blue hair) and B and D are liars (and have green hair).

S57. In the hint we have established that E is a truth-teller. Because the first part of E's statement is false, the second part must be true, so either both C and D are liars or both are truth-tellers.

If D is a liar then A's hair is blue and B's is green, which implies that C is telling the truth. This contradicts the fact that C and D come from the same community. It follows that D is telling the truth, and therefore so is C. Combining the truth of the statements by C and D, we conclude that B is a truth-teller. Therefore, the path through the mountains leads to the doctor.

> **Q57.** Try to work through the logic of this problem by using truth tables (see the previous problem; in this example, however, there are 32 cases to consider).

S58. It's helpful to draw a tree to show the flow of possibilities.

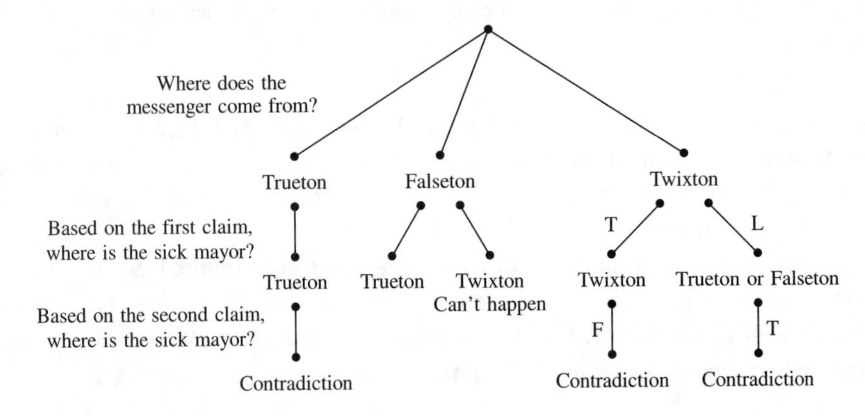

With the diagram as an aid, we first establish that it is impossible for the messenger to have come from Twixton. For suppose so. If his first statement is true then his second, "In Twixton," is not true, which is contrary to our assumption. On the other hand, assuming that one of the mayors is sick, suppose his first statement is false. Then the mayor must not be from Twixton. But this is contrary to his second claim which, by supposition, is true.

The messenger could not have come from Trueton either. For if so, his answer to my question should have been "In Trueton." The only remaining alternative is that the messenger comes from Falseton. The information that *his* mayor is sick is false, so the sick mayor must be from Trueton or Twixton.

If the sick mayor is from Twixton then we could have deduced, with some reflection, that the second statement from the messenger was false. Therefore, I went to Trueton. (My musings took such a long time that by the time I got there I could scarcely help the mayor.)

S59. (a) Yes, it is possible. For example, $1\underline{100}1 \to 10\underline{11} \to \underline{10}000 \to \underline{11}1000 \to$ 0001000.

(b) No, it is not possible. The string 0110 has two 1's in it whereas the string 0110010 has three 1's. However, none of the opertions changes the parity of the number of 1's, so it is not possible to transform 0110 into 0110010.

> **Q59** What strings can be generated with these rules? The following steps are useful.
>
> (i) We can reach 10 from 01, and vice-versa.
> (ii) We can reach 1110 from 111, and vice-versa.
> (iii) We can reach 11111 from 111, and vice-versa.
> (iv) Use the preceding parts (along with the given rules) to show that starting from the string 10, you can construct any string with an odd number of 1's, provided that the string has length greater than one.
> (v) Show that starting from 110 you can construct any string with an even number of 1's (perhaps zero), provided that the string has length greater than three.

S60. The number in the center square must be 11. When we add the numbers from the middle row, the middle column, and both diagonals, we get $4 \times 33 = 132$. On the other hand, the numbers in this sum include all the numbers in the grid exactly once, except for the number in the center square which is counted four times. The sum of all the numbers in the grid is $3 \times 33 = 99$ (for example, add together the three rows, or the three columns). It follows that the number in the center square is $(132 - 99)/3 = 11$.

S61. When Margaret counts $70 + 3n$, Bruce counts $1996 - 7n$, and the difference is $1926 - 10n$. Thus, at each count they are 10 closer together. When $n = 192$ they are 6 apart, and when $n = 193$ they are -4 apart, and that is as close as they'll be. That is, the closest they will be is on the 193rd step when Bruce counts $1996 - 7 \times 193 = 645$ and Margaret counts $70 + 3 \times 193 = 649$.

S62. Yes; there are two solutions. To find them, we continue along the lines discussed in the hint. Suppose we delete the middle digit in A. Then we have $100a + 10b + c = 7(10a + c)$, which simplifies to $5(3a + b) = 3c$. It follows that c must be divisible by 5 and because c cannot be 0 ($a > 0$), we must have $c = 5$. This in turn implies that $a = 1$ and $b = 0$, so we get a solution for A, namely $A = 105$.

In the last case, suppose we leave off the first digit in A. Then $100a + 10b + c = 7(10b + c)$, which simplifies to $50a - 30b = 3c$. This shows that c must be divisible by 10, so $c = 0$, and $5a = 3b$. It follows that $a = 3$ and $b = 5$; that is, $A = 350$.

S63. It is apparent that the multiplier must be 101 and each row must begin with 1.

$$\begin{array}{ccccccc} & & & 1 & * & * & * \\ & & \times & & 1 & 0 & 1 \\ \hline & 1 & * & * & * & & \\ & & & 1 & * & * & * \\ \hline 1 & * & * & 1 & * & * & * \end{array}$$

To get the initial 1 in the final product there must have been carries from each of the previous two columns, and to get the central 1 in the final product there must have been a carry from the column before that. So the multiplicand and each of its products are 1111 and the final product is 1001011.

S64. Assume that Agassi served in four games and Becker had k service breaks. Then Agassi won $4 - k$ service games and made $5 - k$ service breaks. Thus, he won $(4 - k) + (5 - k) = 9 - 2k$ games, an odd number. But Agassi won 6 games, an even number. This contradiction implies that Agassi served five games, and therefore served the first game.

S65. Here's one solution.

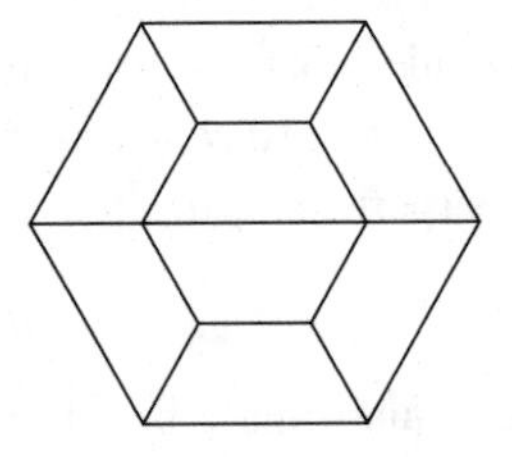

Q65. Is there another way of doing it?

S66. In each case, the difference $A - a$, $B - b$, ... is odd if and only if one of the terms is odd and one is even.

Assume that we can find numbers $A, B, \dots, H, J$, and a rearrangement $a, b, \dots, h, j$ so that the given product is odd. Then, each of the differences $A - a$, $B - b$, ... is odd, and thus one of the numbers in each difference is odd and one is even. If m of the numbers among $A, B, \dots,$ are odd, then m of the numbers among $a, b, \dots$ are even. Because the same nine numbers are in the second list, we must have $m = 9 - m$. However, this equation has no solution in whole numbers. This contradiction shows that the product can never be odd.

S67. In the hint we established that if there were no draws then K, the number of participants in the tournament, must be an even number. Now assume that Jill had n wins and no draws. Then she had $K-1-n$ losses, which gives $n-(K-1-n) = 12$ points; that is, $2n - K = 11$. This equation shows that K must be an odd number. We therefore have a contradiction, and we conclude that either Jack or Jill must have had at least one draw among their matches.

Jack's and Jill's scores could have occurred with as few as 13 students, if Jill wins all her matches, and Jack wins 10, loses one (to Jill), and draws one.

S68. In a knockout tournament the number of players decreases by 1 after each match. Because there is only one winner in a knockout tournament, the number of matches in the tournament is $64-1 = 63$. (Alternatively, the number of matches is $32+16+8+4+2+1 = 63$.) If n is the number of matches that ended with the score 3-0, then $2n$ and $4n$ are the numbers of matches that ended 3-1 and 3-2 respectively. Consequently, $n+2n+4n = 63$, which gives $n = 9$. Therefore, the number of sets is $9 \times 3 + 18 \times 4 + 36 \times 5 = 279$.

S69. As in the hint, let $d = s - t$ be the updated difference after each game, where s is the number of games John has won at that point, and t is the number of games Jane has yet to win. If John wins, the value of d increases by 1, from $s-t$ to $(s+1)-t$, and if Jane wins, the value of d increases by 1, from $s-t$ to $s-(t-1)$. At the start of play, $d = -24$, and therefore, after 24 games (that is, 2 hours), $d = 0$. At that time, the number of games won by John will equal the number of games Jane will yet win. Presumably this happens before the midafternoon break because they play 60 games and that takes 5 hours.

S70. The construction is possible. For example, you can take three blocks and build the component shown in (i); then with six blocks you can build three copies of the component shown in (ii). These four components can be put together to form the cube (iii).

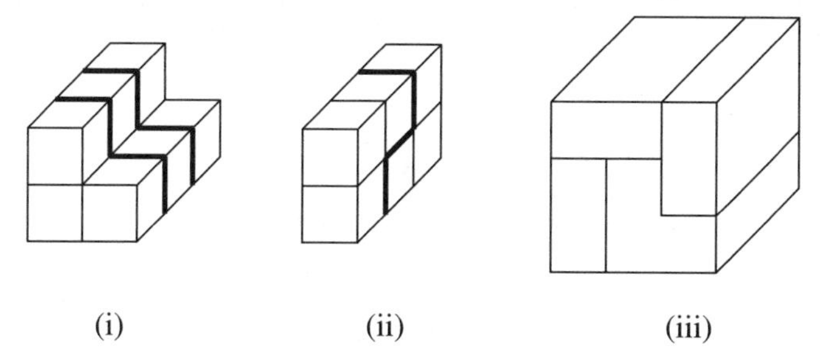

(i) (ii) (iii)

Q70. In how many different ways can you make the cube?

S71. Using the notation introduced in the hint, let $A = 10a+b$. We wish to find a digit x such that

$$9(10a+b) = 100a + 10x + b$$

$$8b = 10a + 10x$$

$$4b = 5(a+x)$$

This shows that 5 divides b, and because $a > 0$ and $b \neq 0$, we must have $b = 5$. Thus $a + x = 4$, and we have four solutions depending upon whether $a = 1, 2, 3$ or 4, namely, $9 \times 15 = 135$, $9 \times 25 = 225$, $9 \times 35 = 315$ and $9 \times 45 = 405$.

S72. The sum of all the numbers in the grid, that is, the numbers from 1 to 30, is 465 which is not divisible by 6. The column sums, therefore, cannot all be equal. The sum is divisible by 5, however, so if it is possible to have equal row sums, each must sum to $465/5 = 93$. There are several ways to do this, one of which is shown below.

30	29	28	3	2	1
27	26	25	6	5	4
24	23	22	9	8	7
21	20	19	12	11	10
18	17	16	15	14	13

Q72. Find a solution with equal row sums having column sums within one of each other.

S73. (a) The smallest amount that can be made up in two different ways from \$4 and \$7 coins is \$28 (see the hint). The smallest amount that can be made up in three different ways is \$56: it can be made from eight \$7 coins, or fourteen \$4 coins, or four of one and seven of the other.

(b) In the hint we showed that every amount greater than or equal to \$18 can be made from \$4 and \$7 coins. It is easy to check that every amount greater than or equal to \$18 + \$28 = \$46 can be made from \$4 and \$7 coins in at least two ways, and every amount greater than or equal to \$18 + \$56 = \$74 in at least three different ways.

Alternatively, let x be the number of \$4 coins and y the number of \$7 coins, and consider the Diophantine equation

$$4x + 7y = n.$$

The general solution (see **Diophantine equations** in the Treasury) is

$$x = 2n - 7t, \qquad y = 4t - n.$$

In our case, we want $x \geq 0$ and $y \geq 0$, so $n/4 \leq t \leq 2n/7$. This interval has length $n/28$, so there cannot be two solutions if $n < 28$ and there cannot be three solutions if $n < 56$. Also, $n = 28$ is the least n for which there exist two solutions: $(x, y) = (7, 0)$ or $(0, 4)$, and $n = 56$ is the least n for which there exist three solutions: $(x, y) = (14, 0), (7, 4)$, or $(0, 8)$. There is only one solution for $n = 45$, but any larger n has two solutions: $n = 46, 47, 48$, $t = 12, 13$; $n = 49, 50, 51, 52$, $t = 13, 14$; $n = 53, 54, 55$, $t = 14, 15$; and so forth. Similarly, any $n > 73$ has at least three solutions.

S74. If you push the U-button seven times ($7 \times U$) you can go from floor 1 to floor 57. If you then push the D-button five times ($5 \times D$) you can go back down to floor 2. In the same way ($7 \times U$ followed by $5 \times D$) you can get from 2 to 3, from 3 to 4, ..., from 10 to 11 (to get from 10 to 11 takes you to floor 66). If you push $4 \times U$ and then $3 \times D$, you can get from floor 2 to floor 1, and similarly, from 3 to $2, \ldots, 11$ to 10. Thus, it is possible to take the elevator between any two arbitrary floors among the first eleven.

We will now show how to get from floor k to floor $k+1$ for any k. Write $k = 11t+r$ and $k+1 = 8m+s$, where the remainder r is larger than zero but smaller than 12 and the remainder s is larger than zero but smaller than 9. First, push $t \times D$ to get down to floor r. Then, from above, take the elevator by the described procedure to floor s. Finally, push $m \times U$ to get from floor s to floor $k+1$. If we write $k = 11t + r$ and $k - 1 = 8n + u$, where $1 \le r \le 11$ and $1 \le u \le 8$, we can, in the same manner, get from floor k to floor $k - 1$.

The preceding instructions show that it is possible to get between any two floors in the building.

S75. The building must have more than 11 floors. In fact, it must have at least 19 floors, for otherwise it would not be possible to go anywhere from floor 11. It is not difficult to show that in a building with 19 floors such an elevator will allow a person to get from any one floor to any other. It's enough to show that one can get from floor k to floors $k+1$ and $k-1$. We can argue as in the preceding problem, with minor modifications, or alternatively, we can follow the following algorithm: go up as far as possible with the U-button, and then down as far as possible with the D-button, and continue this until you arrive at the desired floor. For example, with 19 floors, to get from floor 1 to floor 2, push the buttons, $U, U, D, U, D, U, U, D, U, D, U, D$ which will take you to the following floors: $1 \to 9 \to 17 \to 6 \to 14 \to 3 \to 11 \to 19 \to 8 \to 16 \to 5 \to 13 \to 2$.

If we continue with U, U, D, U, D, U, D we have $2 \to 10 \to 18 \to 7 \to 15 \to 4 \to 12 \to 1$. The combination of these sequences takes us to each of the floors. The answer, therefore, is that the building has 19 floors.

The heart of the solution is that the Diophantine equation $8x - 11y = 1$ has a solution in integers. Also, start at 1 and repeatedly add 8 modulo 19, and visit all the floors: 1, 9, 17, 6, 14, 3, 11, 19, 8, 16, 5, 13, 2, 10, 18, 7, 15, 4, 12, 1, ...

S76. The edge-length of the square is $2\sqrt{5}$, so the edges are each covered by two hypotenuses. These eight triangles must be arranged cyclically round the border, as shown in the diagram, in order that the remaining space, which forms a cross, has rectangular corners. There are 8 ways of filling the cross with 6 dominoes, each domino being a pair of triangles in one of two ways (draw either of its two diagonals). So, by the **Fundamental counting principle** (see the Treasury), noting that we can reflect the whole picture, the total number of solutions is $2 \times 8 \times 2^6 = 1024$.

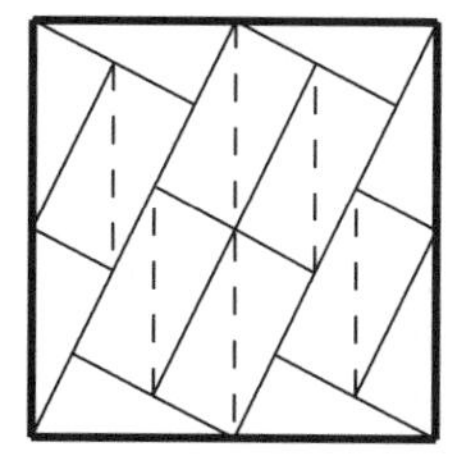

S77. For the sum of the 18 products to be 0, nine of them must be 1 and nine must be -1, so the product, P, of all 18 numbers is -1. But P is also the product of the eleven row products *and* the seven column products, and this is 1 because it is the *square* of the product of all the numbers in the grid. This contradiction shows that such an arrangement of numbers is impossible.

S78. Because four of the non-American, non-Swedish employees are women, there are 20 women who are either American or Swedish. If A denotes the number of Swedish women in the company, then $20 - A$ is the number of American women in the agency. The remaining American employees, equal in number to $15 - (20 - A) = A - 5$, are male. The number of American men working for the agency is therefore five fewer than the number of Swedish women employees. If at least one American man works for the company, then A must be at least 6. Therefore, the number of employees must be at least $15 + 8 + 6 = 29$.

Here's a more algebraic approach. Suppose the number of employees are labeled as in the following table.

	American	Swedish	Other
Males	a_1	b_1	c_1
Females	a_2	b_2	c_2

We are told that

$$a_1 + a_2 = 15 \tag{1}$$

$$a_2 + b_2 + c_2 = 24 \tag{2}$$

$$c_1 = c_2 = 4 \tag{3}$$

Equations (2) and (3) give $a_2 + b_2 = 20$, and substituting (1) gives $b_2 - a_1 = 5$, so the answer to the first question is "No", because there are 5 more Swedish women than American men.

If $a_1 \geq 1$, then $b_2 \geq 6$, $b_1 + b_2 \geq 6$, and (1) and (3) give

$$(a_1 + a_2) + (b_1 + b_2) + (c_1 + c_2) \geq 15 + 6 + 8 = 29.$$

If there is only one American man and no Swedish men, there could be as few as 29 employees.

S79. Suppose the original piece of paper is divided into $a \times a$ small squares, and that we cut out a $b \times b$ square, where a and b are positive numbers. Then $a^2 - b^2 = 148$. Because $a^2 - b^2 = (a - b) \times (a + b)$, $a - b < a + b$, and $148 = 2 \cdot 2 \cdot 37$, it must be the case that $a - b$ and $a + b$ are either equal to 1 and 148 respectively, or 2 and 74, or 4 and 37. If one of the numbers is even and the other is odd, then $a - b$ and $a + b$ would both be odd and their product would also be odd. Because 148 is even, either a and b are both even or both odd. It follows that $a - b$ and $a + b$ are even, and consequently, $a - b = 2$ and $a + b = 74$. These equations imply that $a = 38$ (and $b = 36$), so the original piece of paper measured $9\frac{1}{2}$ inches by $9\frac{1}{2}$ inches.

Q79. A matte for a square picture is made from a rectangular sheet of $\frac{1}{4}$-inch ruled paper by cutting along the lines. The top, left, and right borders are all the same width, but the bottom border is wider, but not more than four times as wide. If the number of $\frac{1}{4}$-inch squares in the four borders between the picture and the frame is 652, what are the dimensions of the picture and the frame?

S80. Each yellow hexagon is adjacent to three black pentagons. If you go around each hexagon and count the number of pentagons adjacent to it you will get a total of $20 \times 3 = 60$ pentagons. But because each pentagon is adjacent to five hexagons, each pentagon will have been counted five times. Thus, the number of pentagons must be $60/5 = 12$. (See **Euler's formula** in the Treasury for a solution where you don't need to know the number of hexagons.)

The soccer ball was not always made in this truncated isohedral shape, and in fact the article, J. M. Goethals and J. J. Seidel, The Football, *Nieuw Archief voor Wiskunde* **29** (1981) pp. 50–58, proposes an even better design. The official soccer ball is what combinatorists call a spherical 5-design, but they find a 9-design which is even more spherical.

S81. Clicker could start at A and follow the path shown in the figure, until he reaches B. The lights are now off everywhere except in A and the room south of B. If he returns to A along the same path all lights will be on except A and B. Next a similar trip from A, but only as far as C and return, will leave all lights on except for A, B, and C. Then a trip from A to D and return leaves A, B, C, D in the dark. If Clicker continues in this way, making 31 return trips in all, the last one being from A to X and back, he leaves alternate rooms in the dark, including his own.

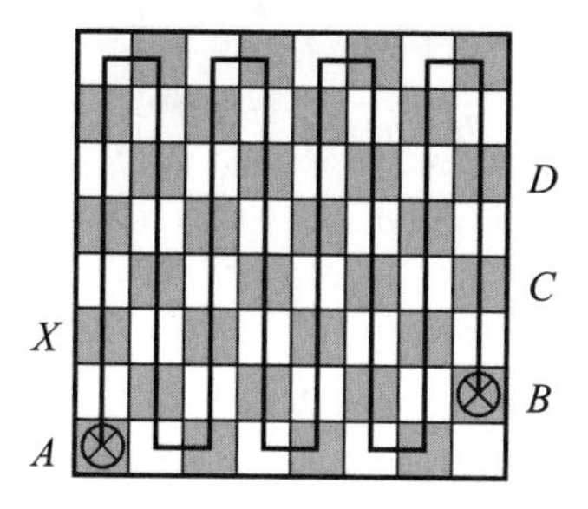

Q81. Can Clicker do it by passing through as few as 64 doorways?

S82. Yes, it is possible. For example, divide both your wires into three pieces of lengths 50, 25, and 25 centimeters. You can make one triangle with your two 50 centimeter pieces and my longest piece (which is less than 100 centimeters). Both my other pieces are shorter than 50 centimeters, so you can use your other two pieces to make two triangles with different colored sides.

S83. The possible scores are $1, 2, \ldots, 14$, 16 or 17 points. Suppose that some child scores 16 points. This child's list agrees with the teacher's list in sixteen places. But then the seventeenth number would have to agree with the teacher's number as well. Therefore, no child could have scored exactly 16 points. So only fifteen scores are possible, and because no two children have the same score, the highest score must have been 17 and the next highest, Mark's score, was 14.

S84. There are several solutions to each part. For example, for the first part, $159 \times 48 = 7632$, and for second part, $402 \times 39 = 15678$.

Q84. How many different solutions can you find?

S85. Of the 97 people who answered "yes" to at least one of the questions, $97 - 78 = 19 = b + y + c$ answered "no" to the first question. Of the 70 who answered "yes" to question B, at least $70 - 19 = 51 = x + w$ also answered "yes" to question A. Among the 97 who answered "yes" to one of the questions, at most $97 - 51 = 46 = y + c + z$ answered "no" to question A or question B (or both). Of the 63 who answered "yes" to question C, at least $63 - 46 = 17 = w$ answered "yes" to all three questions.

Those who answered "yes" to all three of the questions is largest if no one answered "yes" to exactly two of the questions, that is, if $x = y = z = 0$. Then $78 - w$, $70 - w$, and $63 - w$ people answered "yes" only to question A, B, and C respectively. This leads to the equation

$$w + (78 - w) + (70 - w) + (63 - w) = 97,$$

which yields $w = 57$.

The number of people who answered "yes" to all three questions is therefore at least 17 and at most 57.

Here's a more algebraic approach. Using the notation from the hint,

$$a + x + z + w = 78$$

$$b + x + y + w = 70$$

$$c + y + z + w = 53.$$

Adding these gives

$$(a + b + c) + 2(x + y + z) + 3w = 211. \tag{1}$$

Also, we know that

$$(a + b + c) + (x + y + z) + w = 97. \tag{2}$$

Eliminating $a + b + c$ from these equations (by subtracting (2) from (1)) yields

$$(x + y + z) + 2w = 114, \tag{3}$$

and eliminating $x + y + z$ (by subtracting (twice (2) from (1)) yields

$$-(a + b + c) + w = 17. \tag{4}$$

From (3) we see that $w \leq 57$, with equality when $x = y = z = 0$, and from (4), $w \geq 17$, with equality when $a = b = c = 0$. Thus, as before, the number of people saying "yes" to all three questions is between 17 and 57, inclusive.

S86. Continuing from the hint, $b = 12$, $c = 13$. If $a = 14$, then $h = 15$ and 10 is cut off from 7, so $d = 14$, $e = 15$, $f = 16$. To link 7 with 10, $a = 8$, $h = 9$ and finally, $i = 6$, $g = 5$, $j = 4$, $k = 3$, $l = 2$, $m = 1$.

12	11	10	1
13	8	9	2
14	7	6	3
15	16	5	4

Q86. In how many different ways can you fit a sixteen segment numberworm into a 4×4 grid?

S87. (a) Continuing from the hint, b–c–f must be 3–4–5. Then $u = 23$, $t = 22$, and $s = 21$ (not $q = 21$ by the *corner principle* that a corner will be a cul-de-sac if it is not occupied by three consecutive numbers, although this can happen if the corner is one end of the worm, which is not the case here). Thus, $r = 20$. Then, not $q = 19$, $p = 18$, since o or m would become a cul-de-sac, so $q = 1$, $o = 19$. Again, not $p = 18$ else m is a cul-de-sac, so

$$n\text{–}m\text{–}l \quad \text{is} \quad 18\text{–}17\text{–}16$$

and

$$p\text{–}d\text{–}e\text{–}h\text{–}g\text{–}k \quad \text{is} \quad 14\text{–}13\text{–}12\text{–}11\text{–}10\text{–}9.$$

(b) Continuing from the hint, if $d = 18$ then i–j–k–g is 16–15–14–13 and 18 is cut off. Therefore, $d = 16$, $i = 18$ and j–k–g is 15–14–13. Then $l = 19$, and, in order to connect with 26, m–n–o is 20–21–22. Also, in order to use those squares, p–q–r must be 23–24–25. The ends of the worm are

$$s\text{–}t\text{–}u\text{–}v\text{–}w \quad \text{and} \quad x\text{–}y\text{–}z\text{–}*\text{–}\cdots,$$

and there's only one way to fill in these squares.

17	16	9	8	7
18	15	10	11	6
19	14	13	12	5
20	1	2	3	4
21	22	23	24	25

4	5	6	7	8	9
3	26	25	12	11	10
2	27	24	13	16	17
1	28	23	14	15	18
30	29	22	21	20	19
31	32	33	34	35	36

S88. There is such a circuit. Begin by filling in the path through those squares where there are only two outlets, such as the corner squares. By proceeding with this single principle we find that the circuit is completely determined, so there is only one solution.

Q88. (Marc Paulhus) Find such a circuit (of horizontal and vertical segments) for the 10×15 board with guideposts as shown.

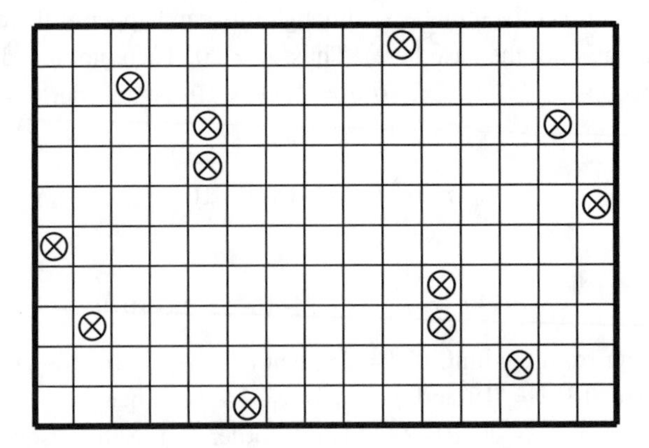

S89. The line segments in the circuit alternate between horizontal and vertical. Moreover, if the circuit begins with a horizontal (vertical) segment, it must end with a vertical (horizontal) segment. This means that the number of segments in a circuit is even, so there is no circuit with 25 segments.

There are many ways to construct a circuit with 26 segments. One idea is to extend the "spikes" in the 16-segment loop shown at the left.

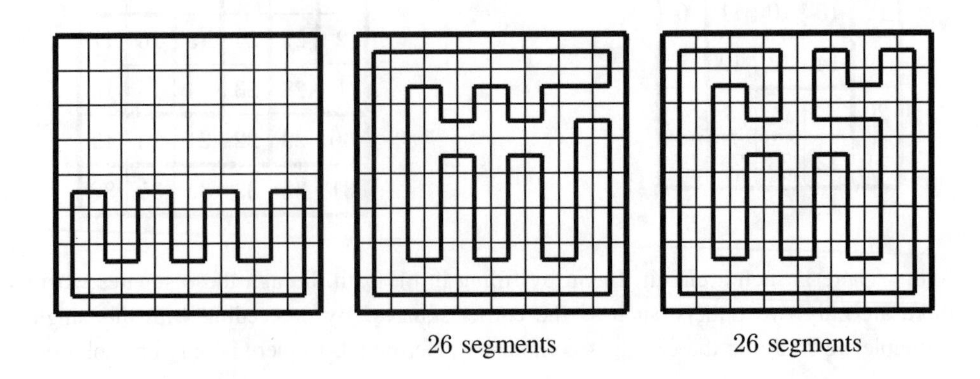

26 segments 26 segments

Q89. What is the fewest number of segments in such a circuit of the 8×8 chessboard? Find as many different circuits as you can with this minimal number of segments.

S90. (a.) No, it is not possible. The reason is that after each move, the number of tails will be an even number, so that there can never be seven tails face-up. To see this, suppose that t go from tails-up to heads-up. Then, at the same time, $4-t$ go from heads-up to tails-up. It follows that the number of tails showing after the move changes by $(4-t)-t = 4-2t = 2(2-t)$, an even number. At the beginning 0 tails are showing, and therefore, after an arbitrary number of even changes, the number of tails showing will remain even.

(b.) Yes, it is possible, and it can be done in three moves. If we write H for heads and T for tails, we can proceed as follows:

$$\text{HHHHHHH} \to \text{TTTTTHH} \to \text{HHHHTTH} \to \text{TTTTTTT}.$$

The following graph provides information that allows you to go from i heads-up to j heads-up in the minimal number of moves, for any pair i and j. The vertices in the graph represent the number of coins that are placed heads-up. Two vertices are connected by an edge if it is possible to go from the one to the other by turning over five coins.

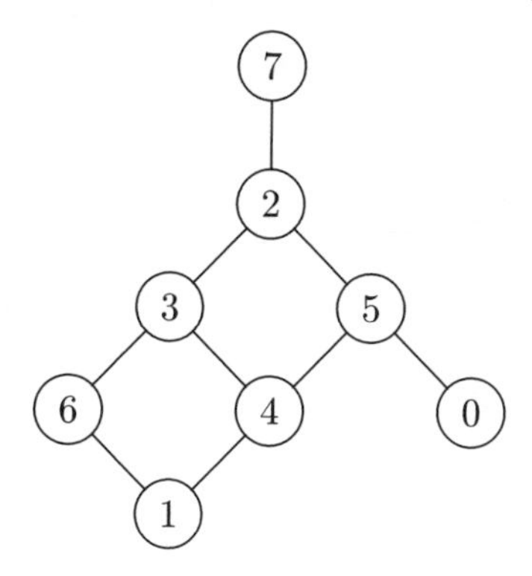

For example, it takes a minimum of three moves to go from six heads-up coins to five heads-up coins, and it can be done in three different ways: (i) from six heads-up to three, to two, to five, or (ii) from six to three, to four, to five, or (iii) from six to one, to four, to five. The greatest "distance" between two vertices is four; from seven to one, or from six to zero.

Q90. Seven pennies lie in a row in the order HHTHHTH. In a single move you are allowed to turn over any five coins, but you are not allowed to change the order of the coins on the table. Using a sequence of such moves, can you get the arrangement THHHHHH?

S91. Let's investigate the possibility (see the hint) that $P = 20$. This implies that $w = 2$. Therefore, the maximum number of points awarded under this system is 9 (Sarah received 9 points in the first week), which means that Camilla could not have acquired 22 points in two weeks. Therefore P must be 10 or 8.

If $P = 10$, then $w = 4$. The three-point award system, therefore, must be $7, 2, 1$, or $6, 3, 1$, or $5, 4, 1$, or $5, 3, 2$. The first alternative would mean that Sarah should have accumulated at least $7 + 1 + 1 + 1 = 10$ points in four weeks. In each of the three other cases, Camilla could not have accumulated 22 points (for example, 3 points in the first week and 6 points in each of the next three weeks yields only 21 points altogether).

This means that $P = 8$ and $w = 5$. The number 8 can be written as a sum of three different positive numbers in two ways, as $5+2+1$ and as $4+3+1$. But the latter would allow Camilla to accumulate at most $3+4+4+4+4 = 19$ points, so the scores must be 5, 2, 1. From this it is easy to deduce that the five weekly awards to Sarah, Camilla, and Daniella were $(5, 1, 1, 1, 1)$, $(2, 5, 5, 5, 5)$, and $(1, 2, 2, 2, 2)$ respectively. Thus, Daniella came in second place in the second week.

S92. In the hint we argued that $4b$ couples danced during the evening. We also know that $7g$ couples danced during the evening (every girl danced with seven boys). Therefore $4b = 7g$, so we may write $b = 7t$ and $g = 4t$, for some integer t. Then $22 < b + g < 44$ becomes $22 < 11t < 44$ so that $t = 3$ and the number of young people at the dance was $12 + 21 = 33$.

S93. In the hint we found that 0 must be placed in circle f. Then 4 goes in circle e. Furthermore, because $4 \to 3$, and $3 = 2 + 1 + 0$, 1 and 2 must go into circles d and c. It is easy to see that 1 must go in d and not in c because $1 \to 12$. Now it is easy to place the remaining numbers.

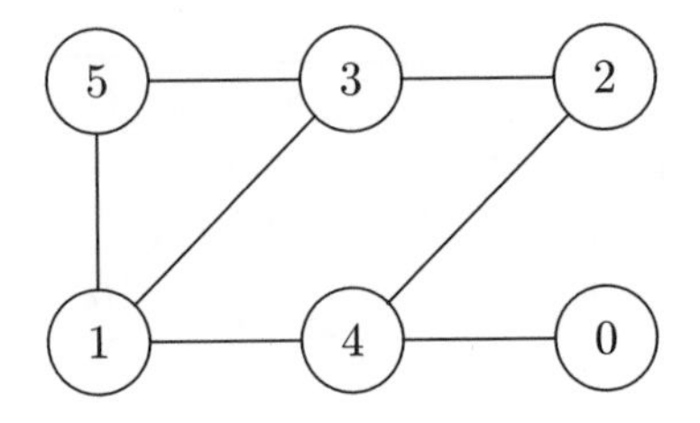

Here's a more algebraic solution. We have $a \to b + d$, $b \to a + c + d$, $c \to b + e$, $d \to a + b + e$, $e \to c + d + f$, and $f \to e$. We know that

$$a + b + c + d + e + f = 15, \tag{1}$$

and adding the right-sides of the arrow relations we have

$$2a + 3b + 2c + 3d + 3e + f = 4 + 12 + 7 + 8 + 3 + 4 = 38. \tag{2}$$

Taking three times (1) and subtracting (2) gives

$$a + c + 2f = 7.$$

This equation implies that $f = 0, 1, 2$, or 3. If $f = 1, 2$, or 3 then $f \to e$ implies that $e = 12, 7$, or 8, which can't happen. Therefore $f = 0$. Then $e = 4$, and $e \to c + d + f$ implies that c and d are 1 and 2, but not in that order because of $1 \to 12$. Therefore, $c = 2$, $d = 1$, $b + e = 7$, $b = 3$ and $a = 5$.

S94. As we saw in the hint, vertex k is 1-valent, so (k, g) is $(2, 8)$ or $(7, 4)$. Vertex 1 is at least 4-valent, and so is vertex 9, since $24 = 7 + 8 + 9$ would duplicate 9. So a and f are 1 and 9 in some order, since g is not 1 or 9. Then $h \to g + f$ which is $4 + 1 = 5$ or $8 + 1 = 9$ or $4 + 9 = 13$ or $8 + 9 = 17$. Only the last two are available. And not $4 \to 13 = 9 + 4$, which duplicates 4, so $h = 5 \to 17 = 8 + 9$, $g = 8$ (and $k = 2$), $f = 9$ and $a = 1$. Now $g = 8 \to 19 = e + f + h + k = e + 16$, and $e = 3$. $a + e = 1 + 3 = 4 \leftarrow b = 7$. Finally, $a + e + f = 1 + 3 + 9 = 13 \leftarrow 4 = d$, leaving $c = 6$.

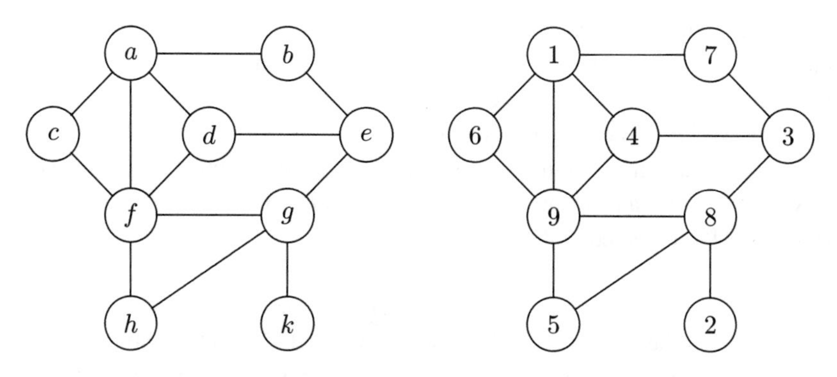

In part (b) of the hint, we established that $t = 5, r = 0, p = 2, s = 7, q = 10$ and $u = 6$. From $6 \to s + x$ we conclude that $x = 3$; from $v \to 8$ we get $v = 4$, and from $3 \to u + v + y$ we get $y = 8$. Then $8 \to t + x + z$ implies that $z = 1$, which means that $w = 9$.

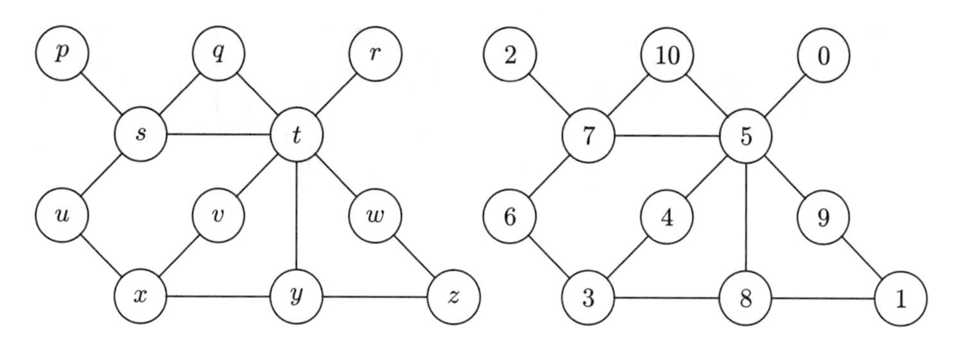

S95. Let t denote the number of times Romeo and Juliet have met when they meet at the entrance for the first time. Then $20t$ must be a multiple of 36 and a multiple of 45. The smallest such common multiple is 180, namely, $20t = 36 \times 5 = 45 \times 4$. That is to say, after five rounds for Romeo and four rounds for Juliet, they will meet at the entrance for the first time, and this will happen after 180 seconds, disregarding the waltzes. They will meet nine times in their circuits, and therefore, their appearance takes $180 + 9 \times 10 = 270$ seconds, or equivalently, four and one-half minutes.

S96. In the hint we found that S can be at most 17. For this to happen, $C + E + G = 23$, and this can happen only for the digits 6, 8, and 9. The pair C and E cannot be 8 and 9 because then $C + D + E$ would be more than 17. Similarly, E and G cannot be 8 and 9 because $E+F+G$ would be more than 17. Therefore C and G must be 8 and 9, leaving E to be 6. This yields a solution: $1+7+9=9+2+6=6+3+8=8+5+4$.

Q96. Is E uniquely determined if the sum of these four numbers is as small as possible?

S97. In the hint we have seen that none of the letters can denote the number 0. The same argument shows that none can equal 5 or 7 either. (If one of the numbers is 5, then P is divisible by 5. But 5 can occur in at most two of the triples, so the third triple could not be divisible by 5.) The remaining seven digits, therefore, must each occur just once. Because 1 occurs as one of the factors in one of the triples, P can be at most $1 \times 8 \times 9 = 72$. On the other hand, P must be divisible by both 8 and 9, and therefore $P = 72$. The only ways to express 72 as the product of three single digits are $1 \times 8 \times 9$ or $2 \times 4 \times 9$ or $3 \times 4 \times 6$, where the only common digits are 9 and 4, which must be C and E in some order. Apart from rearrangement, the only solution is $8 \times 1 \times 9 = 9 \times 2 \times 4 = 4 \times 6 \times 3$. In each case, D must be equal to 2.

S98. In the hint we established that the first letter in the top row is A and the last letter in the top row and the last letter in the first column is B. The other two letters in the top row must be different, and the same is true for the other two letters in the first column. So the first row and column are AABB and ABAB in some order. The conditions on the words now forces all but the last row and column; then to avoid duplicates, the third row and second column are forced, and then so is the last row and column. It follows that there are just four solutions.

(i)

A	A	B	B
B	B	A	B
A	A	B	A
B	A	B	A

(ii)

A	B	A	B
A	B	A	A
B	A	B	B
B	B	A	A

(iii)

B	B	A	A
A	A	B	A
B	B	A	B
A	B	A	B

(iv)

B	A	B	A
B	A	B	B
A	B	A	A
A	A	B	B

Note that (ii) is a reflexion of (i) about the falling diagonal, and (iii) and (iv) are obtained from (i) and (ii) by swapping A and B.

S99. If k denotes the number of men who left without galoshes, then another k men went out with the galoshes of these men, so $2k+3 \leq 16$, and consequently k cannot be larger than 6.

Q99. Is it possible for six men to leave without galoshes (and three to leave with galoshes)?

S100. Suppose $s = 15$ and $m = 0$. The only triples containing $m = 0$ are 960 and 870, whereas we need three such triangles for B, C, E. Therefore $s = 14$ and $m = 3$. There are exactly three possibilities for triangles adding to 14 containing 3: 923, 743, and 653. The possibilities for the corner triangles are 950, 941, 860, 851, 842, 761, and 752; we need three disjoint triangles from this list, and there are just two sets of these:

$\{950, 842, 761\}$ and $\{941, 860, 752\}$. Only the first of these connects with $m = 3$ in the middle. So, except for symmetries, there are just two solutions: the one at the left, and the one at the right with numbers subtracted from 9 and triangle sums of 13.

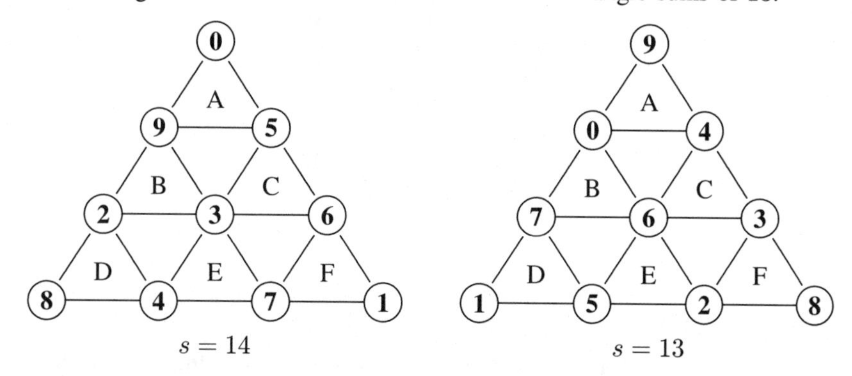

S101. Let a, b, c, d be the four costs in increasing order, and let $D = a + b + c$, $C = a + b + d$, $B = a + c + d$ and $A = b + c + d$. Then $A > B > C > D$, and $A{+}B{+}C{+}D = 3(a{+}b{+}c{+}d)$. If $A = 46$, then $A{+}B{+}C{+}D = 43{+}44{+}45{+}46 = 178$, which is not a multiple of 3. If $B = 46$, then $A + B + C + D = 182$, which is not a multiple of 3. If $D = 46$, then $A + B + C + D = 190$, which is not a multiple of 3. Therefore $C = 46$ and $A + B + C + D = 186$, so $a + b + c + d = 62$. The cost of the saucer is $a = 62 - A = 14$ dollars, and the other items cost 15, 16, and 17 dollars.

S102. If there were six or more countries, we could divide the coins according to their country, and then take at most two coins from each country until we have a total of eleven coins. We then will have 11 coins without 3 from any one country, so there are at most five countries. If there were no more than 19 coins from each country, there would be at most 95 coins altogether, and therefore there are at least 20 coins from some country.

> **Q102.** Could Peppi conclude that at least 21 coins in her backpack are from the same country?

S103. Continuing, using the notation introduced in the hint, the possibilities for a, b, c, d and e, are given in the following table:

16		d	
f	e	11	8
g		c	
a	15	14	b

From the hint

	i	ii	iii	iv
a	1	2	3	4
b	4	3	2	1
c	2	5	4	7
d	7	4	5	2
e	12	10	12	10

The second and third columns are ruled out when we calculate $f = 15 - e$, and the first when we calculate $g = 18 - f - a$. The missing numbers 3, 6, 12, 13 fit into the remaining squares in just one way, giving the famous magic square from Albrecht

Dürer's *Melencolia I* (now in the UCLA Grunwald Center for the Graphic Arts) with its date, 1514, in the middle of the bottom row.

16	3	2	13
5	10	11	8
9	6	7	12
4	15	14	1

S104. Color the original grid in a checkerboard fashion (white in the lower right corner) and note that the difference between the sum of the numbers in the white squares and the sum of the numbers in the black squares is equal to 3. This difference is unchanged by the operation, and this means that one can never obtain the position required in line (a) (all zeros would imply that the difference would be 0), or in line (b) (all zeros except for a single 1 would imply that the difference would be 1 or -1), or in line (d) (all 1's would imply a difference of 1).

One can obtain the configuration described in (c). One can proceed, for example, as shown in the following figure (the two adjacent squares are marked with a heavy border).

4	6	0
8	7	2
5	0	3

$\longrightarrow$

1	3	0
5	4	0
5	0	1

$\longrightarrow$

1	0	0
0	1	0
0	0	1

Q104a. Describe the grids that can be attained from the original starting position.

Q104b. Suppose you are allowed only to *decrease* (by a positive number) the two numbers in two adjacent squares. Some grids that can be attained using this operation are *minimal* in the sense that no further subtractions are possible without producing negative numbers (for example, the grid in (c), with 1's on the main diagonal and zeros elsewhere). Describe the set of attainable minimal grids.

S105. (a) We can get rid of any four consecutive numbers by the sign pattern $+--+$. Because 2001 is 1 more than a multiple of 4, we can make the last 2000 terms add to 0, and get the sum to add to 1:

$$1 = 1 + \underbrace{(2-3-4+5)}_{=0} + \cdots + \underbrace{(1998-1999-2000+2001)}_{=0}.$$

(b) Consider four consecutive numbers: $n, n+1, n+2$, and $n+3$. We see that

$$n^2-(n+1)^2-(n+2)^2+(n+3)^2 = n^2-n^2-2n-1-n^2-4n-4+n^2+6n+9 = 4.$$

It we take the next four consecutive numbers, but exchange the signs on their squares, we have $-(n+4)^2+(n+5^2+(n+6)^2-(n+7)^2=-4$. Consequently, we can get rid of any eight consecutive numbers by the sign pattern $+--+-++-$. Because 2001 is one more than a multiple of 8, we can get the sum to add to 1.

S106. On Monday, Skip catches Harry for the first time when his father has run one-third of a lap. At this point, Skip has run one and one-third laps. Therefore, Skip's pace is four times faster than Harry's. On Tuesday, Harry and Skip meet for the first time after Harry runs one-fifth of a lap and Skip runs four-fifths. They meet again when Harry completes two-, three-, and four-fifths of a lap, and then again when Harry finishes the lap, which means that they meet five times after the start.

S107. Every number with digits a and b in this grid can be written as $10a+b$. Therefore, instead of one grid of numbers, we can consider two grids, one for ten times the tens digit and the other for the units digit.

0	0	0	0	0	0	0	0	0	0
10	10	10	10	10	10	10	10	10	10
20	20	20	20	20	20	20	20	20	20
30	30	30	30	30	30	30	30	30	30
40	40	40	40	40	40	40	40	40	40
50	50	50	50	50	50	50	50	50	50
60	60	60	60	60	60	60	60	60	60
70	70	70	70	70	70	70	70	70	70
80	80	80	80	80	80	80	80	80	80
90	90	90	90	90	90	90	90	90	90

0	1	2	3	4	5	6	7	8	9
0	1	2	3	4	5	6	7	8	9
0	1	2	3	4	5	6	7	8	9
0	1	2	3	4	5	6	7	8	9
0	1	2	3	4	5	6	7	8	9
0	1	2	3	4	5	6	7	8	9
0	1	2	3	4	5	6	7	8	9
0	1	2	3	4	5	6	7	8	9
0	1	2	3	4	5	6	7	8	9
0	1	2	3	4	5	6	7	8	9

Now distribute 50 minus-signs among the numbers in the original grid, and use the same distribution of minus-signs in each of the two grids shown above. If we denote the sum of the numbers of these two grids by S_1 and S_2 respectively, then it is clear that $S=S_1+S_2$.

Because five of the numbers in each row of the original grid are negative, the sum of each row of the first of the two grids is equal to 0. Because S_1 is the sum of the row sums of the first grid, it also must be 0. In the same way, we find that $S_2=0$ (because five of the numbers in each column of the original grid are negative, the column sums of the second grid are 0). It follows, therefore, that $S=0$ regardless of how the minus-signs are distributed, as long as each row and column contains exactly five of them.

S108. There are $7\times 7\times 7=343$ such numbers less than 1000 (each digit can be chosen in 7 different ways). The largest three-digit number left is 888 and it is the 343rd number in the sequence. The next number that remains is 1000 and $7\times 7=49$ places after 888 is the number 1088. In the same way, 49 places further in the sequence is the number

1188, and so forth, $4 \times 49 = 196$ places after 888, in the $343 + 196 = 539$-th place, is the number 1488. Seven places further is 1508, and then 5×7 places further, in the $(539+7+35)$-th place, is the number 1578. The next six numbers are 1580, 1581, 1582, 1584, 1585, and 1587. We find the 587th number is 1587, a number that does not contain the digit 2.

There's an easier way to do it. Note that in this deleted set we're counting in base 7, but using the symbols $0, 1, 2, 4, 5, 7, 8$ in place of the usual digits $0, 1, 2, 3, 4, 5, 6$, respectively. As we are counting from 0, the 587th number will be the equivalent of 586_{10} (that is, 586 base 10), which is 1465_7 (keep dividing by 7), which is 1587 when translated into our symbols!

Q108. What is the millionth number in the sequence?

S109. If we think of the friends as living along the same street, then their locations are shown below.

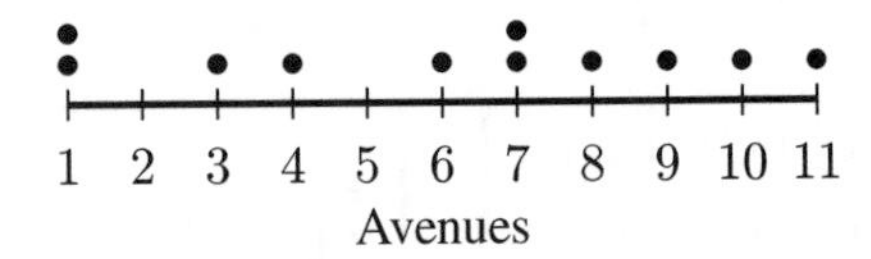

We find that the friends should meet on 7th Avenue, where the "middle" person lives (the *median* of the set of avenue numbers). This is because if we move the meeting place one block to the left (or to the right) the distance would decrease by one for five of the friends (or four of them, respectively) but would increase by one for six of the friends (or seven of them, respectively). The total distance would therefore increase. This means that 7th Avenue is optimal.

Now disregard the avenues and think of the friends as living on the same avenue. Their locations along such an avenue are shown below.

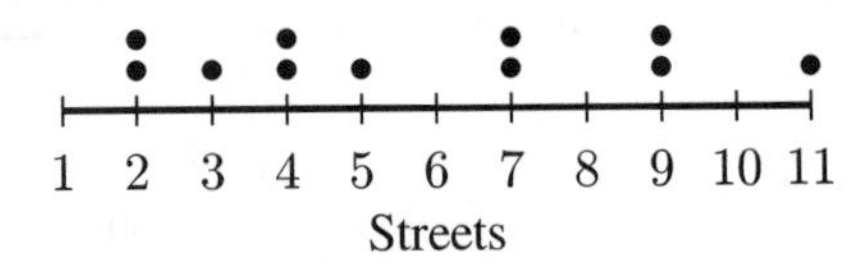

Reasoning as before, we conclude that the friends should meet on 5th Street. Putting these results together, the friends should meet at the intersection of 5th Street and 7th Avenue.

S110. In the hint we have seen that there could not have been 17 containers in the laboratory on May 1. But 553 can also be written as $553 = 16 \times 31 + 57$ (that is, we could have put out 16 containers on May 1). Because $57 = 3 \times 19$, the first new container was placed in the incubator 19 days prior to June 1, on May 13.

S111. We have seen from the hint that the number of circles required for such arrangements, if they exist, are six and twelve respectively. Arrangements with these minimal numbers of circles are possible.

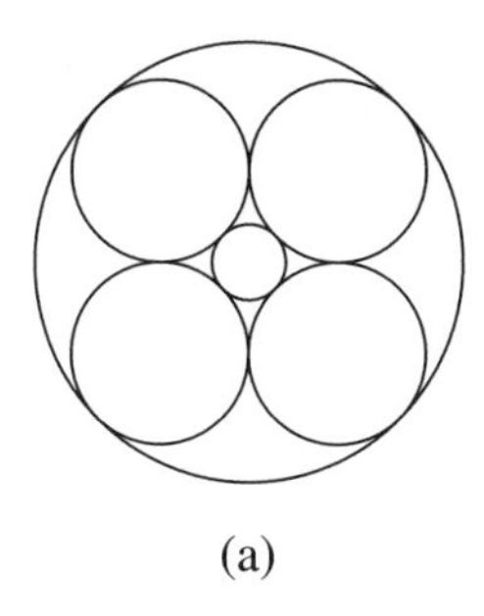

(a)

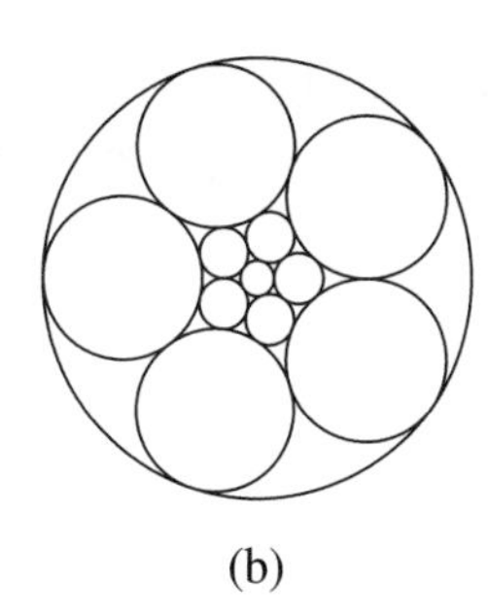

(b)

Q111. Can you arrange a collection of circles (not necessarily of the same radius) so that each circle is tangent to exactly six others and no three circles pass through the same point?

S112. The following figures show that it is possible to construct a closed polygon with six edges in which each edge intersects exactly one other, and a closed polygon with seven edges in which each edge intersects exactly two others.

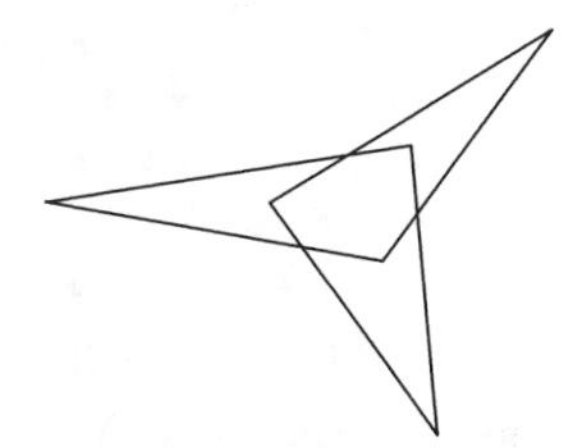

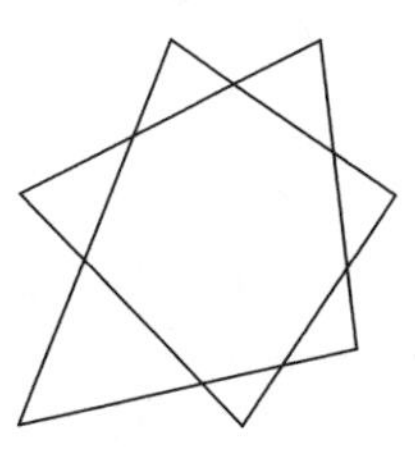

It is not possible to construct a closed polygon with six edges in which each edge intersects exactly two others. For suppose there is such a polygon. The intersection points divide the edges into three parts. If we consider only the middle parts (see below) we see that they form a hexagon. If we lengthen the edges of this hexagon we get two overlapping triangles, which is not a closed polygon.

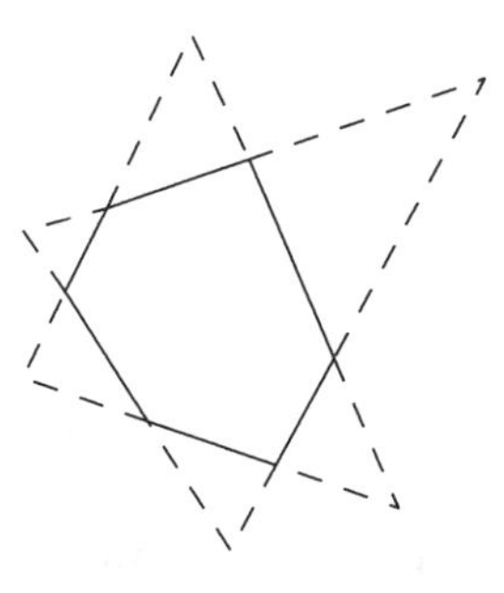

Likewise, it is not possible to construct a closed polygon with seven edges so that each will intersect exactly one other. For each edge, seven of them, uniquely determines the edge that it intersects, which in turn determines the intersection point. But each intersection point is counted twice in this way, which implies that there are $7/2$ intersection points, an impossibility because $7/2$ is not an integer.

S113. Yes, there is a polygon with 16 corners.

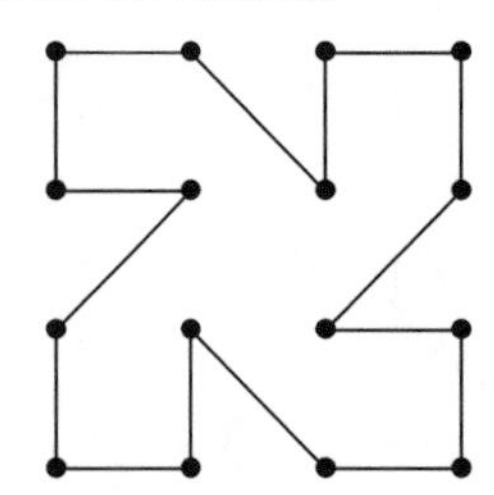

Q113. What is the maximum number of vertices of such a polygon drawn on an $n \times n$ grid of points?

S114. The polygon on the triangular array can't have more than 13 corners. In the figure at the left, fourteen points are visited, but one of them isn't a corner. On the other hand, the figure on the right shows how you can make a corner at every one of the 37 points of the hexagonal array.

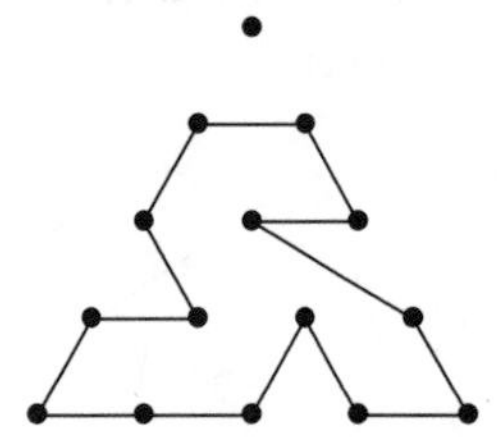

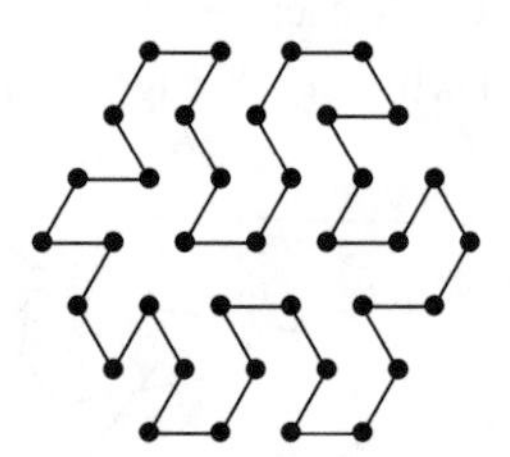

Q114. These two problems used a triangular grid with $n = 5$ points on each side, and a hexagonal grid with $n = 4$ points on each side. What about other values of n?

S115. The answer to the first question is "Yes." Let E denote an even number and O an odd number. Then consider the sequence EOO EOO EOO EOO EOO EOO EOO EOO. Any three successive terms are comprised of two odds and one even, so their sum is even. Furthermore, 17 of the numbers in the sequence are odd and so the sum of all 25 numbers is odd.

It is not possible to construct a sequence of numbers to satisfy the conditions of the second problem. For suppose we have constructed a sequence of 25 numbers so that the sum of any five successive terms is even but the sum of all 25 numbers is odd. Partition the sequence into five sets of five successive terms. The sum of the numbers in the sequence is equal to the sum of the totals from the five sets, and therefore is a sum of five even numbers. It follows that the sum of all the numbers in the sequence is even.

S116. (a) The number 2 must appear in the down diagonal. It can only be placed on the middle square. After this, it is easy to figure out where the other twos must go. Continue next with the fives, and then with the other digits.

(b) There must be a 1 in the down diagonal, and it has to be in the fourth place. Then the 1 in the other diagonal is in the lower left corner, and this determines the last of the ones. Now the up diagonal reads $1, 2, 3, 4, 5$, and the top row is $2, 4, 1, \ldots$, and so forth.

1	2	3	4	5
2	5	4	1	3
4	3	2	5	1
5	4	1	3	2
3	1	5	2	4

(a)

2	4	1	3	5
3	5	2	4	1
4	1	3	5	2
5	2	4	1	3
1	3	5	2	4

(b)

S117. If the number A has n digits, then A^2 has either $2n$ or $2n-1$ digits, and $2A$ has either n or $n+1$ digits. Thus A^2 and $2A$ have the same number of digits if and only if n is equal to 1 or 2. If A has two digits then A^2 must have three digits. If $2A$ has three digits then A must be at least 50, but then A^2 would have four digits (since A^2 would be at least 2500). Thus A must have just one digit. A check shows that only 0, 2, and 9, satisfy the conditions of the problem, so there is just one other such number, namely 9.

Q117. What if we only ask for the *sum* of the digits of $2A$ and A^2 to be equal?

S118. Think of the $3 \times 3 \times 3$ cube as made up of small black and white $1 \times 1 \times 1$ cubes arranged in a checkerboard fashion: two small cubes with a common face have different colors. Let us suppose that the $1 \times 1 \times 1$ space in the center is black. Then the remainder of the $3 \times 3 \times 3$ cube will consist of 12 black and 14 white cubes. Each $1 \times 1 \times 2$ block consists of two small cubes, one of each color. Thirteen such blocks, therefore, contribute 13 white and 13 black cubes, and therefore the desired construction is not possible. (You can pack the thirteen blocks if you leave a corner empty.)

S119. Two sets of pieces will tile the 5×8 grid; here's one solution.

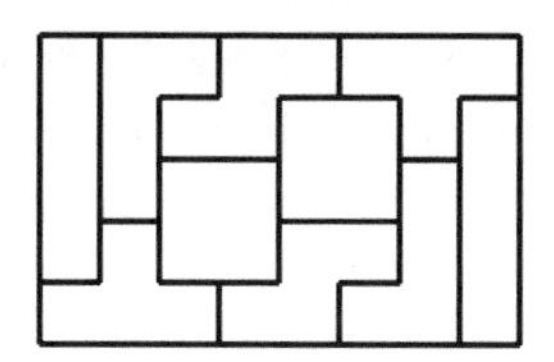

A computer search (using a backtracking algorthim) showed that there are 783 solutions, 13 of which have symmetry with respect to a 180° rotation about the center, such as the one above.

Three sets of five tetrominoes will not tile a 12×5 grid. For if the squares of the grid are colored in a checkerboard fashion, half of them are white and half are black. On the other hand, the S, L, I, and Z pieces have an equal number of white and black squares, whereas the T has an unequal number (1 and 3). Thus, the number of white and black squares among the five pieces is not equal, as it would have to be in order to tile the grid.

Q119. Can you tile a 4×10 grid with two sets of tetrominoes?

S120. Using the notation introduced in the hint, let us suppose that some line in the first set, say for example line t_1, is also a line in the second set, say s_1. If none of the lines

t_2, t_3, t_4 are parallel to any of the lines s_2, s_3, s_4, s_5, these lines will intersect pairwise in $3 \times 4 = 12$ points. There is one other line (the ninth), and if it is not parallel to any line in either set, it will intersect these lines in eight points, which gives a total of $2 + 12 + 8 = 22$ intersection points. But this contradicts our assumption that we have only 21 intersection points.

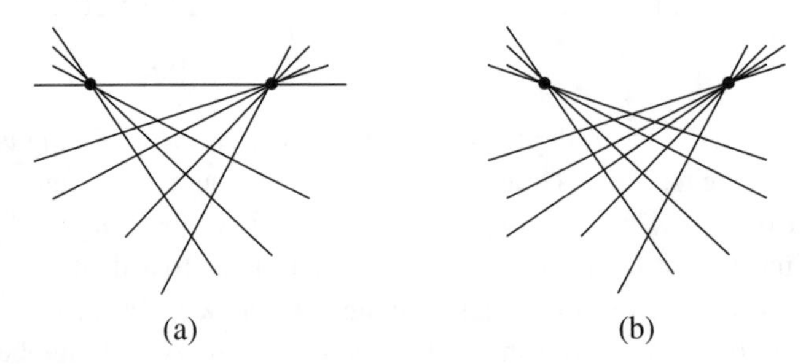

(a) (b)

Now assume that no line in the first set is in the second set. If no line in first set is parallel to any line in the second set, then they intersect in pairs to give $4 \times 5 = 20$ intersection points, for a total of $2 + 20$ intersection points all together. Again, this is too many, and therefore we must conclude that at least two of the nine lines are parallel.

Q120. Can you say, in each case, how many pairs of parallel lines there are among the nine lines?

S121. Color the grid as a checkerboard (white in the lower right corner). We observe that four of the five deleted squares are black (B) and one is white (W). Each domino covers one white and one black square. The deleted board will have three more white squares than black, and therefore at least three white squares cannot be covered by the dominoes.

One approach for covering the deleted board is to construct a tour through all but three of the white squares (the three marked with an $\times$ in the following grid), using only horizontal and vertical lines. The dashed line is such a tour. Then lay down dominoes along the path to get the desired covering.

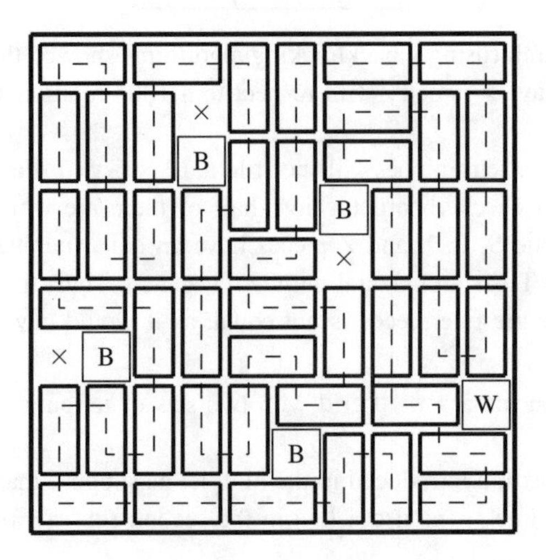

S122. Since at the end there is only one type of item left, one of the sums $C+B$, $B+K$ and $C+K$ (see the hint) must be 0, while the other two are the same. In the beginning, $C+B=73$, $C+K=74$ and $B+K=77$. After each visit one of the sums $C+B$, $C+K$, $B+K$ decreases by 2 while the other two remain unchanged. So, $C+K$ must eventually equal 0, because the other two sums will always be odd. Therefore, at the end, Elizabeth's hiding place will only contain baseball cards.

Q122. How many baseball cards could there have been at the end?

S123. According to the new rule, the chips lying on the black squares will always move to black squares, and those on the white squares will move to white squares. To reorder your chips, you can separately reorder the eight chips that lie on the black squares ($7+6+5+4+3+2+1=28$ exchanges) and then reorder the seven chips that lie on the white squares ($6+5+4+3+2+1=21$ exchanges). The total number of exchanges is therefore $28+21=49$.

I can never reorder my chips according to the rule. If the first chip lies on a black square, it can never occupy the last square which is white.

S124. (a) Yes, it is possible as the following figure demonstrates.

1	1	1	1	4
−1	−1	−1	0	−3
0	1	1	1	3
−1	−1	0	0	−2
−1	0	1	2	

(b) No, it is not possible. First observe that the row and column sums account for eight of the possible nine sums, so either 4 or -4 occurs among these sums. A solution remains a solution if you change the signs of all the entries, so we may assume that 4 is a row or column sum. A solution remains a solution when the grid is reflected about a diagonal, so we may assume that 4 is a row sum.

Given that some row is filled with 1's, no column or diagonal sum can be -4, so suppose that some row sum is -4. Then no column or diagonal sum can be 3 or -3, so one of the row sums is 3 and another is -3. But this implies that two column sums must be equal.

Q124. Is it possible to place $-1, 0, 1$ on the squares of a 2000×2000 grid so that the 4000 row and column sums will all be different?

S125. Yes, it is possible. Here's one way of doing it.

1	1	1	1	0	4
−1	0	−1	−1	−1	−4
1	1	1	0	0	3
1	−1	−1	−1	−1	−3
2	1	0	−1	−2	

Q125. Is it possible to place -1, 0, 1 on the squares of a 2000×2001 grid so that the row and column sums will all be different?

S126. Let a and p denote Anna's and Peter's dollar allowances, respectively. We know that Anna is older than Peter, say $a = p + k$, where k is at least 1. If we double Anna's allowance, $2(p+k)+p = 3p+2k \leq 31$. If we double Peter's allowance, $2p+(p+k) = 3p+k \geq 29$. Twice the latter inequality minus the former gives $2(3p+k)-(3p+2k) \geq 27$, or equivalently, $3p \geq 27$, $p \geq 9$ (see **Inequalities** in the Treasury). Because $k \geq 1$, the first inequality gives $3p \leq 31 - 2$, so $p = 9$. But now the first and second inequalities give $k \leq 2$ and $k \geq 2$ so $k = 2$ and Anna is 11 and Peter is 9.

S127. Here are three solutions from which all others can be generated by rotation, reflexion, or cycling, as we shall see.

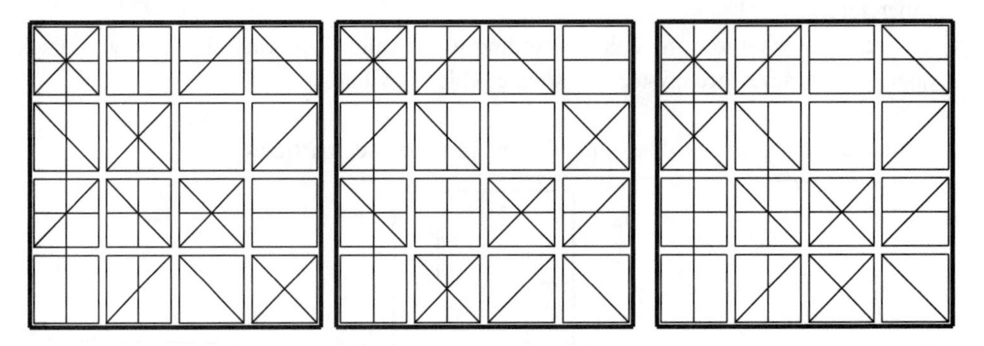

Specifically, you can generate other solutions by reflecting the board about a line through the center (horizontal, vertical, or diagonal), and by rotating the board 90°. In addition, you can produce other solutions by cycling the topmost row to the bottom or the rightmost row to the left. This is because these solutions have the toroidal property: when a line leaves one edge it re-enters at the corresponding point on the opposite edge. Thus, rotation, reflexion, and cycling, or any combination of these, will generate new solutions.

For a proof that these give all possible solutions, see Barry Cipra, The Sol LeWitt Puzzle, *Puzzlers' Tribute: A Feast for the Mind*, Tom Rodgers, David Wolfe, editors, AK Peters, Natick MA, 2001.

S128. (a) With each move you add two disks, so there is always an odd number of disks on the wheel. Should you be able to get the same number of disks on each space, the number of disks on the wheel would be even. Therefore, the answer to the question is "No!"

(b) It is not possible to get the same number of disks on each space in this case either. To see this, color the spaces alternatively black and white in such a way that the spaces occupied by the two disks in the starting position are black. When you add two disks to two neighboring spaces, one of them will be on a white space and one on a black space. Therefore the number of disks on the four black spaces will always be two more than the number of disks on the four white spaces.

S129. Color every other space red. The number of disks on the red squares increases by 2, decreases by 2, or remains unchanged, after each move. At the start, there are three disks on the red squares, so there will always be an odd number of disks on the red spaces. Therefore, six disks on the same space can never occur.

Here's an alternative solution: Let S denote the sum of the products of the space number and the number of disks on that space. In the beginning, $S = 1\times1+2\times1+3\times1+4\times1+5\times1+6\times1 = 21$. If a disk moves one space counterclockwise from positions 2, 3, 4, 5, or 6, the value of S decreases by 1, and if the disk comes from position 1 it increases by 5. If a disk moves one space clockwise from positions 1, 2, 3, 4, or 5, the value of S increases by 1, and if the disk comes from position 6 it decreases by 5. In each move, two disks move one space (clockwise or counterclockwise, independently), and the result is that S decreases by 6, 4, 2, or 0, or increases by 0, 2, 4, or 6. In each case, S changes by an even number. Because at the beginning S is an odd number, it will always be an odd number. Suppose it is possible to move all the disks to space k. Then S would equal $6k$, an even number. This shows that the position cannot be attained.

S130. The sum of the lengths of fencing is $1+2+\cdots+43 = 946$. If a square has a side whose length is a whole number, then its perimeter must be divisible by 4. But 946 is not divisible by 4, so it is not possible to construct a square with the given line segments.

For the second part, divide the line segments into 22 sets as follows: $\{1,42\},\{2,41\}$, $\{3,40\},\ldots,\{21,22\},\{43\}$. Twenty-one of the sets have two elements and one has just one element, but in each case, the sum of the lengths of the segments in the set is the same: 43 meters. It is now an easy matter to construct a rectangle having one side $43k$ meters and the other $43(11-k)$, for each integer k between 1 and 5 (43×430, 86×387, 129×344, 172×301, and 215×258).

Q130. What other rectangles can be constructed?

S131. The answers are uniquely determined by the conditions.

1	2	5	3	7	9	4	6	8
9	6	4	2	8	1	5	3	7
8	7	3	4	6	5	1	2	9
3	5	1	6	2	8	9	7	4
6	4	2	9	1	7	3	8	5
7	8	9	5	4	3	2	1	6
5	9	8	7	3	2	6	4	1
2	1	6	8	9	4	7	5	3
4	3	7	1	5	6	8	9	2

(a)

2	3	5	8	7	6	9	4	1
7	8	1	5	4	9	3	6	2
9	4	6	1	3	2	8	7	5
3	9	7	6	5	8	1	2	4
1	6	4	9	2	7	5	8	3
8	5	2	3	1	4	6	9	7
5	2	9	7	8	3	4	1	6
4	1	8	2	6	5	7	3	9
6	7	3	4	9	1	2	5	8

(b)

S132. The first question is answered in the hint. The smallest number which satisfies the second question has 31 digits: 1213121412131215121312141213121.

If you're patient, you can carry the pattern as far as you like. For example, to get from $1, 2, 3, 4, 5$, to $1, 2, 3, 4, 5, 6$, copy the 31-digit number for 5, adjoin 6, and then copy the 31-digit number again. For $1, 2, \ldots, 9$, the number will have $2^9 - 1 = 511$ digits.

S133. (a) Using the notation from the hint, we can write

$$\begin{aligned} D = 10b + d &= 10(A - 10a) + (B - 10c) \\ &= (10A + B) - 10(10a + c) \\ &= 10A + B - 10C. \end{aligned}$$

Since K divides A, B, and C, it must also divide D.

(b) This can be done using the same method.

a	b	c	A
d	e	f	B
g	h	i	C
D	E	F	

Using the notation in the figure,

$$\begin{aligned} A &= 100a + 10b + c \\ B &= 100d + 10e + f \\ C &= 100g + 10h + i \\ D &= 100a + 10d + g \\ E &= 100b + 10e + h \\ F &= 100c + 10f + i. \end{aligned}$$

It follows that

$$\begin{aligned} F &= 100(A - 100a - 10b) + 10(B - 100d - 10e) + (C - 100g - 10h) \\ &= 100A + 10B + C - 100D - 10E. \end{aligned}$$

Because A, B, C, D, E are divisible by K, so is F.

(c) From the last equation in part (b),

$$100A + 10B + C - 100D - 10E - F = 0.$$

Because K divides five of the six among A, B, C, D, E, F on the left side of this equation, it must also divide the sixth. (Do you see where we used the hypothesis that K is a prime larger than 5?)

S134. We have established (see the hint) that a Neptune has one head and a Siren has two. Let n denote the number of Neptunes by the Great Volcano, and m the number of

Sirens, and let y be the number of legs on each Siren. Then $n+2m=23$ (the number of heads) and $3n+ym=134$ (the number of legs). The first equation multiplied by 3 gives $3n+6m=69$, and subtracting this from the second equation yields $(y-6)m=65$. But $65=5\times 13$, and since m cannot be larger than 11 (eleven Sirens account for 22 heads), and m must be larger than 1 (implied in the problem, *some* Sirens), we must have $m=5$. This yields $y-6=13$, so there are 5 Sirens near the volcano, each Siren has 19 legs, and there are 13 Neptunes in the group.

S135. Let m and w denote, respectively, the number of men and women who took part in the contest. If each of the women had solved five problems correctly and each of the men had solved four problems correctly, the combined total score would be $4m+5w$. In the other case, their score would have been $5m+4w$. But the first result is 4 percent better than the second, so

$$(5m+4w)\frac{104}{100}=4m+5w.$$

Multiplying each side by 100 gives $520m+416w=400m+500w$, or equivalently, $10m=7w$. It follows that the number of men is divisible by 7. If $m=7$ then $w=10$. If $m\geq 14$, then $w\geq 20$. But the number of contestants is at most 33 (35 minus *several*) so the alternative $m\geq 14$ must be rejected. Therefore 7 men and 10 women took part in the contest.

S136. We've seen from the hint that if there is a solution, the magic product will have at least three prime divisors. It can't be of the form p^2qr with p,q,r distinct primes, because it has 12 divisors, only 8 of which are not divisible by p^2. So at least one entry is divisible by p^2; in fact, it must be p^2 with the other entries in its row and column being $1,q,r$ and qr in some order. But there is no way to complete the other line containing qr with two entries whose product is p^2. So we must try p^3qr, and there is a solution with a magic product $120=2^3\times 3\times 5$.

4	10	3	120
15	1	8	120
2	12	5	120
120	120	120	

S137. As for most relative velocity problems, bring something to rest! An observer in a boat floating by the goggles (at rest relative to the goggles) need not know how fast she is moving. She sees John swimming away for 10 minutes, and knows, if he maintains his pace, that he'll be back in another 10 minutes. The goggles go half a mile in 20 minutes, or $3/2$ mile per hour.

S138. Yes, it can be done. First, fold the left third down, then the bottom up, and finally, fold the left third up and tuck it into the pocket.

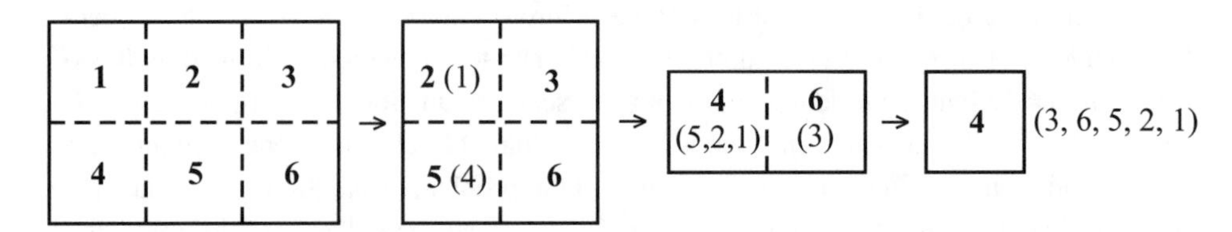

Q138. In how many different orders can you arrange the squares by making three folds about these three dotted lines?

S139. Using the notation introduced in the hint, we find that

$$abcduvwt = (av)(bu)(cw)(dt) = 12 \times 20 \times 15 \times 21,$$

and at the same time

$$abcduvwt = (aw)(bt)(cu)(dv) = 8 \times 25 \times 14 \times x.$$

These two equations imply that $12 \times 20 \times 15 \times 21 = 8 \times 25 \times 14 \times x$, which gives the answer: $x = 27$.

The acreage of each of the plots is now easy to compute; the size of the farm (including the farmhouse plot) is $287\frac{6}{35}$ acres. (The actual dimensions of the farm can't be determined.)

$7\frac{7}{15}$	20	14	$16\frac{4}{5}$
12	$32\frac{1}{7}$	$22\frac{1}{2}$	27
8	$21\frac{3}{7}$	15	18
$9\frac{1}{3}$	25	$17\frac{1}{2}$	21

S140. It is not difficult simply to count those numbers less than 1000 that contain the digit 9. There is one between 1 and 10, 19 between 1 and 100, and 271 between 1 and 1000 (which represents about 27% of these numbers). It is more complicated to continue with this method for the larger set of numbers from 1 to 10,000,000. But it is possible to determine an explicit formula for each positive integer n, which will indicate how many numbers between 1 and 10^n contain 9 among its digits.

Consider those numbers which have n or fewer digits which do not contain 9 among its digits. The first digit (which can be 0 if the number has fewer than n digits) can be chosen in 9 different ways, the second digit can also be chosen in 9 different ways, and so forth. Therefore, there are 9^n numbers with n or fewer digits that do not contain a 9 among its digits. (Indeed, this is equivalent to counting in base 9.) It follows that $10^n - 9^n$ of the numbers between 1 and 10^n contain at least one 9 among its digits. For $n = 3$ we get $1000 - 729 = 271$ as we've already seen, and for $n = 7$, we get

$$10000000 - 4782969 = 5217031,$$

which is approximately 52% of all the numbers between 1 and 10000000.

Q140. Consider the set of all eight-digit positive integers (that is, those integers n such that $10^7 \leq n < 10^8$). Which is larger: the sum of those which do not contain a zero among its digits or the sum of those which do contain a zero among its digits?

S141. Because the number must have at least one even digit, the number must be even. No digit can be 0 because no number is divisible by 0. Consequently, the number is not divisible by 5, and therefore 5 cannot be one of the digits. We are left with the eight digits 1, 2, 3, 4, 6, 7, 8, and 9. If 9 is left out, the sum of the remaining seven digits is 31, so no number constructed with these seven digits will be divisible by 3. Therefore 9 must be included. The sum of the eight digits is 40, so if there is such a seven-digit number, the sum of its digits must be 36 (to be divisible by 9). This means that 4 must be omitted.

So we must make a seven-digit number from the digits 1, 2, 3, 6, 7, 8, 9 that is divisible by each of them. To be divisible by 8, the last three digits must be divisible by 8, and the only possibilities are d**1**2, e**3**2, e**1**6, d**3**6, d28 and d68, where d means "odd", e means "even", **1** means "1 or 9", and **3** means "3 or 7", respectively.

Now if $abcdxyz$ is a seven-digit number, then

$$abcdxyz = abcd \times 1001 - abcd + xyz \equiv xyz - abcd \bmod 7 \equiv 0 \bmod 7$$

provided $abcd \equiv xyz \bmod 7$.

So, to get 7-digit examples, suppose the last three digits are 216. This gives a remainder of 6 when divided by 7, and the permutations 7398, 8973, 7839 of 3, 7, 8, 9, also give remainder 6, so 7398216, 8973216 and 7839216 are each divisible by each of 1, 2, 3, 6, 7, 8, and 9.

We've already seen that no eight-digit number can have the property, because the digits would have to be 1, 2, 3, 4, 6, 7, 8, 9, and any number constructed from these digits would give a remainder of 4 when divided by 9, since their sum, 40, gives this remainder.

There are 105 solutions to the 7-digit problem. If $abcd$ contains 1 & 8 and/or 2 & 9, then they may be swapped without changing the residue class of the number mod 7, so solutions can come in pairs or fours.

(a) Two sets of four are represented by 1289736 and 2189376, i.e., 8 solutions. There are no solutions with $xyz = 672$ or 936. The other 24 possibilities for xyz are

(b) 392, 792, 632, 296, 976 which yield twelve 1-8 pairs, i.e., 24 solutions:

1687392	1863792	7198632	1738296	1382976
1876	3186	9718	3187	1823
7168				2138

swap 1 and 8 in each of these

(c) 816, 136, 176, 168, 368, 768 which yield twelve 2-9 pairs; 24 more solutions,

2937816	2789136	2983176	2793168	7291368	1293768
	2978	3298	3297		2931
	7829		7329		

swap 2 and 9 in each

and

(d) 312, 712, 912, 192, 832, 872, 216, 896, 128, 328, 728, 928, 968 which give a further 49 solutions, $8 + 24 + 24 + 49 = 105$.

8796312	3869712	3768912	3678192	1679832	1369872	7398216
9678	9389	7863	3867	6719	1936	7839
9867	6893	8367	6387	6971	3196	8973
			7836	7916	6913	
1679328				9176		
6719				9617		
6971	3619728	1376928				
7916	6139	6731	1372896		3796128	2137968
9176	6391	7361	2317		6379	3271
9617	9163	7613	7231		7639	3712

The largest and smallest solutions are 9867312 and 1289736.

Q141a. Can the ten digits $0, 1, 2, \ldots, 9$ be arranged to form a ten-digit number $a_9a_8 \ldots a_1a_0$ with the property that for each $k \geq 1$, $a_ka_{k-1} \ldots a_1a_0$ is divisible by k?

Q141b. Can nine of the ten digits be arranged to form a nine-digit number $b_1b_2 \ldots b_8b_9$ so that for each $k \geq 1$, $b_1b_2 \ldots b_k$ is divisible by k?

Q141c. Can nine of the ten digits be arranged to form a nine-digit number $c_1c_2 \ldots c_8c_9$ such that for each $k \geq 1$, $C_k = c_1 \ldots c_{k-1}c_{k+1} \ldots c_9$, the number formed by removing the kth digit, is divisibe by k?

S142. (a) The hint enables us to find all but three of the numbers, and these we have labeled x, y, and z in the following figure.

x	×	12	÷	x	=	12
+		+		+		×
9	+	15	÷	3	=	8
÷		÷		−		÷
y	+	9	÷	z	=	6
=		=		=		=
4	×	3	+	4	=	16

(a)

The first and third columns together with the third row give the following system of equations:

$$x + 9 = 4y$$

$$x - z = 1$$

$$y + 9 = 6z$$

From this we see that $x = 1 + z$ and $y = 6z - 9$. Substituting into $x + 9 = 4y$ gives

$$(1 + z) + 9 = 4(6z - 9)$$
$$-23z = -46$$
$$z = 2,$$

and from this we find $x = 3$ and $y = 3$ (so the solution is unique).

(b) The figure below shows what we have established from the hint.

x	+		÷	1	=	7
+		+		+		−
	×		−	9	=	y
÷		−		÷		×
1	×	3	−	2	=	1
=		=		=		=
8	+		−	5	=	

(b)

Next we will express each of the missing numbers in terms of x and y. The empty square in the top row must be $7 - x$; the empty square in the first column is $8 - x$; the empty square in the last column is $7 - y$; the other empty square in the last row is therefore $4 - y$; the remaining empty square in the second column is therefore $x - y$.

It then follows that the second row is $(8 - x)(x - y) - 9 = y$, which may be written $(9 - x)(x - y + 1) = 18$, and since x lies between 1 and 6, $9 - x = 3$ and $x - y + 1 = 6$, or $9 - x = 6$ and $x - y + 1 = 3$, giving two solutions labeled (b1) and (b2) as shown.

3	×	12	÷	3	=	12
+		+		+		×
9	+	15	÷	3	=	8
÷		÷		−		÷
3	+	9	÷	2	=	6
=		=		=		=
4	×	3	+	4	=	16

(a)

6	+	1	÷	1	=	7
+		+		+		−
2	×	5	−	9	=	1
÷		−		÷		×
1	×	3	−	2	=	1
=		=		=		=
8	+	3	−	5	=	6

(b1)

3	+	4	÷	1	=	7
+		+		+		−
5	×	2	−	9	=	1
÷		−		÷		×
1	×	3	−	2	=	1
=		=		=		=
8	+	3	−	5	=	6

(b2)

S143. (a) In the hint we observed that each boy played at least once in each pair of two successive games. Because Victor played in 17 games there couldn't have been more than $2 \times 17 + 1 = 35$ games altogether. On the other hand, Dan played 35 times, which means that the boys played 35 games. Therefore, Dan must have won all of these except possibly the last game. In particular, Dan won the fifteenth game. Christopher, who only played against Dan, evidently played $35 - 17 = 18$ times.

(b) No, you can't be certain about who lost the sixth game or who played in the antepenultimate game. For example, suppose the outcomes on the second day were as shown (W for winner, L for loser):

Victor	L		L		W	L			W	L		L
Christopher		L		W	L		…	W	L		L	
Dan	W	W	W	L		W		L		W	W	W

10 sets of 3

Here, Victor lost the sixth game, and the antepenultimate game was between Victor and Dan. However, it might have been different, as in the next example, in which Dan lost the sixth game and the antepenultimate game was between Christopher and Dan.

Victor	L		L		L		W		L		W	W
Christopher		L		L		W	L	…		W	L	
Dan	W	W	W	W	W	L			W	L		L

10 sets of 3

Q143. On the third day the boys played once again. This time Victor played 15 games, Christopher played 22 games, and Dan played 25 games. Can you say who lost the sixth game and who played in the antepenultimate game?

S144. Let P_1, P_2, P_3, P_4 be the products of the numbers in each of the four 2×2 squares. If each of these numbers is divisible by $16 = 2^4$, the product $P_1P_2P_3P_4$ is divisible by $16^4 = (2^4)^4 = 2^{16}$. However, $P_1P_2P_3P_4$ is equal to the product of all the numbers in the grid, namely, the product of the integers from 1 to 16. The number of factors of 2 in this product is 8 (the number of even numbers) $+4$ (a second 2 in the multiples of 4) $+2$ (a third 2 in the multiplies of 8) $+1$ (a fourth 2 in 16) $= 15$. Therefore, $P_1P_2P_3P_4$ is not divisible by 2^{16}, so the numbers cannot be placed in the grid as prescribed.

S145. If the numbers are $a < b < c < d < e < f < g$ and $e+f+g < 51$, then $e < 16$ (because $e \geq 16, f \geq 17, g \geq 18$ imply $e+f+g \geq 51$). So $e \leq 15$, $d \leq 14$, $c \leq 13$, $b \leq 12$, $a \leq 11$, and $101 = 11+12+13+14+51 > a+b+c+d+(e+f+g) = 101$ is a contradiction.

S146. Suppose the dimes are placed so that four or more of them are in the same row. Then choose that row. Afterwards you can choose two rows and three columns and take all the remaining (at most five) coins. The same reasoning holds, of course, if more than four dimes are placed in a column.

Consider now the case where exactly three coins are in one row and no more than three in any other row. One of the remaining rows has more than one coin in it (otherwise there would be fewer than nine coins on the board). Choose these two rows and you will have at least five coins. You now have four choices (one row and three columns) so you can get the remaining (at most four) coins.

Finally, consider the case in which no row or column contains three or more coins. In this case at least three of the rows must contain two coins each. Choose these three rows, and then choose the columns to obtain the last three coins.

This shows that regardless of how I distribute the nine coins, you can choose three rows and three columns in such a way as to obtain all the coins.

Q146. Can you get all ten coins if I place *ten* dimes on the grid?

S147. Continuing along the lines of the hint, we want to assign numbers 0, 1, 2 to the edges so that the three edges at each vertex will have different numbers assigned to them. One easy way to do this is to think of the graph as a hexagonal prism, and draw a Hamilton circuit on this prism (it's easy to find lots of ways to do this).

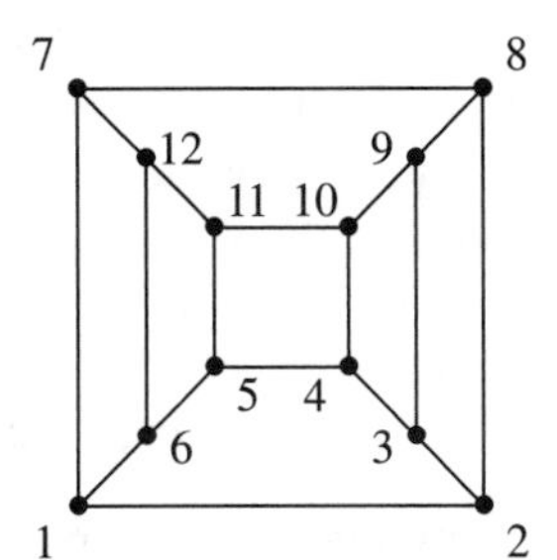

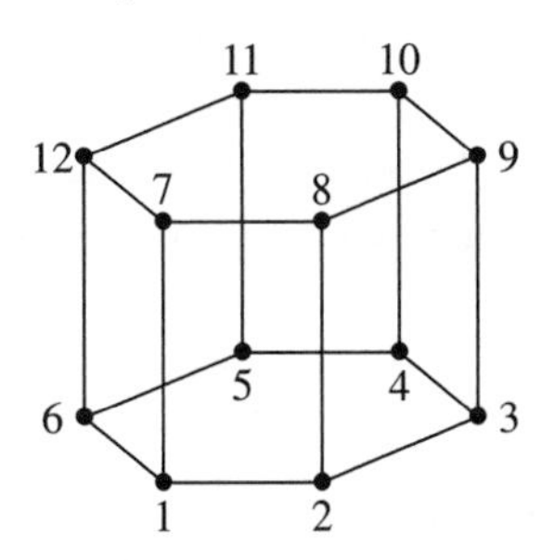

Now assign 0 and 1 alternately to the edges of the Hamilton circuit and assign 2 to the unused edge at each vertex. For the circuit $1, 2, 3, 4, 5, 6, 12, 11, 10, 9, 8, 7, 1$ we get the assignment on the left. The figure at the right gives one of the $(6!)^3$ ways of completing the diagram.

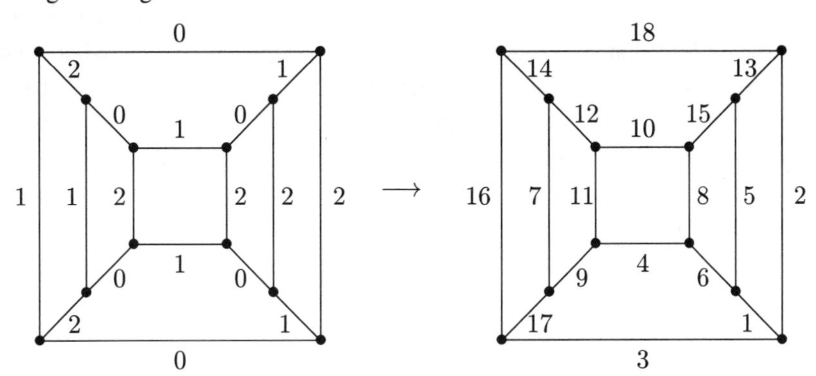

S148. Observe that the number of white pieces on one half of the board is equal to the number of black pieces on the other half of the board. Thus, the hypothesis of the statement is false, so the statement itself (by convention) is (vacuously) true.

S149. Yes, there will always be three gold points which form the vertices of an equilateral triangle. Number the points, clockwise in order, from 1 to 27. Partition these into nine subsets of size three: $\{1, 10, 19\}, \{2, 11, 20\}, \{3, 12, 21\}$, and so forth. Each subset forms the vertices of an equilateral triangle. Suppose none of these are all gold vertices. In that case, each subset contains at least one green vertex, so there must be at least nine green vertices around the circle. But between two green points there are at least two gold points, so the number of gold points is at least $9 \times 2 = 18$. On the other hand, there are just 27 vertices altogether. Thus, we must conclude that every third vertex is green. Without loss of generality, we may assume that we have labeled the points so that point 1 is green. Then, all the points in $\{2, 11, 20\}$ are gold. In fact there are six gold equilateral triangles and three green ones.

S150. (a) The sum of the page numbers of the torn-out sheets is at least

$$1 + 2 + 3 + \cdots + 50 = 1275.$$

Therefore, the sum can never equal 1273.

(b) We observe that the sum of the page numbers for a single torn-out sheet is an odd number: this is because the two numbers are consecutive, so one is even and one is odd. When we compute the sum of all page numbers from the torn-out sheets, we will be adding 25 odd numbers. The sum is therefore odd, so can never equal 2450.

(c) Suppose the two sides of a torn-out sheet are numbered $2k - 1$ and $2k$ with sum $4k - 1$. When we add the page numbers of all 24 torn-out sheets we get

$$(4k_1 - 1) + (4k_2 - 1) + \cdots + (4k_{24} - 1) = 4(k_1 + k_2 + \cdots + k_{24}) - 24,$$

which is a multiple of 4, so cannot equal 2450. If the convention that recto is odd and verso is even is not adhered to, the argument is the same, with "+" in place of "−".

S151. In the hint we showed that the sum of the numbers around each face must be 26, and with this knowledge it is not difficult to find a satisfactory assignment. However the assignment is made, the cube can be rotated so that 12 is on the top-front edge. Then, if necessary, by reflexion (about the plane through the top-front and back-bottom edges), the label on the bottom front edge can be made less than the label on the top back edge, and, again, if necessary, by reflexion (about the vertical plane through the center of the front and back faces), the label on the left front edge can be made less than the label on the right front edge. Given this standard form, there are nineteen essentially different solutions. Each of these generate 47 others (by rotation and reflexion), for a total of $48 \times 19 = 912$ different solutions.

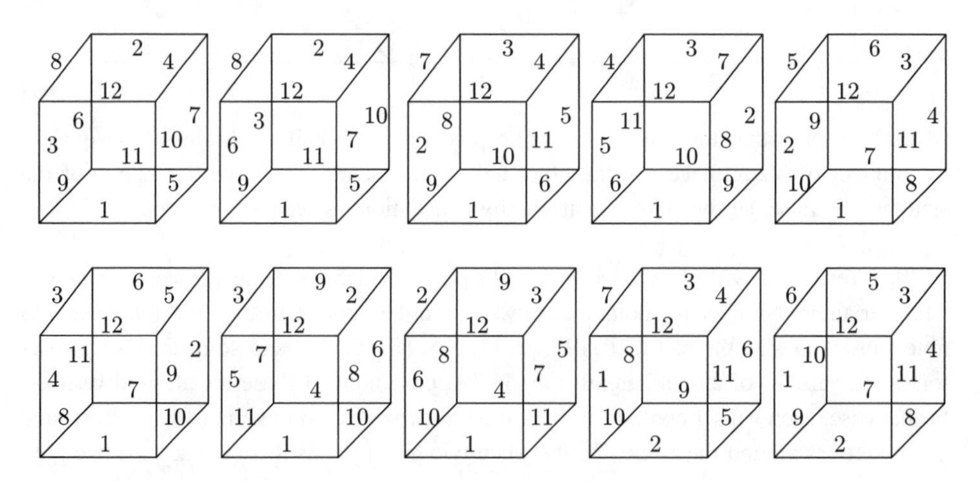

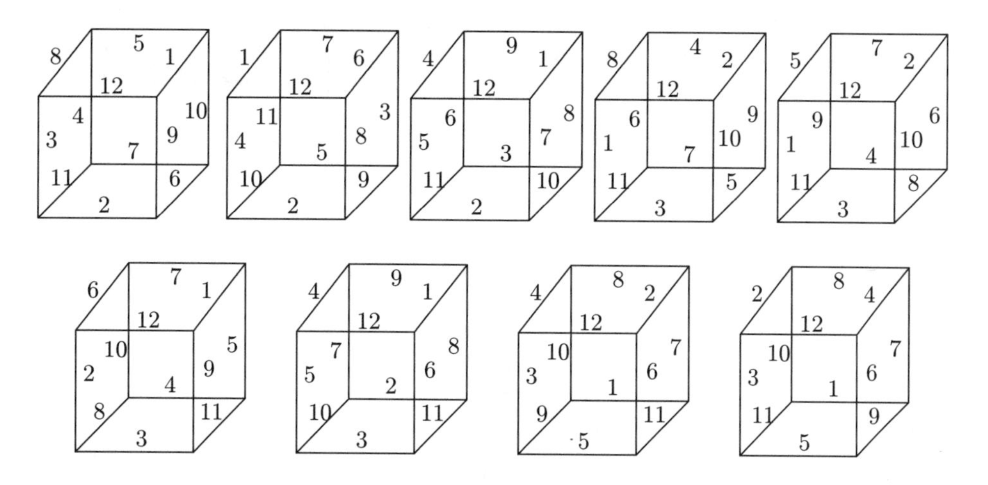

S152. The first figure cannot be packed because it contains an odd number of small triangles, in fact, only 117.

The other boxes each contain 120 small triangles, but the second one has 62 of its triangles pointing up (△) and only 58 of them pointing down (▽), whereas each calisson, whichever way you place it, covers just one of each kind of triangle.

The third box has 60 triangles pointing up and 60 pointing down, but the top and bottom edges have an angle of 60 degrees at each end, and if you start from either one and proceed along the row, the positions are forced and finally cut off a single triangle from the other end that cannot be used.

The last box can be packed, and there are many ways of doing it, but the positions of 24 of the calissons are forced and don't allow one to be placed where we've put it.

(See **Tilings** in the Treasury.)

S153. Yes, it is possible to carry out the King's order. The dances can be organized in the following way: Number the knights and the ladies from 1 to n. The knight numbered k is to dance with, and only with, the ladies numbered from 1 to k. The result of this schedule is that the lady numbered m will dance with, and only with, the knights numbered from m to n. It is easy to check that this scheme will satisfy the King's wish.

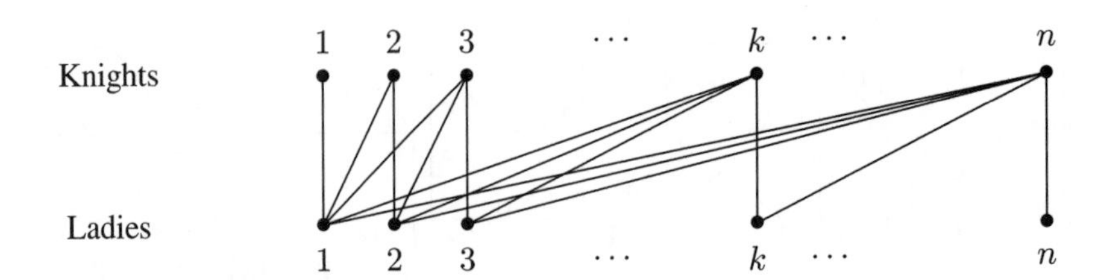

S154. We let a, b, c, d, e, f be the numbers of shoes in each of the categories shown in the following table.

	Color 1	Color 2	Color 3
Left	a	b	c
Right	d	e	f

We will show that Asa can be assured of finding 40 pairs of shoes.

Let P be the number of pairs of shoes that Asa can be assured of finding, and suppose that P is less than 40. Observe that after Asa has taken these P pairs out of the box, at least one of the numbers in each column is 0. It cannot happen that both numbers in the same column are 0 because the original sum of numbers in each column was 80 and we have taken out P pairs, $P < 40$. In the same manner, the three 0's in the table cannot be in the same row. Thus, after Asa has taken out the P pairs of shoes, two of the numbers in one of the rows and one of the numbers in the other row must be 0. Say, for example, that a, b, and f are 0. Then the number denoted by c must be $120 - P$ (P pairs in this row have already been taken out). But, because this number cannot be larger than 80, we must have $120 - P \leq 80$, or equivalently, $P \geq 40$. But this contradicts our assumption that P was less than 40.

Can Asa always find more than 40 pairs of shoes? No, because the shoes might be distributed in the following way:

	Color 1	Color 2	Color 3
Left	40	80	0
Right	40	0	80

S155. (a) Yes; an example is shown in figure (a).

(b) Yes; figure (b) shows one way of doing it.

(c) No. Make 20 squares of the chessboard black as shown in figure (c). Regardless of how each piece is put onto the board (to cover three squares), it will cover exactly one black square. But we need 21 pieces to cover the deleted chessboard, so such a covering is impossible.

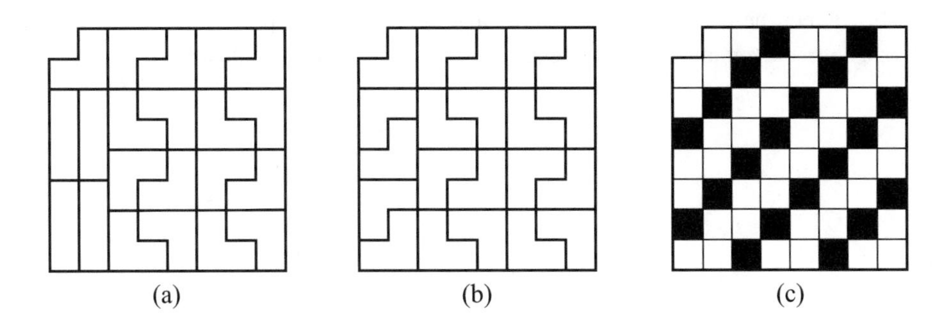

S156. Color every other square of every other row of the board black, as shown below. Each 2×2 square covers exactly one black square, whereas each 1×4 piece covers either 0 or 2 black squares.

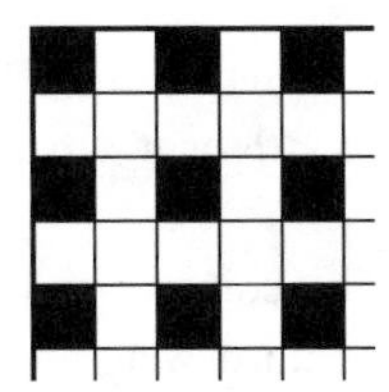

If the number of black squares in the grid is odd, then the number of 2×2 squares in the original tiling must have been odd. On the other hand, if the number of black squares had been even, the number of 2×2 squares would also have been even. If we replace a 2×2 square with a 1×4 rectangle, the number of 2×2 squares goes from odd to even, or vice-versa, while the number of black squares remains unchanged. This shows that no tiling is possible after the substitution.

S157. As you walk around the boardroom table, at each step you either change gender, or stay the same. When you get back to where you have started, there have been an even number of changes (if you start with a woman, say, then you finish at that woman). But we are told that there is the same number of non-changes (people of the same gender seated together). So the total number of steps is a multiple of four.

> **Q157.** Under the conditions of the problem, must there be the same number of each gender?

S158. Here are a few different solutions.

For readers interested in puzzles of this sort, see Major P. A. MacMahon, *New Mathematical Pastimes*, Cambridge University Press, 1921.

S159. Label the coins A, B, C, D, E, and denote their respective weights by a, b, c, d, e. The tree diagram shows the sequence of decisions and conclusions.

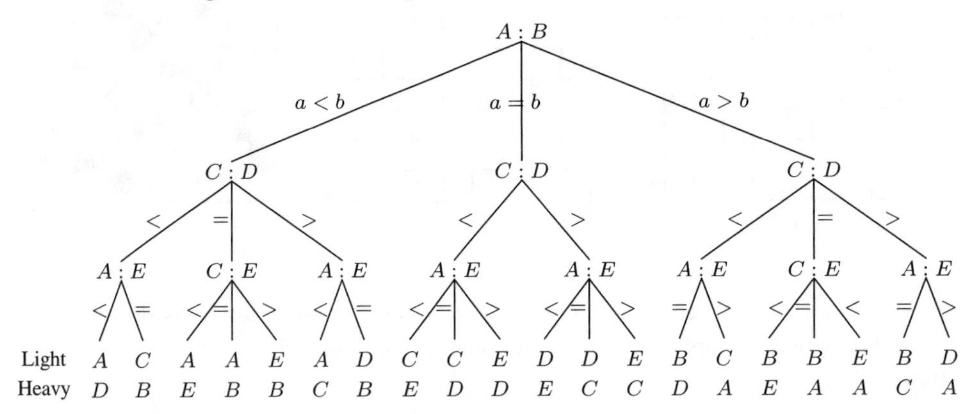

In the first weighing, compare coins A and B. If (Case 1) $a = b$, then the two coins are true, whereas, if (Case 2) $a < b$ or $b < a$, then at least one of the two coins is counterfeit.

In Case 1, continue by comparing coins C and D. These coins will not balance. It suffices to suppose that $c < d$. In the third weighing, compare coins A and E. If $a < e$ then E is the heavier coin and C is the lighter. If $e < a$ then E is the lighter and D is the heavier. If $a = e$ then C is the lighter and D is the heavier.

In Case 2, suppose that $a < b$ (the case $a > b$ is completely analogous). In the second weighing, compare coins C and D. If (Case 2.1) $c = d$, then C and D are true coins. If (Case 2.2) $c < d$ or $d < c$, then just one of these two coins is counterfeit.

In Case 2.1, continue in the third weighing by comparing coins C and E. If $c < e$ then E is the heavier and A is the lighter coin. If $e < c$, then E is the lighter and B is the heavier. If $c = e$ then A is the lighter and B is the heavier.

In Case 2.2, you can conclude that E is a true coin. Suppose that $c < d$ (the case $d < c$ is similar). In the third weighing, compare coins A and E. If $a < e$ then A is the lighter and D is the heavier. If $a = e$, then B is the heavier and C the lighter. The case $a > e$ cannot happen.

> **Q159.** If the total weight of the two duds is equal to that of two good coins, can you specify all three weighings *before* you start?

S160. Yes, it is possible. Take the delegates into the hall one at a time, and suitably place them in the following way: Suppose that person P is to be seated. There are at most 18 tables where a jury member might be seated who voted for the same studio as P in the first category, and there are at most 18 tables where a delegate might be seated who voted for the same studio as P in the second cagegory, and there are at most 18 tables where a delegate might be seated who voted for the same studio as P in the third category. We are left with at least $55 - 3 \times 18 = 1$ table, and P can sit there.

S161. Let A be a team with the most wins, which, as we argued in the hint, is at least four, and let X, Y, Z, T, denote four of the teams who lost to A. These four teams played six games between them, and at least one of the teams must have won at least two of them. Let that team be labeled B, and label the two teams who lost to B by C and D, where C beat D. Then teams A, B, C, D, satisfy the conditions of the problem.

Q161. Will there necessarily be four such teams if only seven teams are in the tournament?

S162. The sum of the numbers which each pentomino must cover is equal to the sum of all the numbers in the rectangle divided by 9; that is $153/9 = 17$. The twelve pentominoes are shown below, arranged for easy memory, in the order T, U, V, W, X, Y, Z, together with the mnemonic FILiPiNo. The nine pentominoes indicated in the problem are the ones depicted, but omitting F, P, and N.

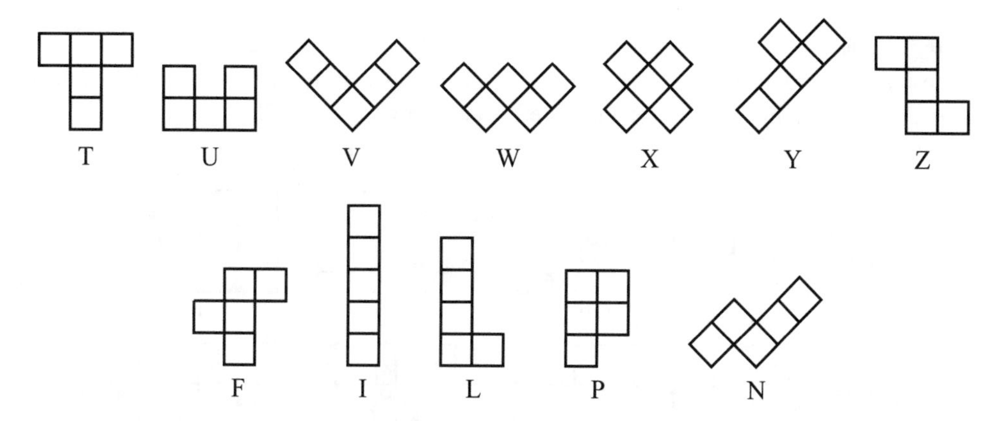

There is only one way to place the nine pentominoes indicated in the problem in the 5×9 rectangle so that the sums of the numbers covered by each is 17. One way to arrive at the solution is to begin in the top-right corner. That square can be covered in the prescribed manner by the I, or by the T. However, if the T is placed there, there is no way to cover the 1 (under the 3). Therefore I fits along the right column. Next, consider the 3 next to the top-right corner. It must be covered either by the V or the L. It can't be covered by the L because then there is no way to cover the 5 next to it. So it must be covered by the L. Then the 8 (under the 5) can only be covered by the T, and this in turn means that the 4 under the 8 must be covered by the W. The 7 (under the 6) can be covered by the I or the Y; but I has been used, so we must try Y. (It is not hard to show that placing the I there does not lead to a solution.) Then the 5 (above the 1) must be covered by the Z, the 5 (above the 6) must be covered by the L, the 2 (under the 8) must be covered by the X, and the 7 in the corner must be covered by the U.

3	8	1	5	2	3	5	3	3
0	2	3	6	0	6	8	0	1
4	0	2	2	1	5	4	6	4
3	9	6	1	5	0	6	1	0
7	1	0	5	1	4	7	1	9

Q162. Will the twelve different pentominoes cover the 6×10 rectangle in such a way that the sums of the numbers covered by each pentomino are all the same?

5	2	3	2	0	4	0	5	4	2
2	3	4	1	9	1	4	3	3	1
4	6	6	1	2	5	2	1	5	6
7	0	4	2	1	3	2	4	0	3
2	0	2	4	7	4	0	8	1	4
2	4	6	2	1	2	1	6	0	2

S163. Not counting rotations and reflexions as different, there are 23 different solutions. These can be grouped into 9 different classes, with those in the same class distinguished from each other by internal rotations, reflexions, or swaps (the shaded areas show the region that has been changed from a preceding solution in its class).

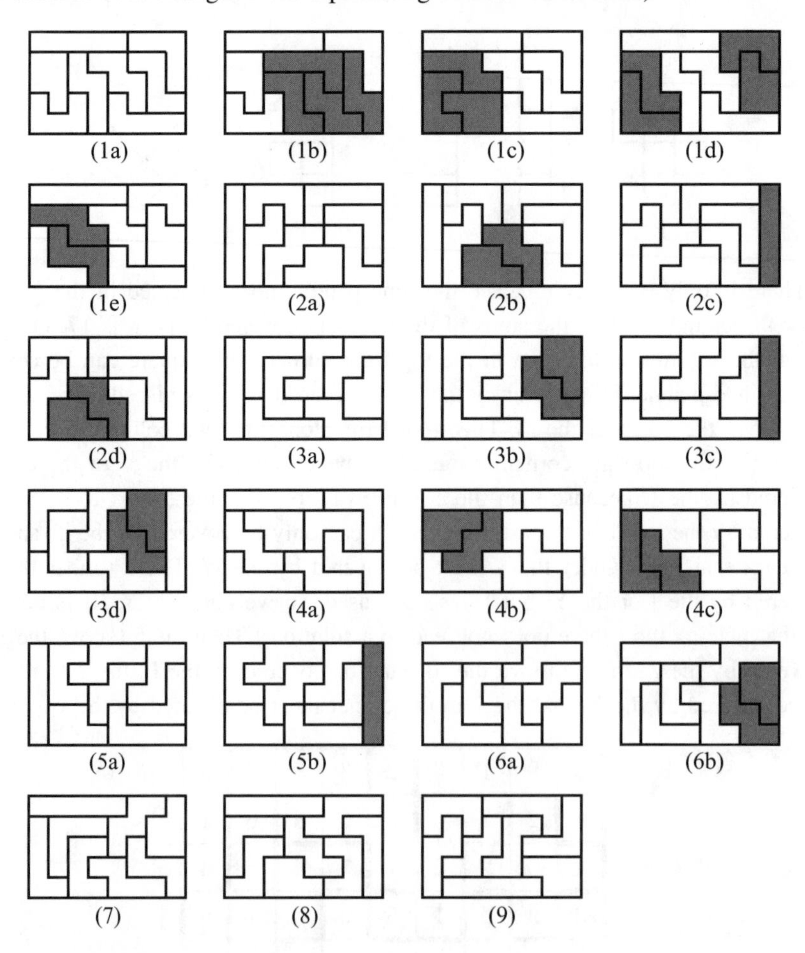

Q163. Will these eight pentominoes tile a 4×10 rectangle?

S164. We have placed six points on the perimeter of the equilateral triangle as shown in the hint, and the distance between any two of them is at least 10 yards. If the plot can be covered with five tarps (each congruent equilateral triangles) then at least one of the tarps must cover at least two of the six points. Thus, the sides of each of the tarps must be at least 10 yards in length. If we therefore divide the garden plot into four triangular regions as shown below, then a single tarp will cover one of these smaller triangular regions. This shows that the plot can be covered by four tarps.

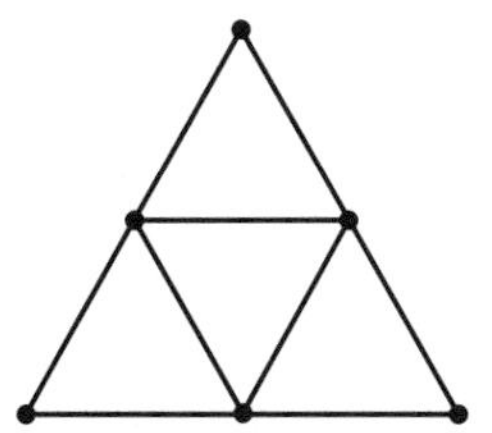

S165. Let S denote the common sum of the numbers in the stack at each vertex. Suppose that Selma makes K triangles. The sum of all the numbers at the corners of the triangles is $K(1+2+3) = 6K$. Clearly, $3S = 6K$, so that $S = 2K$. Since this is an even number, S cannot be equal to 99.

We've seen that the sum (the equal sum at each vertex) has to be even. It's clear that S can't be 2, but it can be 4; just stack $(1, 2, 3)$ on top of $(3, 2, 1)$. We also can make $S = 6$ by stacking $(1, 2, 3)$, $(2, 3, 1)$, and $(3, 1, 2)$. And S can be any larger even number because any such number can be made of fours and sixes; for example, $98 = 23 \times 4 + 6$.

> **Q165.** What triples of positive integers A, B, C are attainable as the corner sums of a stack of such triangles?

S166. If a person, A, is in a room with two or three others who are incompatible with A, move A to the other room. The number of "bad" relationships in the first room thereby decreases by 2 or 3, and increases by 1 or 0 in the second room. Therefore, the value of S (see the hint) decreases by 1, 2, or 3. Furthermore, since $S \geq 0$, there will be a point at which S will reach a minimal value. At this point, no delegate will be in a room with more than one other incompatible person.

S167. Using the notation introduced in the hint, we find that square (3) has side length $2 + s$, square (4) has side length $2 + (2 + s) = 4 + s$, square (5) has side length $(2 + s) + (4 + s) = 6 + 2s$, square (6) has side length $(4 + s) + (6 + 2s) = 10 + 3s$, and square (7) has side length $(10 + 3s) + (4 + s) + 2 = 16 + 4s$. At the same time we find that square (8) has side length $s + (2 + s) + (6 + 2s) = 8 + 4s$, and square (9) has side length $s + (8 + 4s) = 8 + 5s$. But squares (2) and (9) fit against (1) and (7), so $s + (8 + 5s) = 2 + (16 + 4s)$, and it follows that $s = 5$. The dimensions of the rectangle are

$$(10 + 3s) + (16 + 4s) = 61 \quad \text{by} \quad (16 + 4s) + (8 + 5s) = 69,$$

and its area is $69 \times 61 = 4209$ square inches.

S168. We can write $9 = 10 - 1$, $99 = 10^2 - 1$, $\ldots$, $\underbrace{999\cdots 9}_{99 \text{ digits}} = 10^{99} - 1$. It follows that

$$\begin{aligned} A &= (10-1) + (10^2 - 1) + \cdots + (10^{99} - 1) \\ &= 10 + 10^2 + 10^3 + \cdots + 10^{99} - 99 = \underbrace{111\cdots 1}_{99 \text{ digits}}0 - 99 \\ &= (\underbrace{111\cdots 1}_{97 \text{ digits}}000 + 110) - 99 = \underbrace{111\cdots 1}_{97 \text{ digits}}000 + 11 \\ &= \underbrace{111\cdots 1}_{97 \text{ digits}}011 \end{aligned}$$

which contains 99 ones.

S169. Suppose there are $n+1$ players. Each person plays n matches, and therefore $W_k = n - L_k$, for each player k. It follows that

$$\begin{aligned} W_1^2 + W_2^2 + \cdots &= (n - L_1)^2 + (n - L_2)^2 + \cdots \\ &= (n^2 - 2nL_1 + L_1^2) + (n^2 - 2nL_2 + L_2^2) + \cdots \\ &= (n+1)n^2 - 2n(L_1 + L_2 + \cdots) + (L_1^2 + L_2^2 + \cdots). \end{aligned}$$

Now, $L_1 + L_2 + \cdots$ is equal to the number of matches in the tournament, namely, $n(n+1)/2$. Substituting this into the last equation, we find that

$$\begin{aligned} W_1^2 + W_2^2 + \cdots &= (n+1)n^2 - 2n\Big(n(n+1)/2\Big) + (L_1^2 + L_2^2 + \cdots) \\ &= L_1^2 + L_2^2 + \cdots . \end{aligned}$$

S170. There are many solutions. To write some down, notice that, using the notation introduced in the hint, $2S + b_1 + b_3 - c = 91$ (consider the top and bottom triangles), and similarly, $2S + b_2 + b_4 - c = 91$. Equating these we can conclude that $b_1 + b_3 = b_2 + b_4$, and B is even. Also, any apex a_i can be swapped with its corresponding diagonal d_i, so that solutions come in sets of 16. To save writing them all out you could take $a_i > d_i$. Put $p_i = a_i + d_i$ for the sum of an apex-diagonal pair. Then $p_1 + b_1 = p_3 + b_2$ (compare the top and right triangles), and similarly, $p_1 + b_4 = p_3 + b_3$, $p_2 + b_1 = p_4 + b_4$, and $p_2 + b_2 = p_4 + b_3$ (visualize all of these in geometrical terms). From these, we find that $p_1 - p_3 = b_2 - b_1 = b_3 - b_4$, and $p_2 - p_4 = b_4 - b_1 = b_3 - b_2$.

For examples of the largest sum, $= 49$, take $c = 13$ and $b_1 = 1$, $b_2 = 2$, $b_3 = 5$, and $b_4 = 4$. Then $p_1 - p_3 = 1$ and $p_2 - p_4 = 3$. Thus, $p_1 + p_3$ and $p_2 + p_4$ are odd, and $(p_1 + p_3) + (p_2 + p_4) = 91 - 13 - (1 + 2 + 5 + 4) = 66$.

So, working backwards, we must divide 66 into two odd parts, say $p_1 + p_3 = 33$ and $p_2 + p_4 = 33$, giving, combining with the previous equations, $p_1 = 17$, $p_2 = 18$, $p_3 = 16$, and $p_4 = 15$. Can we arrange the missing numbers $3, 6, 7, 8, 9, 10, 11, 12$ in pairs with these four sums? Yes: $6 + 11, 8 + 10, 7 + 9, 3 + 12$, or $8 + 9, 7 + 11, 6 + 10, 3 + 12$. With the possible apex-diagonal switches, these two together yield 32 solutions! Or, we could have divided 66 into 31 and 35, with apex-diagonal pairs $p_1 = 16$, $p_2 = 19$, $p_3 = 15$,

$p_4 = 16$, which can be written as $6 + 10, 8 + 11, 3 + 12, 7 + 9$, or with the outside pairs swapped, giving 32 more solutions! Here are some samples:

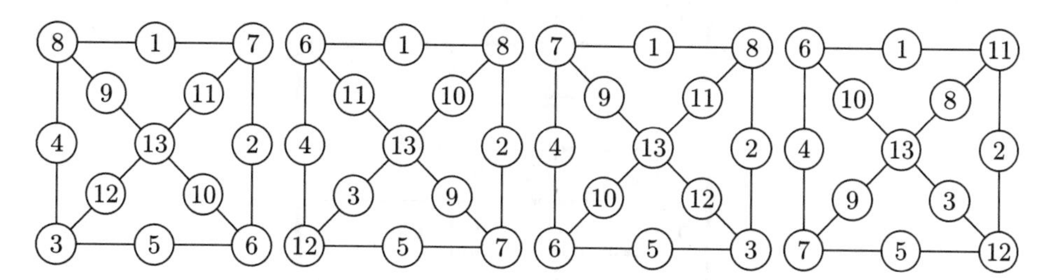

You can find another collection of solutions for $S = 49$ by taking $c = 12$ and $b_1 = 1$, $b_2 = 2$, $b_3 = 4$, $b_4 = 3$. And to get solutions with the smallest sum subtract everything from 14 in the preceding solutions. There are lots of solutions in between as well; here are some:

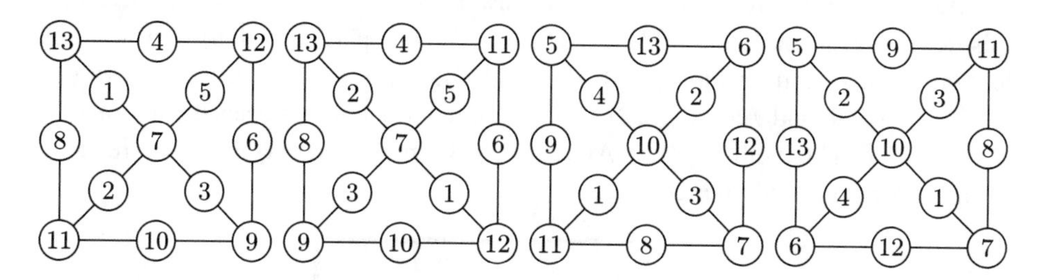

S171. The prime factorization of 195 is $3 \times 5 \times 13$. The hypotheses of the problem imply that the number of visitors is more than 5 (with 5 visitors, a distribution of one from the north (20%), two from the east (40%), and two from the west (40%), contradicts the hypothesis that more than 40% are from the west continent). Therefore, there are either 13 or 15 visitors from Tralfadom.

We also know that the proportion of delegates from the north, east, and west continents is more than 16%, 27%, and 40%, respectively. We can summarize this information in a table.

	16%	27%	40%
15	2.4	4.05	6.0
13	2.08	3.51	5.2

Because the number of visitors from each continent is strictly larger than these respective percentages, there are at least 3, 5, and 7 representatives from these continents (in the case of 15 visitors), or at least 3, 4, and 6 representatives from these respective continents (in the case of 13 visitors). But $3 + 5 + 7 = 15$ and $3 + 4 + 6 = 13$, which means that in each case, there is no visitor from the south continent, so the answer is "No."

S172. The hint tells us where eight of the dominoes are:

G	4	0	2	2	6	2	3	4
F	2	1	6	0	0	5	5	3
E	1	1	0	5	4	6	2	3
D	4	0	4	4	3	0	5	1
C	5	1	5	6	1	6	3	1
B	4	2	6	3	6	5	0	5
A	0	2	6	4	1	3	2	3
	a	b	c	d	e	f	g	h

As we know where $(2,6)$ is, the 2 in row E must be covered by a vertical $(2,5)$. This determines the neighboring $(1,1)$. There's only one place for $(4,6)$ (in row E), and this forces $(0,5)$, and, in turn, $(1,6)$, $(0,4)$, $(1,2)$, $(0,1)$, $(4,5)$, and $(1,5)$, with $(2,4)$ and $(0,2)$ at Bab and Aab. If $(2,3)$ is placed at Agh it forces a second copy of $(0,5)$, so $(3,5)$ is at hAB with $(2,3)$ at Afg, and $(0,3)$ (not $(0,5)$) at gBC. The remaining dominoes are now easily placed:

4	0	2	2	6	2	3	4
2	1	6	0	0	5	5	3
1	1	0	5	4	6	2	3
4	0	4	4	3	0	5	1
5	1	5	6	1	6	3	1
4	2	6	3	6	5	0	5
0	2	6	4	1	3	2	3

S173. (a) Continuing from the hint, the 6 under $(4,4)$ is covered by $(4,6)$ at aCD, because $(3,6)$ at Fba is forced. The 3 at dF is covered by $(3,4)$, either vertically or horizontally, so that $(3,3)$ is at dDE. We draw thick lines between squares Efg, gEF, gJG and Fef. Then $(0,1)$, $(1,3)$ are forced. If $(1,4)$ and $(3,4)$ are in either orientation in row G, then $(1,5)$ is fixed in G, $(4,5)$ in column h, $(0,6)$ in row F, and $(1,2)$ in column g. In addition $(0,4)$ is forced in A, and then $(5,5)$ in h, $(0,5)$ in e, and now we have a contradiction with $(0,5)$ in the g. The upshot is that $(0,4)$ is vertical in column e. We now place $(3,4)$, $(1,5)$, and $(1,4)$, and place $(0,0)$ and $(4,5)$ in the lower corners. $(5,5)$ is near the middle, and then we can place $(0,6)$ (vertically, not horizontally, because $(0,1)$ is already placed), $(0,5)$; $(1,2)$ is in column g (not e, since $(1,6)$ covers that 1). To finish up, we locate $(3,5)$, $(2,5)$, $(0,3)$, $(2,2)$, $(5,6)$, $(1,6)$, $(2,4)$, $(2,3)$, $(2,6)$, and $(0,2)$.

(b) After the hint, we continue by placing $(1,1)$, $(4,5)$. $(1,2)$ is in column h; then $(1,5)$, $(5,6)$, $(1,3)$, $(5,5)$ are forced, as well as $(2,2)$, $(0,1)$, $(0,2)$, $(4,4)$, $(3,5)$, $(3,3)$, $(2,3)$, and $(1,4)$. If we now place $(0,0)$ in the middle, it separates the board into two odd numbers of squares, so $(0,0)$ goes into G with $(3,6)$ below it, and finish up with $(3,4)$, $(0,6)$, $(0,4)$, $(6,6)$, and $(4,6)$.

	a	b	c	d	e	f	g	h
G	1	1	6	4	1	5	1	4
F	6	3	6	3	4	6	0	5
E	4	4	1	3	0	0	1	5
D	6	3	0	3	5	5	2	3
C	4	1	2	6	1	6	2	5
B	0	2	4	2	2	5	3	5
A	0	3	2	6	0	2	0	4

(a)

	a	b	c	d	e	f	g	h
G	2	0	0	1	1	5	0	4
F	2	3	6	5	5	2	6	2
E	1	3	6	4	2	0	4	2
D	4	4	0	3	3	2	4	1
C	2	3	0	6	5	3	6	5
B	0	0	4	6	6	1	3	5
A	1	3	4	6	5	5	1	1

(b)

S174. (a) Because an odd number (65) of guests are sitting around the table, at least 33 of them are of one gender and at most 32 are of the other. Consequently, at least two people of the first gender are sitting next to each other.

(b) Suppose this were not so. Then every man is sitting between a man and a woman, and every woman is sitting between a woman and a man. Consequently, the guests are sitting in pairs with respect to genders: two men, two women, two men, and so forth. Because the number of people around the table is odd, this arrangement is not possible. Therefore the statement must be true.

(c) and (d) We shall show that for every number $2 < k < 33$, there will either be two men or two women with exactly k people between them. For let $K = k + 1$ and suppose the claim is not true. Label the places around the table $0, 1, 2, \ldots, 64$, and suppose a man is sitting at the table in position 0. Now go around the table, say clockwise, and alternately place a man or a woman at every Kth position. Then, because we are assuming the statement of the problem is not true, our labels will be assigned M, W, M, W, ... in positions $0, K, 2K, 3K, \ldots$, each number taken modulo 65. Men are sitting in positions $0, 2K, 4K, \ldots$, modulo 65. Let s be the smallest positive number such that $sK \equiv 0$ modulo 65. No number ≤ 33 is a multiple of 13 *and* 5, and so s must be an odd number. But this means that a woman is placed in position 0, contrary to assumption. (Alternatively, $s - 1$ is even, so a man is placed in position $(s - 1)K$ modulo 65, which means the men in positions $(s - 1)K$ and 0 have $K - 1 = k$ people between them.) We conclude that our assumption is false, and the proof is complete.

S175. (a) In the hint we established a notation and showed there could not be a coin on squares Bb, Db, Dc, Bd, or Dd.

	a	b	c	d	e	f	g
G			2				
F					1		1
E			3				
D	1	–	–	–	4		1
C			1				
B	3	–	2	–		1	
A			1				

(a)

Now, if we add the equations (Fe) $=1$, (Bf) $=1$, (Dg) $=1$, (Fg) $=1$, and subtract (De) $=4$, we get $-\mathrm{Cd}+\mathrm{Fd}+\mathrm{Gd}+\mathrm{Ae}+\mathrm{Be}+\mathrm{Ge}+\mathrm{Af}+\mathrm{Cf}+2\mathrm{Ef}+2\mathrm{Ff}+2\mathrm{Gf}+\mathrm{Ag}+\mathrm{Bg}+2\mathrm{Cg}+2\mathrm{Eg}+\mathrm{Gg}=0$. The only negative term is the first term, so $\mathrm{Ef}=\mathrm{Ff}=\mathrm{Gf}=\mathrm{Cg}=\mathrm{Eg}=0$, as they have coefficient 2. Equation (Fg) $=1$ is now $\mathrm{Gg}=1$, and the last long equation now requires $\mathrm{Cd}=1$ and all other variables 0. We established in the hint that $\mathrm{Cb}+\mathrm{Cd}=1$, so $\mathrm{Cb}=0$, and (Ba) $=3$ now gives $\mathrm{Aa}=\mathrm{Ca}=\mathrm{Ab}=1$, (Da) $=1$ implies $\mathrm{Ea}=\mathrm{Eb}=0$, (Bf) $=1$ implies $\mathrm{Ce}=1$, (Dg) $=1$ implies $\mathrm{Df}=1$, while (Ec) $=3$ requires $\mathrm{Fb}=\mathrm{Gc}=\mathrm{Ed}=1$, and we've located all ten coins.

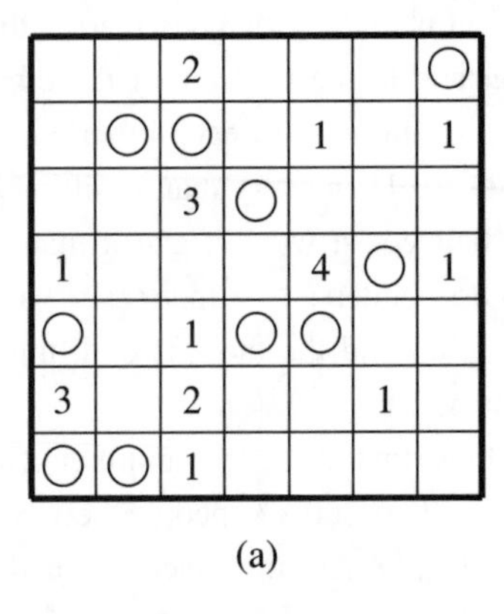

(a)

(b) We have shown the following (see the hint):

	a	b	c	d	e	f	g
7							
6			2	1	3		
5		2					
4				3		1	
3	2	–	–	–			
2	○		1	–	3	3	
1	2		–	–			

(b)

Now, from (Be) $= 3$ and (Bf) $= 3$ we find that Ad + Bd + Cd = Ag + Bg + Cg, so these are all zero. Subtracting (Be) $= 3$ from (Df) $= 1$ gives $-$Ae + De + Ee $-$ Af + Ef + Cg + Dg + Eg $= -2$, so that Ae = Af = 1, and the rest are zero. From (Fd) $= 1$, Ec + Ed is at most 1, so (Dd) $= 3$ implies that Dc = Ce = 1, Ec + Ed = 1, and (Fd) $= 1$ gives Gc = Gd = Ee = Ge = 0. (Bf) $= 3$ is now Cf = 0, and (Fe) $= 3$ implies that Ed = Ff = Gf = 1. Therefore Ec = 0. We have found 8 of the 10 coins, so Da + Ea + Fa + Ab + Bb + Db + Fb + Gb = 2. But Ab + Bb = 1 and Fb + Gb = 1, so Da = Ea = Fa = Db = 0, and the last two coins are at Bb and Fb.

					○	
	○	2	1	3	○	
	2		○			
		○	3		1	
2				○		
○	○	1		3	3	
2				○	○	

(b)

S176. Yes, Toby can ride the elevator for more than 8 minutes whether he starts from the basement or from the first floor. The diagrams below show how to go for 9 minutes from the basement, or $8\frac{1}{2}$ minutes from the first floor.

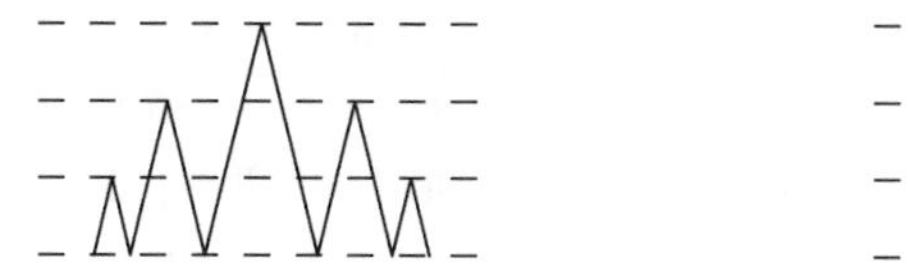

ABAABBAAABBBAABBAB BAABBAAABBBAABBAB

Q176. If Toby starts in the basement and rides until the elevator stalls, how many different rides could he make?

S177. Suppose we have a board with five or more black squares. If three or more are in any one row or column we operate on that row or column and reduce the number of black squares.

Suppose two of the rows, say A and B, contain two black squares apiece. Consider any black square, c, in one of the other rows, and let C denote the column which contains c. Let a and b denote the squares in column C that lie in rows A and B, respectively. By applying the operation to row A if necessary, and to row B if necessary, we can transform our board into one with the same number of black squares as before but with three black squares, a, b, c, in column C. But now, the operation applied to column C will reduce the number of black squares on the board.

A similar argument shows that if the if two columns contain two black squares, we could transform the board with our operation to one with fewer black squares.

Finally, if there is only one row and one column with two black squares, operate on the column, producing two rows with two black squares, and operate on these as above. Thus, the operation can always be used to reduce the number of black squares to 0, 1, 2, 3, or 4. For example:

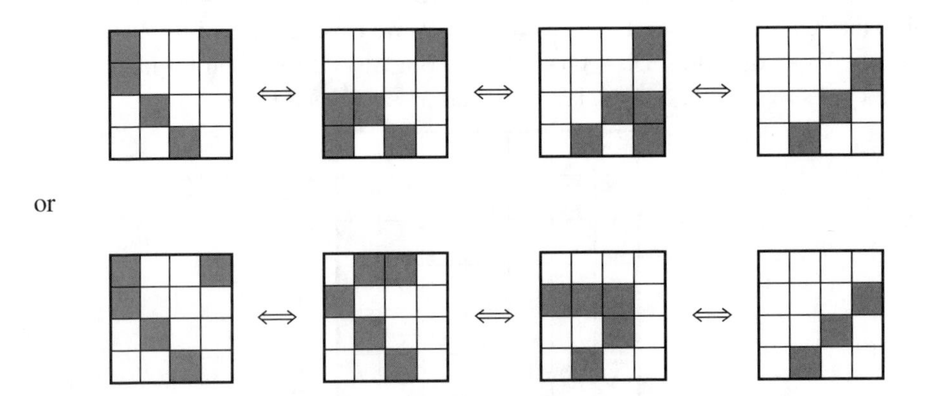

> **Q177.** We say that two grids are equivalent if one can be transformed into the other by the prescribed row and column operations. Describe a set of grids $\mathcal{S}$ which have the property that no two grids in $\mathcal{S}$ are equivalent, but yet every grid not in $\mathcal{S}$ is equivalent to some grid in $\mathcal{S}$.

S178. After each operation the number of numbers on the paper goes down by one. Therefore, after 2000 operations, there is only one number left on the paper. If you choose two even numbers, or two odd numbers, the replacement number is even. If you choose an even number and an odd number, the difference between their squares is odd. In the first case, the number of odd numbers on the paper decreases by two (if the chosen numbers are both odd) or remains the same (if the chosen numbers are both even). In the second case, the number of odd numbers on the list remains the same. Because we begin with 1001 odd numbers, the number of odd numbers on the list will decrease through the sequence 999, 997, 995, and so forth. That is, the number of odd numbers is always odd. So when there is only one number left on the paper, it will be an odd number, and therefore cannot be 0.

> **Q178.** If you start with the numbers $1, 2, \ldots, 9$, and successively replace pairs $a \geq b$ by the single integer $a^2 - b^2$, what is the smallest possible final integer? What is the answer for the first 10 integers? For the first 11? For the first 12?

S179. As in the preceding problem, we see that the number of men in the restaurant decreases by 0 (if two women or a man and a woman leave) or by 2 (if two men leave). When the four of them first arrive at the restaurant, there is an odd number of men in the room, and consequently there is always an odd number of men in the room. At the end, when only their party of four remains in the room, there is an odd number of men present. Since there are at least two men (Tom and Jerry) we must conclude that one of their friends is a man and the other is a woman.

S180. All four figures can be constructed with 18 notched blocks. The best way to see this is to see how the box is constructed out of smaller units.

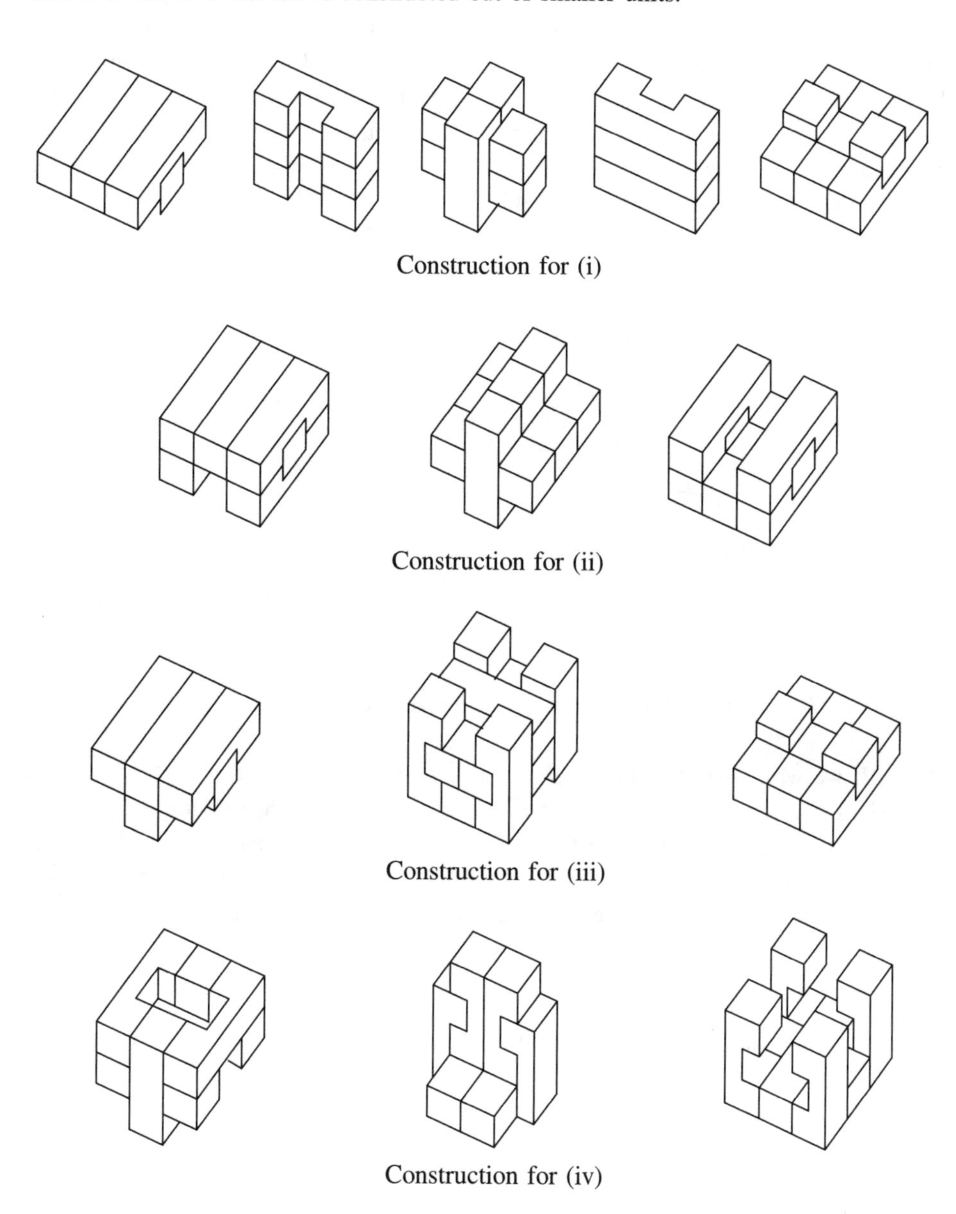

Construction for (i)

Construction for (ii)

Construction for (iii)

Construction for (iv)

Except for minor variations, in addition to rotations and reflexions, these are the only ways to pack these 18 notched blocks into a $3 \times 3 \times 5$ box. The variations have to do with the two different ways of orienting two blocks that interlock to form a + (one of these a 90° rotation of the other). Taking combinations of these orientations into account, construction (ii) can be carried out in four ways, and (iv) in six ways, making a total of twelve essentially different solutions. (This problem was posed by Loren Larson and dedicated to the memory of David Klarner, a mathematician who contributed much to the literature on packing problems.)

S181. Yes, it is always possible. The following figure shows how the plane can be tiled with copies of the quadrilateral from the figure in the statement of the problem. The fundamental region is the hexagon in the upper left corner, made by gluing together two copies of the quadrilateral along a common edge, where one of the copies is rotated 180° about the midpoint of an edge. This hexagon can be translated and pieced together to tile the plane. The fact that the hexagons perfectly fit together is a consequence of the fact that the sum of the angles in the quadrilateral is 360° (a quadrilateral is made up of two triangles, each of whose angles add to 180°).

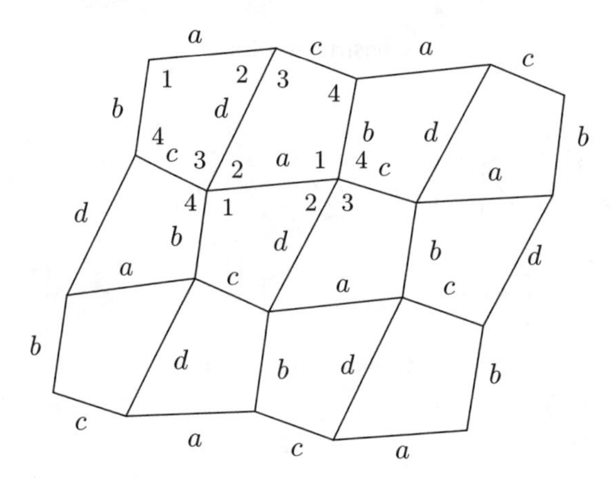

Q181. What happens if an angle at one of the vertices is more than 180 degrees?

S182. If we write A in the form $A = 3a + r$, where the remainder r is 0, 1, or 2, we find that $A^2 = 9a^2 + 6ar + r^2$, where r^2 is 0, 1, or 4. If we then divide A^2 by 3, we get a remainder of 0 or 1 (because $9a^2 + 6ar + r^2 = 3(3a^2 + 2ra) + r^2$, and r^2 gives a remainder of 0 or 1 when it is divided by 3). The number 44 gives a remainder of 2 when it is divided by 3. Therefore, the digits of A^2 can never sum to 44.

Q182. What if A is written in a base other than 10?

S183. Yes, 666 can be written as the sum of numbers constructed by using each of the digits $1, 2, \ldots, 9$ exactly once. One solution is $12 + 85 + 93 + 476 = 666$.

The number 555 cannot be so expressed. For suppose you use the digits $1, 2, \ldots,$ 9, to construct k numbers: $A_1, A_2, \ldots, A_k$. Suppose that their remainders when divided by 9 are $r_1, r_2, \ldots, r_k$, respectively. According to the rule for divisibility by 9, these k remainders are the same as you would get if you were to divide the digit sums of $A_1, A_2, \ldots, A_k$ by 9.

Now consider the number $A = A_1 + A_2 + \cdots + A_k$. When you divide A by 9 you get the same remainder as if you divide the number $r_1 + r_2 + \cdots + r_k$ by 9, which is the same as the remainder one gets when 9 divides the number $B =$ (the sum of the digits of A_1) + (the sum of the digits of A_2) $+ \cdots +$ (the sum of the digits of A_k). Because each digit $0, 1, 2, \ldots, 9$ occurs exactly one time among the numbers $A_1, A_2, \ldots, A_k$, we see that $B = 0+1+2+\cdots+9 = 45$. When this number is divided by 9 we get a remainder of 0. Consequently, A cannot be equal to 555, because $555 = 61 \times 9 + 6$.

S184. (a) If one of the chosen digits is 3, 6, or 9, we would have a one-digit number divisible by 3. If one of the digits has a remainder of 1 when divided by 3 and another has a remainder of 2, then one can use these two to make a two-digit number divisible by 3. Finally, if all three numbers have the same remainder when divided by 3 (either 1 or 2), the three of them taken together will make a three-digit number divisible by 3.

(b) First note that you can't do it for four, because the digits 1, 3, 4, and 7 cannot be used to make a number divisible by 9. Suppose we choose five different digits. If one of the five digits is 0 or 9, then we have a one-digit number divisible by 9. Otherwise, consider the four two-digit sets:

$$\{1,8\},\ \{2,7\},\ \{3,6\},\ \text{and}\ \{4,5\}.$$

Because we have chosen five different digits, two of them must be in the same two-element set. These two digits, which sum to 9, can be used to make a two-digit number divisible by 9. Therefore, the answer to the question is "Yes."

S185. Suppose one could find two such numbers, A and B, with $AB = C$, where C is the number of question. Then, because C is divisible by 3, either A or B must be divisible by 3. But the digit-sums of A and B are the same, because they are formed from the same digits. Because one of the numbers is divisible by 3, and their digit-sums are equal, they both must be divisible by 3. But this would imply that their product AB is divisible by 9, which it isn't. Therefore, such numbers A and B cannot be found.

S186. The number A_1 is divisible by 9. Consequently, the sum of the digits of A_1, which we've denoted by A_2, is divisible by 9. In a similar way, $A_2, A_3, A_4, \ldots$ are all divisible by 9. The last single digit number must also be divisible by 9, and because it cannot be 0, it must be 9. The same is true if you started with the product of $1, 2, 3, \ldots, n$, provided $n \geq 6$.

S187. As we've seen in the hint, the problem is equivalent to the sliding block puzzle

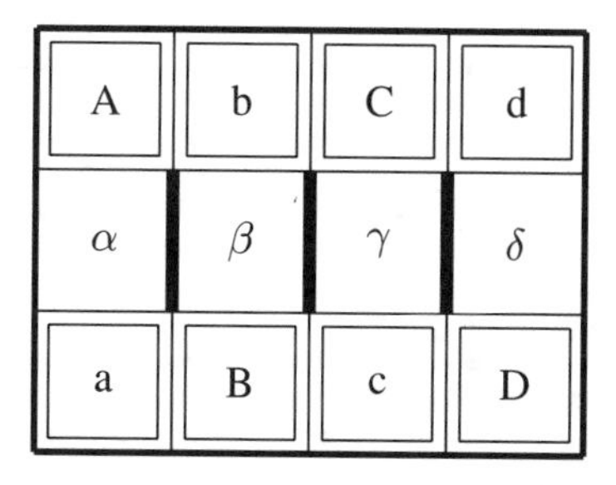

Sliding Block Problem

in which the pieces can slide to the left or right in the top and bottom rows or up and down in each of the four columns, and the object is to swap each roman lettered piece with that of the other roman lettered piece (upper or lower case).

There are several ways of doing it. Here is a particularly nice symmetrical solution with 36 moves.

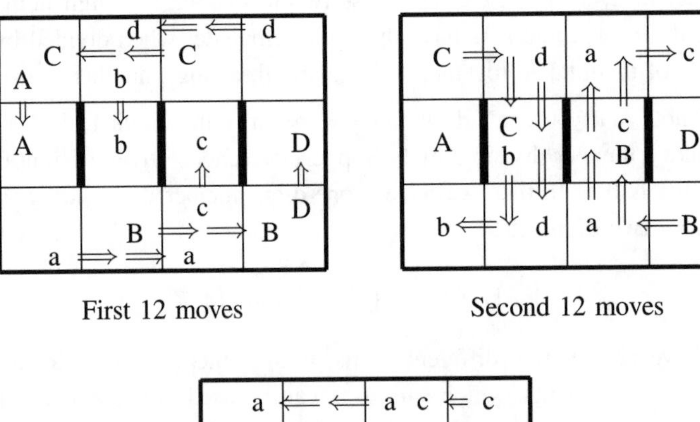

First 12 moves

Second 12 moves

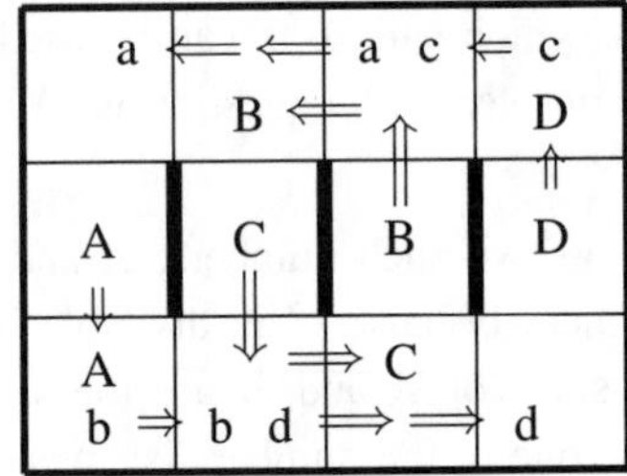

Third 12 moves

Q187. What is the fewest number of moves required to solve the puzzle?

S188. From the hint we can fill in the segments shown on the left. In the notation which follows, the first letter of a string indicates the line (capital = horizontal, lower case = vertical) in which the segment is drawn. When there appear to be alternatives, figure out why they are wrong. We must have just one of Ged or dGF and each implies cGH, dHI and IcbaIHabHGbaGFab. Now not Fbc as this would force Fcd and Ged, closing the loop too soon; so bFEbaED. Then CgfCBfg, DjiDEijEF, hDE, iFG, jFGHIji, Gihg, HhiHI, HhgfeHG, dFG, Ggf, Dfe and eBAed. Now Dcd is the only possible single edge of the 1-square, and then dDCdeCD, cDCBcde, AdcbABCbaCD and we're done.

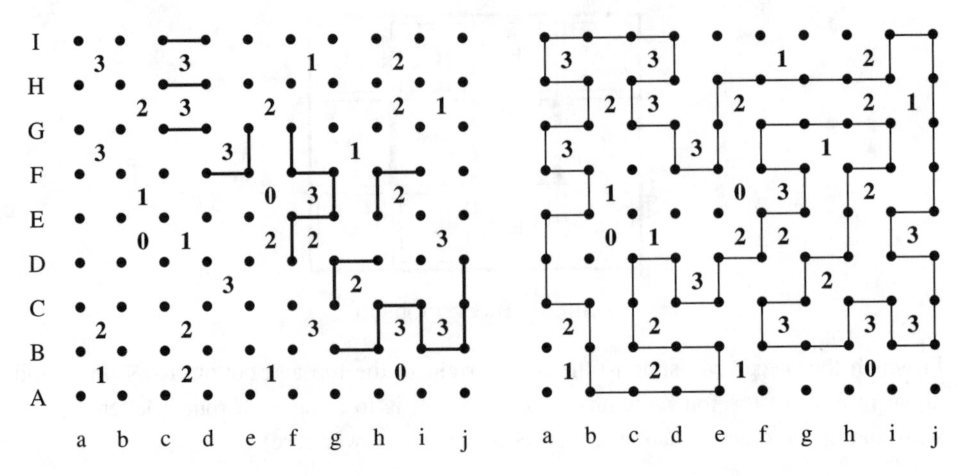

For the second board, using the ideas in the hint, we can fill in the segments on the left. From there, we can fill in hDCBhiBCijCDji. Then GgfGF, eGH, EgfED, Cefg, BfedBAdc and Ffe. The two neighboring 3-squares must be connected by either Ebc and Dcd or by Dbc and Ecd. If the former, then, to bound the 2-square above left, we need FbaFEDC and we have three loose ends in the lower left quarter of the board. So Dbc and Ecd, then dDCdcbaCBAabc and EbaEFGHabcHIcdIHdeHGedcGFcde and we're done.

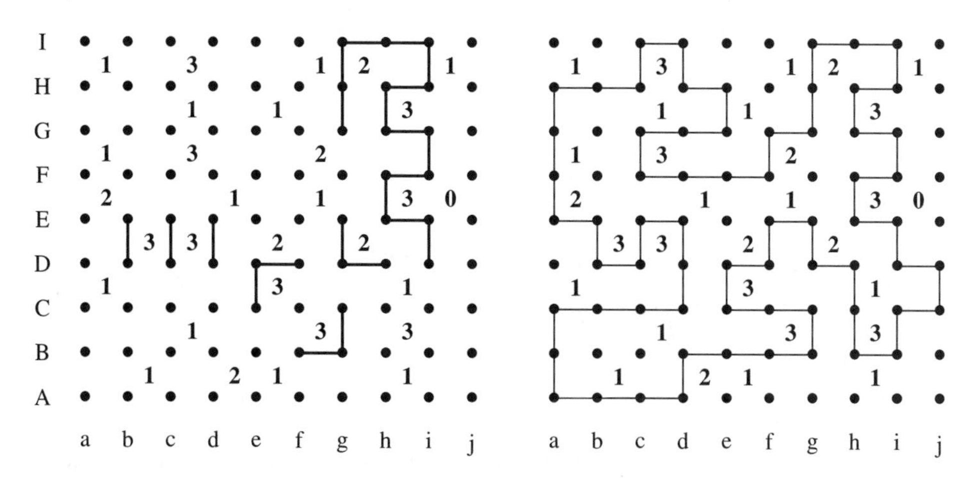

S189. Suppose that $T = abcd$ is a winning number that gives a cart as a prize. Then, $a+b+c+d = 18$. We can write this in the form

$$a+b = 18-(c+d) = (9-c)+(9-d).$$

If we replace the digits c and d with $c' = 9-c$ and $d' = 9-d$, we obtain another lottery number $T' = abc'd'$, which wins a car.

Conversely, if we have a ticket which wins a car, whose last two digits are e and f, we can replace them with $9-e$ and $9-f$ respectively, and obtain a ticket which wins a cart.

We have shown that each ticket that wins a car corresponds to a ticket that wins a cart, and conversely. Moreover, a ticket that wins *both* a car and a cart corresponds to a *different* ticket that wins both a car and a cart (because if $abcd$ wins a car and a cart, then

$$a+b = c+d = 9 \quad \text{and} \quad a+b = c'+d' = 9,$$

so $abc'd'$ wins both a car and a cart, and $c \neq c', d \neq d'$). The organizers, therefore, will give away the same number of cars as carts.

To find the number of tickets $abcd$ such that $a+b+c+d = 18$, let $a+b = 19-k$ and $c+d = k-1$. Then, for $k = 1, 2, \ldots, 10$, there are k possibilities for the pair (a, b), and k possibilities for the pair (c, d), giving k^2 altogether. It follows that the total number is

$$1^2+2^2+\cdots+9^2+10^2+9^2+\cdots+2^2+1^2 = 670.$$

The case $k = 10$ means that the first two digits add to the same as the last two digits, so the middle hundred get nothing, meaning that the organizers will give out 570 cars and 570 carts.

S190. Using the notation from the hint, we can use the following plan. In the first move, families should exchange apartments according to the scheme in figure (i), where the lines show which residents are to exchange apartments. In the next week, the exchanges should be arranged according to the scheme in figure (ii).

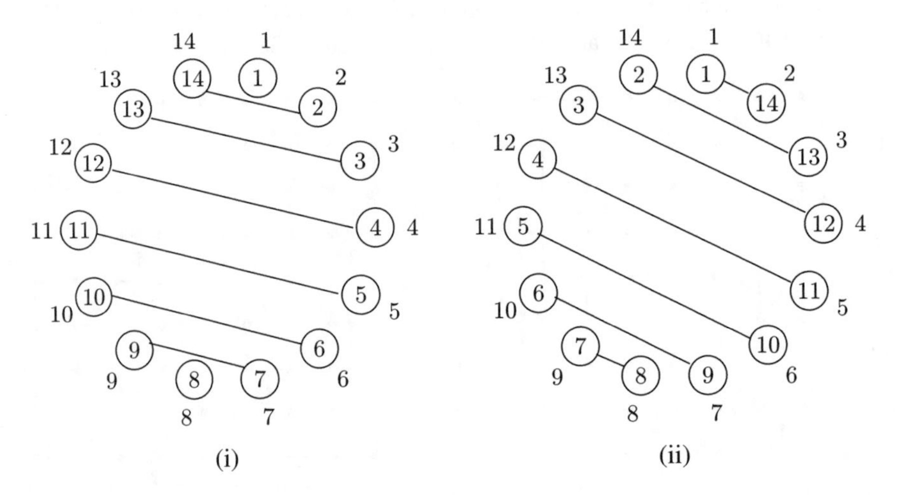

(i) (ii)

The answer does not change when the exchange involves a chain of thirteen apartments; two weeks are necessary, and sufficient. Here is one possible scheme for the exchanges. See **Permutations** in the Treasury.

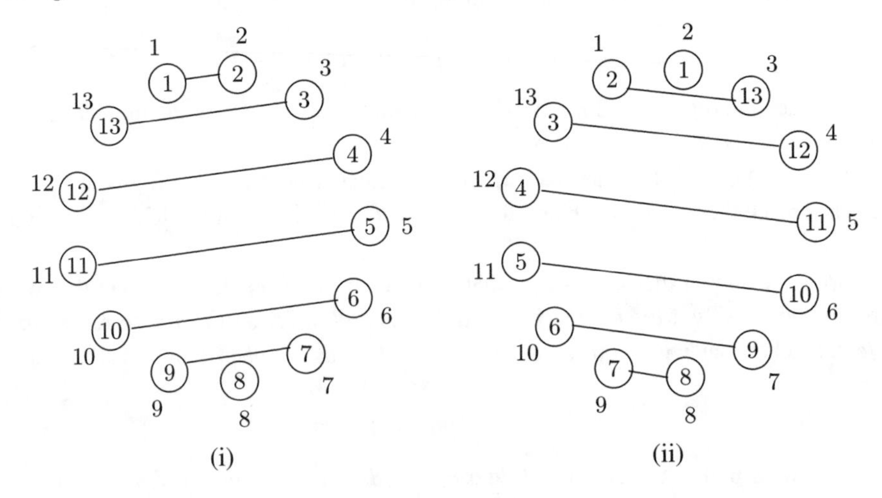

(i) (ii)

S191. We continue with the reasoning begun in the hint.

(a) In the case $E = 3$, B must be 1 (it has to be 1 or 3) and therefore the first digit in the product must be $C = 8$. If $B = 1$, then D must be 0 or 5, but then we cannot obtain $E = 3$ in the product. Therefore, we must have the alternative, $E = 8$. Then $C = 9$ and $B = 3$ (we need a carry in order to get $C = 9$). Again, D must be 0 or 5, but only $D = 5$ satisfies the requirements for the product. Thus, the answer is: $23958 \times 4 = 95832$.

Here's a more algebraic solution. Write $X = 100C + 10D + E$, so that

$$4(10000A + 1000B + X) = 100X + 10B + A$$

and as all the terms in this are even, and A is at most 2, $A = 2$. Division by 6 gives

$$13333 + 665B = 16X$$

Modulo 8, $5 + B \equiv 0$, so $B = 3$ and $X = 958$, $23958 \times 4 = 95832$.

(b) Suppose the last digit of the multiplier is 6. Then the units product is three times as large as the tens product. But this is impossible because the units product is at least 2100 and the tens product is at most 698. Therefore the multiplier is 28. The second digit of the tens product is odd, which means that there is a carry when the last digit of the multiplicand is multiplied by 2; this means that this last digit is 6 or 8. Also, the middle digit of the multiplicand cannot be larger than 4, which means that it can only be 0, 2, or 4. It cannot be 0 for that would make it impossible to get an odd digit in the units product. Therefore we have four possibilities: 326, 328, 346, and 348. When we multiply each of these by 28 we find that only 348 satisfies the requirements: $348 \times 28 = 9744$.

Here's a pretty analysis (can you justify each step?): the units product is ≥ 2100 and the final product is ≤ 9988, so the tens product is ≤ 788 and so is ≤ 698. The ratio of the two even digits of the multiplier is $\geq 2100/698 > 3$, so that the multiplier is 28. Therefore the first digit of the units product is 2, the first digit of the tens product is 6 and the final product lies between 9100 and 9800. The multiplicand must be between 325 and 350 and the units product is between 2600 and 2800, so its first two digits are 27, while the first two digits of the multiplicand are 24. The first two digits of the tens product are 69, and in the final product, the first two digits must be 97. As $9700/28 > 346$, the multiplicand is 348 and the answer is $348 \times 28 = 9744$.

Q191. On page 30 of *The Incredible Dr. Matrix*, Charles Scribner's Sons, New York, 1967, 1976, Martin Gardner is trying to get Miss Toshiyori's telephone number:

"The number has five digits," she said. "If you add a four to the front of it, you make a number that is exactly four times the number you make if you put the 4 at the end instead of the front."

What is Miss Toshiyori's number? What if she hadn't said how many digits there were? What if she had said '5' and 'five times' instead of '4' and 'four times'?

S192. Suppose we begin by placing four coins on the bottom row, in columns a, b, d and e, as shown (we'll use "no" to mean there is not, or cannot be, a coin on that square.

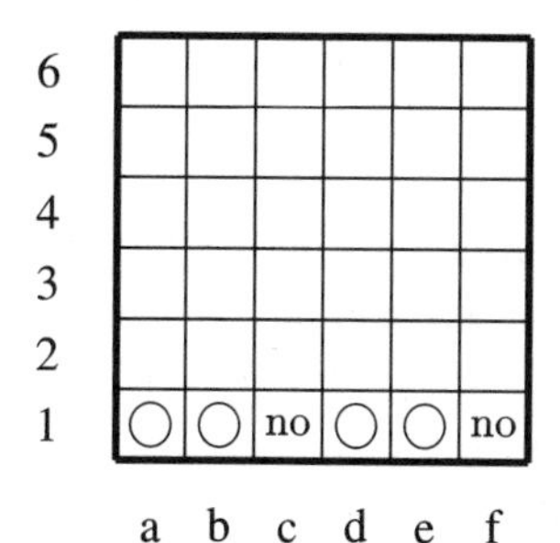

Then in the second row, there cannot be a coin on a2 because that would mean a1 would have two neighbors. Similarly, there can be no coin on b2 otherwise b1 would have two neighbors. There must be a coin on c2 so that c1 will have an odd number of neighbors. Also, no coins are located at d2, e2 or f2, because otherwise d1, e1, f1 would have two neighbors. Thus, the second row is completely determined.

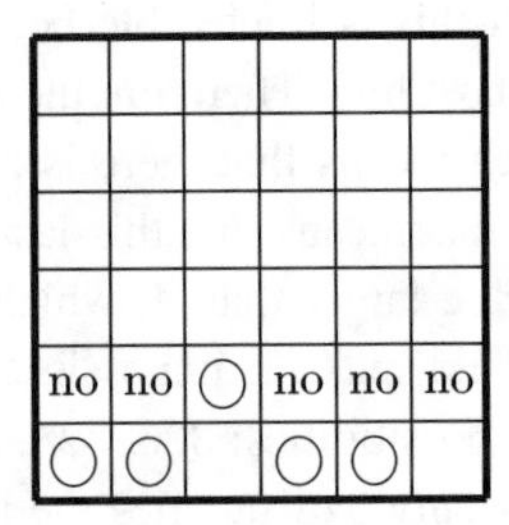

In the same way, subsequent rows are uniquely determined by considering the squares below it.

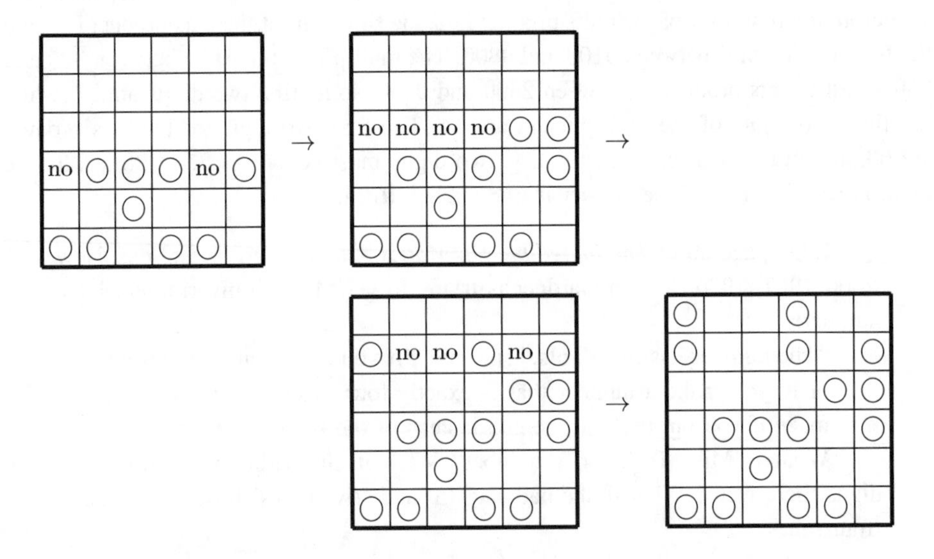

Q192a. Which first row coin placements lead to solutions and what are the fewest and greatest numbers of coins which may be used?

Q192b. What other sizes and shapes of board can be 'coined' in this way, that is, with every square, occupied or not, having just one or three neighboring squares occupied by coins.

S193. 4 raised to any even power ends in 6. 5 raised to any power ends in 5. 6 raised to any power ends in 6, and 8 raised to any power which is multiple of 4, ends in 6. The powers of 7 are a bit more tricky. We need to look at the two last digits. In fact, to make things clearer, we'll look at the last three digits. It will be clearer still if you use modular arithmetic. In any case, here is a table showing the last three digits of the powers of 7, from the zeroth to the nineteenth. It's easy to calculate these if you know your seven times table, and you can verify that from then on the pattern repeats.

001	007	049	343
401	807	649	543
801	607	249	743
201	407	849	943
601	207	449	143

The odd powers of 7 end in 7 or 3, according as the exponent is 1 or 3 more than a multiple of 4.

From the table, the last three digits of 7^7 are 543, which is 3 more than a multiple of 20. So 7^{7^7} ends in 343, which is also 3 more than a multiple of 20, so $7^{7^{7^7}}$ and $7^{7^{7^{7^7}}}$ also end in 343. For another approach to this problem, see **Modular arithmetic** in the Treasury.

S194. Observe first that 631 is a prime number (the only positive numbers that divide it are 1 and 631).

(a) Because $A^2 - B^2 = (A-B)(A+B)$, we must have $A - B = 1$ and $A + B = 631$. These equations have a unique solution: $A = 316$ and $B = 315$.

(b) Because $A^3 - B^3 = (A - B)(A^2 + AB + B^2)$, we have $A - B = 1$ and $A^2 + AB + B^2 = 631$. From the first equation, $A = B + 1$. Substituting this into the second equation yields $(B+1)^2 + (B+1)B + B^2 = 631$, or equivalently, $B^2 + B = 210$. As you can see from **Quadratic equations** in the Treasury, we can multiply by 4 and add 1, giving $4B^2 + 4B + 1 = 841$ or $(2B+1)^2 = 29^2$. $2B + 1 = \pm 29$, $B = 14$ or -15 and correspondingly $A = 15$ or -14. But we were only asked for the positive solution.

(c) Here, $A^4 - B^4 = (A^2 - B^2)(A^2 + B^2) = (A - B)(A + B)(A^2 + B^2)$. Because these three factors are different (neither A nor B can be zero) this product cannot be equal to $631 = 1 \cdot 631$. Therefore, this equation has no solution in nonnegative integers.

S195. For each horse to be unattended for at most one five-minute interval (see the hint), at least five five-minute periods are required (two horses are unattended in each period). The schedule below shows that all ten horses, H1 to H10, can be shod by blacksmiths B1 to B8 in this minimum time interval (twenty-five minutes).

Time Interval \ Horse	H1	H2	H3	H4	H5	H6	H7	H8	H9	H10
0–5 min	B1	B2	B3	B4	B5	B6	B7	B8	–	–
5–10 min	B1	B2	B3	B4	B5	B6	–	–	B7	B8
10–15 min	B1	B2	B3	B4	–	–	B5	B6	B7	B8
15–20 min	B1	B2	–	–	B3	B4	B5	B6	B7	B8
20–25 min	–	–	B1	B2	B3	B4	B5	B6	B7	B8

S196. As explained in the hint, in the top row, 16 across is 7 9, 17 down in the fourth column is 9 8 and also in the first column. 30 across in row two is now 9 7 6 8 and 23 across in row three in 8 6 9. 35 down in column three is 7 6 9 8 5, not 5 8 as there would not be enough for 24 across, which is now 8 9 7 or 8 7 9. But not the latter, else 12 down would be 7 5 and 11 across in the next row would be oversubscribed. 12 down is 9 3 and 15 down in the middle column is 8 7. There are three 3-digit words of sum 23, which each must be 6 8 9 in some order. The digits of 35 across average 7, and so are 5 6 7 8 9 in some order, while 22 down is 6 7 9 in some order, so 14 across has to be 8 6, 22 down is 6 9 7, 23 across in 6 9 8, the top 23 down is 8 6 9, 35 across is 8 9 7 5 6, the other 23 down is 6 9 8 to allow 16 across to be 7 9, and then 29 down is 8 5 7 9. Now 30 across is 7 6 9 8 in order that the second 16 down not grow too big. The first 16 down is 7 9. 13 across is 3 9 1, because we have seen in the hint that the last 4 across is 1 3. We also saw that 3 down is 2 1, so the second 16 down is 6 1 2 4 3 and 7 across and 10 across must be 2 1 4 and 2 4 3 1 in order to make the down entries correct. The other 4 down is now 3 1. 6 down and 7 down in the fourth and third columns are 1 2 3 and 1 2 4 is some order, so the 4 across at the foot which intersects them must be 1 3, and since 15 across is 5 4 3 2 1 is some order, 6 down is 2 1 3 and 7 down is 4 2 1. Now 8 across is 5 2 1, 15 across is 5 3 4 2 1, the remaining 4 across is 3 1 and 11 across is 1 2 5 3, completing the picture.

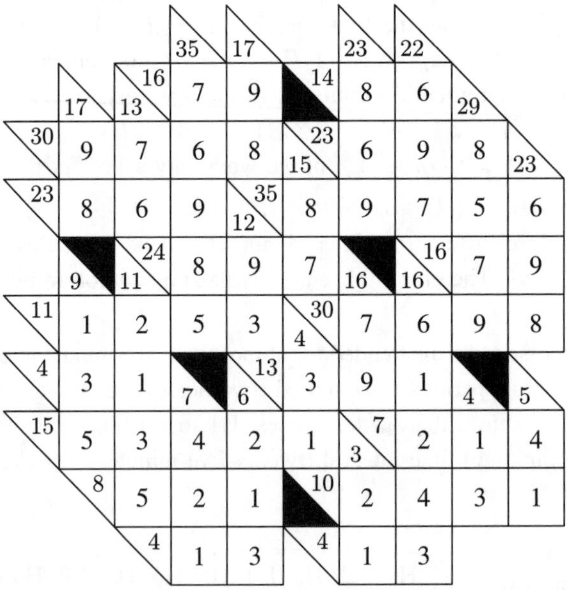

S197. (a) Each child has made between 0 and 197 acquaintances in the first week. If all 198 children have a different number of acquaintances, then there is a child c with exactly 197 acquaintances, and a child d with no acquaintances at all. But this is impossible, because if c has made 197 acquaintances, then c must have become acquainted with d, meaning that d must have at least one acquaintance, contrary to assumption. Consequently, at least two children have the same number of acquaintances at the camp.

(b) Under the assumption, each camper had gotten to know between 132 and 197 other campers, a total of 66 possibilities. Divide the children into distinct groups depending on the number of their acquaintances, with two children in the same group if they have made the same number of acquaintances. If none of these groups consists of four or more children, then each of the 66 possible groups consists of exactly three children, because $66 \times 3 = 198$. Consequently, there are three children with 133 acquaintances, three with 135 acquaintances, three with 137 acquaintances, and so forth. There are 33 odd numbers between 132 and 197, which means that $3 \times 33 = 99$ children at the camp made an odd number of acquaintances. But we saw in the hint, or see the **Handshake lemma** in the Treasury, that the number of such children was necessarily an even number. This contradiction means that one of the 66 groups consists of four or more children.

S198. Three of the nine chosen points must have the same label (two points from each group would only account for eight points altogether). Now the points having the same label form a regular pentagon, and any three points of a regular pentagon form an isosceles triangle (there are only two possible distances between the vertices of a regular pentagon).

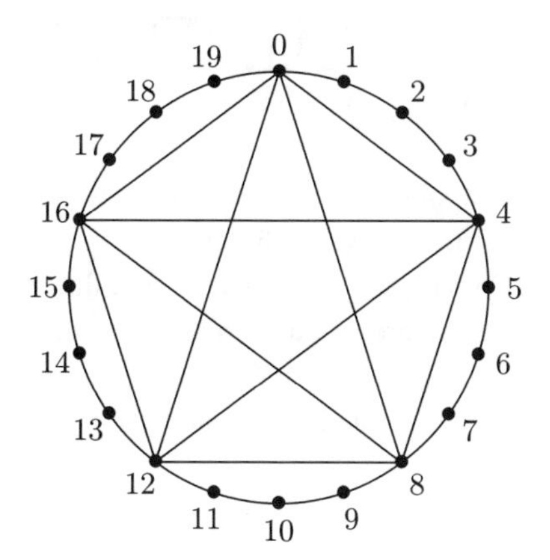

> **Q198a.** How many points must you choose from the 20 to be sure of getting three to form an equilateral triangle?
>
> **Q198b.** If you choose only eight points, will there necessarily be an isosceles triangle?

S199. Since there are evenly many tiles, the difference between the numbers of horizontal and vertical tiles must be even, say $2d$, so that there are $32 - d$ horizontal and $32 + d$ vertical. Color every fourth column of the floor black, so that 64 small squares are black. Each horizontal tile covers just one black square, so

$$64 - (32 - d) = 32 + d$$

are covered by vertical tiles. Each vertical tile covers 0 or 4 black squares, so d must be a multiple of 4, and the difference $2d$ a multiple of 8. Eight is the minimum difference.

S200. The equally spaced triples are

$$1, 3, 5;\ \ 11, 13, 15;\ \ 21, 23, 25;\ \ 1, 11, 21;\ \ 3, 13, 23;\ \ 5, 15, 25;\ \ 1, 13, 25;\ \ 5, 13, 21.$$

The number 13 occurs in four of these, and 1, 5, 21, and 25 each occur in three of them, so the triples are the rows, columns, and diagonals of the square.

1	3	5
11	13	15
21	23	25

So we may put the numbers of the magic square

8	1	6
3	5	7
4	9	2

into the corresponding positions, and each equally spaced triple adds to 15.

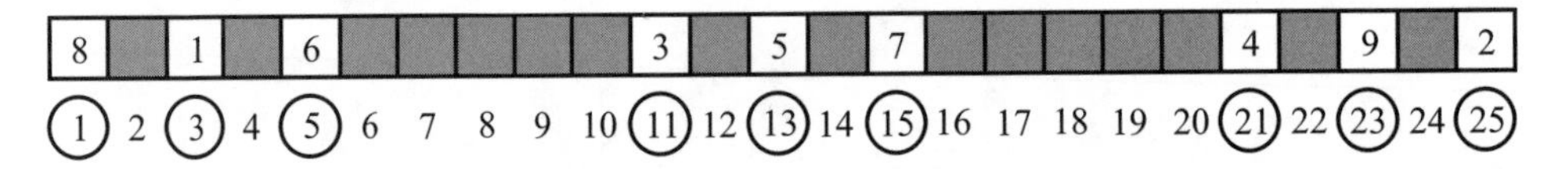

Q200. (Barry Cipra) Suppose Barry starts with a strip of 125 squares, removes the second and fourth fifths (leaving 75 squares), the second and fourth fifths from the remaining strips (leaving 45 squares), and the second and fourth fifths a third time (leaving 27 squares). The remaining squares have labels 1, 3, 5, 11, ..., 25, 51, ..., 75, 101, ..., 125. Can the numbers $1, 2, \ldots, 27$ be placed so that every equally spaced triple has the same sum?

S201. Choosing three numbers that sum to 15 is equivalent to picking three numbers which lie in a single row, column, or diagonal of the 3×3 magic square.

8	1	6
3	5	7
4	9	2

The game looks like Tic-Tac-Toe (which is a tie with best play), but this is a little different because you can make 15 with two numbers, $6 + 9$ or $7 + 8$. Here, the first player can win by starting with 6, 7, 8, or 9 forcing 9, 8, 7, or 6 respectively, and the first player wins with 5 then 2, 6 then 5, 5 then 6, or 2 then 5.

S202. Number the squares of the board as shown.

8	1	6	0
3	5	7	
4	9	2	

The first player can win by playing in square 6. If the opponent replies 8 or 0, then the first player wins by playing 1, which forces 0 or 8, then 5, and 4 or 9 on the final move. If the opponent's first reply is 1, then 5 forces 4, and 2 wins on the next move at 7 or 8. If the opponent's first move is in the second or third row, then 1 forces a win at 8 or 0.

Alternatively, the first player can win by playing in square 7. If the opponent replies 3 or 5, then 9 forces 0, and now 2 wins at 6 or 4. If the opponent's first move is 8, then 0 forces 9 and then 6 leads to a win at 1 or 2. If the opponent's first move is 4, then 9 forces 0 and 5 leads to a win at 1 or 3. If the opponent's first move is 1, 9, or 0, then 5 forces 3 and 2 leads to a win at 6 or 8. If the opponent's first move is 2 or 6, then 5 forces 3 and 9 leads to a win at 1 or 0.

Q202. What is the size and shape of a Tic-Tac-Toe board (not necessarily rectangular), with the smallest possible number of squares, in which the first player has a winning strategy?

S203. Nicole can win if either r or c (or both) is odd. If both r and c are odd then she wins by placing an $\times$ in the middle square (row $(r+1)/2$, column $(c+1)/2$). If one of the numbers, r or c, is odd and the other is even, then she wins by placing an $\times$ in the two middle squares (if, for example, r is odd and c is even, the two middle squares are squares $c/2$ and $c/2+1$ in row $(r+1)/2$. Thereafter, she plays the **Tweedledum and Tweedledee strategy** (see the Treasury): rotate the player's last marks 180° about the center and play in these squares. She will always have a permissible response, and will eventually win the game.

If both r and c are even then Pryor has a winning strategy, which is like that of Nicole in the first case. Namely, in each move, he answers the sequence of consecutive marks from Nicole by placing a mark in those squares obtained by rotating Nicole's last marked squares by 180 degrees about the center point of the grid. In this way, Pryor always has a possible move and will eventually win the game.

Q203. Now consider a slight rule change, and allow players to place their $\times$'s on adjacent squares of a single row, column, *or diagonal.* Find a winning move for each of the boards shown below.

×	×						
×							
×	×	×			×		
			×	×			
			×	×			
		×		×	×	×	×
					×		×
						×	×

(i)

×					×		×
×	×	×	×	×	×	×	×
×	×	×	×	×	×	×	×
×	×	×	×	×	×	×	×
×	×	×	×	×	×	×	×
×	×		×	×	×	×	×
×	×	×	×	×	×	×	×
×	×	×	×	×	×		×

(ii)

×						×	×
×	×	×	×	×	×	×	×
×	×	×	×	×	×	×	×
×	×	×	×	×	×	×	×
×	×	×	×	×	×	×	×
×	×		×	×	×	×	×
×	×	×	×	×	×	×	×
×	×	×			×	×	×

(iii)

×					×		
×	×	×	×	×	×	×	
×	×	×		×	×		
×	×	×	×	×	×	×	×
×	×	×	×	×	×	×	×
×		×	×		×		×
	×	×	×	×	×	×	×
		×					×

(iv)

S204. As Ferdinand starts with $220 = 7 \times 31 + 3$ empty squares in front of him, he can force a win by reducing this number by 3, so his winning move is to move his piece from square 1 to square 4, leaving a multiple of seven empty squares between the pieces.

If Isabella moves k squares to the left, Ferdinand replies by moving $7 - k$ squares to the right, once again leaving a multiple of seven empty squares between the pieces.

S205. Suppose both r and c are odd. Then the second player has a winning strategy. The board minus the lower left corner can be tiled with 1×2 tiles ($(rc - 1)/2$ of them). Once this is done, player 1 will move the piece to a square on one of these 1×2 pieces. Player 2 can now respond by moving the piece to the other square on this 1×2 tile. In each successive move by player 1, player 2 can follow in this same way by moving the piece to the other square of the corresponding tile. Player 2 will always have a move and so will eventually win the game.

If either r or c is even, the first player has a winning strategy. First, cover the board with 1×2 tiles (including the lower left corner). On the first move, player 1 moves the piece to the second square of the 1×2 tile that covers the lower left corner square. Thereafter, player 1 uses the same strategy as player 2 in the preceding case.

> **Q205.** Playing out the winning strategy using 1×2 tiles means that the first player will always move the rook by only one space. Can you design a strategy for the 8×8 board whereby the first player can force a win by moving the rook at least four spaces on each move?

S206. (a) Nicole can win by taking two beans from the heap of 19 and thereafter using the **Tweedledum and Tweedledee strategy**: whenever Pryor takes one, Nicole takes one; if he takes two, so does Nicole. There will always be an odd number of beans after Nicole has played. If it gets down to one bean, Nicole will have 10 and Pryor 8: he has to take the last bean and lose. If it gets down to 3 beans, Nicole will have 9 and Pryor 7: whatever Pryor does, Nicole gets one more bean and wins.

(b) Pryor has a winning strategy in this game with heaps of 10 and 12 beans. However Nicole plays, Pryor responds so that both heaps will have an even number of beans left in them for Nicole to choose from. If she takes one bean from one of the heaps (or one bean from each heap), Pryor takes a bean from the same heap (or a bean from each heap). Because after Nicole's first move, at least one of the heaps will have an odd number of beans, Pryor will always have a move, and therefore will eventually win the game.

If they start with heaps of 10 and 11 beans, then Nicole can take one bean from the odd heap, making it even, and thereafter use Pryor's strategy, since this time he will always have to leave at least one odd heap.

S207. We work backwards as in the hint: we place a P in the lower right corner because the Previous-player has just won, and an N in the three squares nearest to it because from these squares the Next-player can win. Then we can place a P in the two squares from

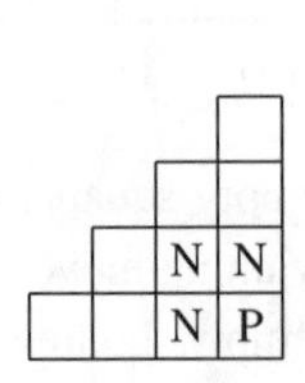

which one can move the piece *only* to an N-square. After that, we can again place an N in those squares (six of them) from which the piece can move to a P-square, and so forth. The rule is: if there is any move to a P-square, then the square is marked with an N, but if all the moves are to N-squares, then it is marked with a P. If we continue marking squares in this way we eventually produce the following board. If it is your turn and you are on an N-square you can force a win; if you are on a P-square, there is nothing you can do to win, for the previous player can force a win. We see that the starting square is a P-square, so Nicole, who starts the game, cannot win, because the Previous-player (Pryor) can force the win.

P	N	P	N	P	N	P	N	P	N	P	N	P	10
N	N	N	N	N	N	N	N	N	N	N	N	N	9
P	N	P	N	P	N	P	N	P	N	P	N	P	8
N	N	N	N	N	N	N	N	N	N	N	N	N	7
P	N	P	N	P	N	P	N	P	N	P	N	P	6
N	N	N	N	N	N	N	N	N	N	N	N	N	5
P	N	P	N	P	N	P	N	P	N	P	N	P	4
N	N	N	N	N	N	N	N	N	N	N	N	N	3
P	N	P	N	P	N	P	N	P	N	P	N	P	2
N	N	N	N	N	N	N	N	N	N	N	N	N	1
P	N	P	N	P	N	P	N	P	N	P	N	P	0
12	11	10	9	8	7	6	5	4	3	2	1	0	

This algorithmic method of working backwards can be applied to other two-person games (see **Winning strategy** in the Treasury), especially in take-away games which can be converted into a chessboard problem such as this one. For example, this problem is identical to **P206b**. To see this, number the rows and columns of the chessboard as shown. A piece on row r and column c corresponds to two heaps of beans, one with r beans and one with c beans. A move to the right or down is analogous to taking a bean from the respective pile, and a diagonal move corresponds to taking a bean from each pile.

> **Q207.** Suppose that the move is to take *any* number of beans from one of the two heaps or an equal number from each of the two heaps. Is the result the same? What about other sizes of heaps?

S208. (a) Adam can prevent Eve from making the number divisible by 9. He can, for example, choose 1 for the first digit (he could choose 2 or 3 as well) and then continue by writing $6 - s$ if Eve chooses s. After Adam has written the 19th digit, the sum of the digits of the 19-digit number is $1 + 9 \times 6 = 55$. But because the last digit must be between 1 and 5, Eve cannot make the sum of the digits add to 63, which would be needed to make the number divisible by 9.

(b) In this case, Adam cannot prevent Eve from making the number divisible by 9. For each digit s played by Adam, Eve will answer $6 - s$. After she has written the 30th digit, the sum of the digits will be $15 \times 6 = 90$. The number is therefore divisible by 9.

S209. Sonja can force a win in the following way. If possible Sonja always chooses an odd number. If Clemens chooses only even digits then Sonja can always choose odd digits, right up to her fifth and final move. If the number after Clemens has made his fourth move is odd then Sonja chooses the last odd number with the plus sign. If the number is even she writes the digit with the multiplication sign. The final number is even because Clemens has only an even digit to write in.

If Clemens chooses at least one odd digit among his first four moves, then Sonja must choose an even digit on her last (fifth) move. She chooses an even digit with the multiplication sign so that the resulting number after her move is even. Clemens must write in an even digit and consequently the final number is even.

S210. As in other touring problems, one way to construct such a tour is to make several loops and then try to connect them into a single loop. The figure on the left shows five loops: four outside loops labeled 1, 2, 3, 4 and one central loop. One way to proceed is to join loop 1 to the inside loop by deleting edges AB and ab and adding Aa and Bb. To keep bilateral symmetry, connect loop 4 to the inside loop by deleting edges CB and cb and then adding Cc and Bb. Similarly, connect loop 2 to the inside loop by deleting DF and df and adding Dd and Ff, and then loop 3 to the inside by deleting DE and de and adding Dd and Ee. The resulting tour on the right is symmetric about the vertical axis.

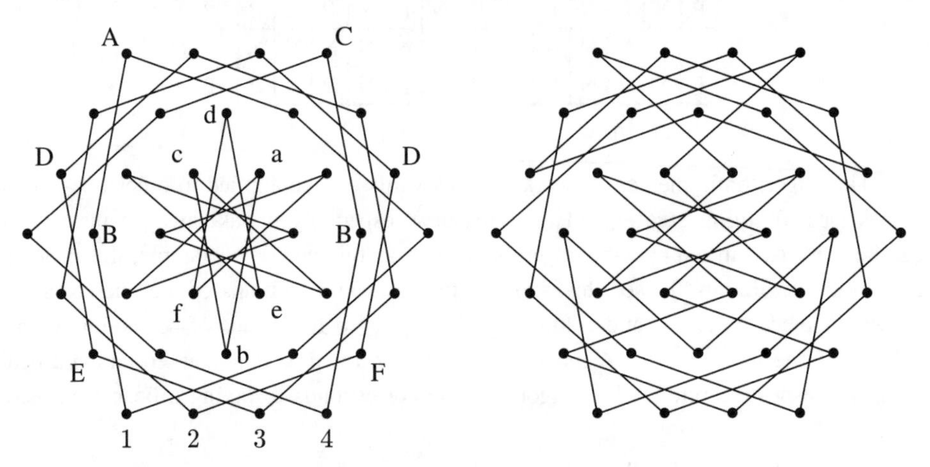

With some extra attention we can construct a tour with six-fold rotational symmetry (invariant under a 60° rotation about the center).

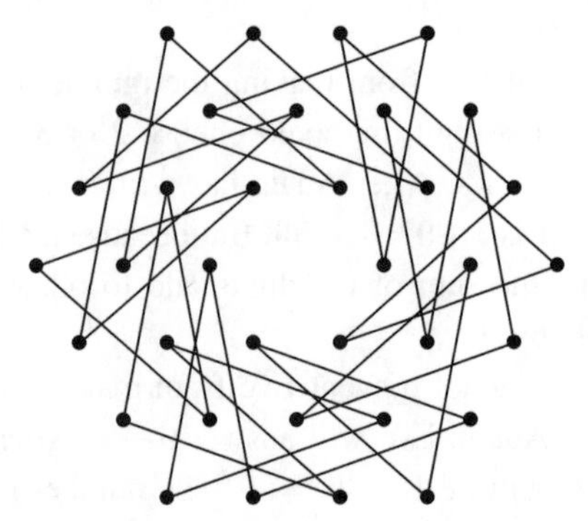

Q210a. Can you construct a tour on a hexagonal grid with five points on a side (instead of four) which has six-fold rotational symmetry?

Q210b. Are there such six-fold symmetric tours on a hexagon with six vertices on a side? seven? eight? ...

S211. From the hint, we begin by disregarding the right boundary and nudge the top row of balls to the right as in the left figure.

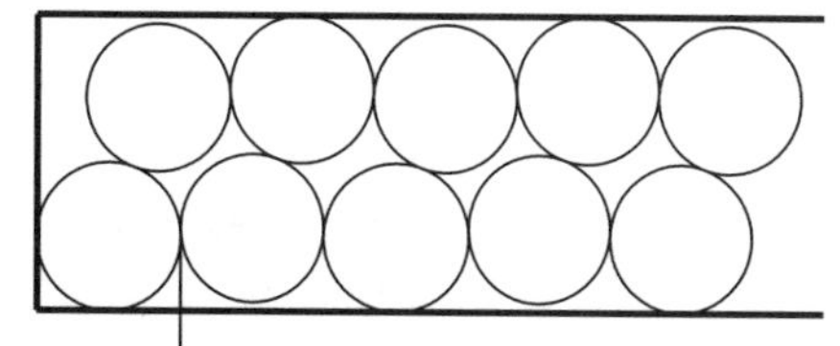

This creates a little room at the top, so now, slide the balls in *every other* column up as far as they will go, as shown on the right (above).

As a consequence the columns of balls are no longer tangent, so there is a very slight space between them, which means the balls can be slid to the left by a small amount. When n is large enough we will have room for another ball in the bottom row!

Q211. How large must n be?

S212. Number the squares of the board as shown.

1	2	3	1	2	3	1	2
3	1	2	3	1	2	3	1
2	3	1	2	3	1	2	3
1	2	3	1	2	3	1	2
3	1	2	3	1	2	3	1
2	3	1	2	3	1	2	3
1	2	3	1	2	3	1	2
3	1	2	3	1	2	3	1

There are 21 squares numbered 2, 21 numbered 3, and 22 numbered 1. Because each 1×3 tile covers three squares with *different* numbers, the board cannot be tiled with these pieces if the 1×1 tile is on a square numbered 2 or 3.

Q212. On which squares can the 1×1 square be placed so that the board can be covered with these tiles?

S213. Because $600 = 85 \times 7 + 5$, there are 85 numbers between 1 and 600 divisible by 7, which means that 600! is divisible by 7^{85}. Moreover, $85 = 12 \times 7 + 1$. Thus, among these 85 numbers divisible by 7, there are 12 divisible by 7^2. This means that 600! is divisible by $7^{85+12} = 7^{97}$. Finally, $12 = 1 \times 7 + 5$, so among the twelve numbers divisible by 7^2, one of them is divisible by 7^3. Consequently, 600! is divisible by $7^{97+1} = 7^{98}$, but not by 7^{99}.

Q213. Can you find 600 consecutive positive integers less than 1000 whose product is divisible by 7^{101}?

S214. In the hint we saw that the digits of A can be no larger than 5. Because $5!+5!+5!=360$, A is less than 360 and its first digit is at most 3. Because $3!+5!+5!=246$, this first digit is at most 2. $2!+5!+5!<255$ and $2!+4!+5!=146$, so the first digit is 1 and $1!+4!+5!=145$ is a solution; in fact the only one, since 135, 125, and 115 are the only other candidates with three digits.

S215. Clearly, f must be 1 or 2, and $fghi$ is a multiple of 111. The multiples of 111 between 1110 and 2997 all have their middle digits equal except $19 \times 111 = 2109$. So $a+b+c=a+d+e=19$, $2<a<9$ (0, 1, 2, 9 are used by f, g, h, i), and since b, c, d, e are all different, $11 \le b+c=d+e \le 13$. Now a short series of trials shows that $a=8$.

$b+c=d+e$	a	Trials for $(b,c),(d,e)$
13	6	(6,7), (5,8)
12	7	(5,7), (4,8)
11	8	(5,6), (4,7), (3,8)

Thus there is essentially just one solution (not counting the order of the summands or the ordering of the two sums): $a=8$, b, c, d, e are chosen from 4, 5, 6, 7 so that the pairs (b,c) and (d,e) are $(4,7)$ and $(5,6)$ in some order, and f, g, h, i are 2, 1, 0, 9, respectively.

S216. (a) Let the first digit of A be a so that $A=10000a+x$, say; $B=10x+a$ and $10A-B=99999a=9\times 41\times 271\times a$ and if A is divisible by 41, so will B be.

In (b), $A=100000a+y$, say; $B=10y+a$ and $10A-B=999999a=999\times 1001a=27\times 37\times 7\times 11\times 13a$ and if A is divisible by 37 (or any other divisor of 999999), so is B. For another approach to these problems, see **Divisibility tests** in the Treasury.

S217. Assume that an odd number of queens are placed on the white squares. Then, from the hint, the sum, S_1, of their row and column coordinates is an odd number. The other queens are on black squares, which have even coordinate sums, and therefore the sum, S_2, of their row and column coordinates is even. Consequently, $S=S_1+S_2$ is an odd number. But the sum S can be computed in another way as we shall see.

Because there is a queen on each row, each row number occurs in S exactly one time. Similarly, each column number occurs in S exactly one time. Therefore, $S=(1+2+\cdots+8)+(1+2+\cdots+8)=72$. This shows that S is an even number, which contradicts our previous conclusion. This means there cannot be an odd number of queens on the white squares.

Virtually the same argument can be used to prove that if every row and column of the chessboard contains an even number of queens (perhaps none) then there must be an even number of pieces on the white squares.

Q217. If you have 8 nonattacking queens on the 8×8 chessboard, must 4 of them be on the white squares?

S218. Triangles abc, dfg, ehi contain the numbers once each, so add to 45, and therefore the "magic number" is 15.

There are eight ways to partition 15 into three distinct parts: $9,5,1$; $9,4,2$; $8,6,1$; $8,5,2$; $8,4,3$; $7,6,2$; $7,5,3$; $6,5,4$.

The corner numbers appear in *three* of the triangles, whereas the other six each appear in just two. From our list of partitions, the corner candidates are 2, 4, 5, 6, and 8. To sum to 15, we must include 5, and the only choices are 8, 5, 2, and 6, 5, 4. By examining the possibilities one finds that there are just four solutions, disregarding reflexions and rotations.

8	8	6	6
1 6	3 4	1 8	7 2
3 4	1 6	7 2	1 8
5 7 9 2	5 9 7 2	5 3 9 4	5 9 3 4

Another way of thinking about the problem is to replace the letters of the original triangle with digits (representing the positions):

1
2 3
4 7
5 6 8 9

In this notation, the relevant equilateral triangles are arithmetical progressions: $1,2,3$; $4,5,6$; $7,8,9$; $1,4,7$; $2,5,8$; $3,6,9$; $1,5,9$. These are the rows, columns, and falling diagonal of the "square of positions" shown on the left.

1	2	3
4	5	6
7	8	9

Positions

8	1	6
3	5	7
4	9	2

Magic Square

So, we will get a solution by placing the numbers of the magic square on the right into the corresponding positions of the triangle prescribed by the square of positions on the left. There are eight 3×3 magic squares gotten by rotating and reflecting the one shown above, and these generate the solutions given above (disregarding triangle reflexions and rotations).

Q218. (Barry Cipra) Can the stars be replaced by $1, 2, 3, \ldots, 27$ so that the vertices of every equilateral triangle with edges parallel to those of the large triangle (there are 37 of them) all have the same sum? (No integer is to be used more than once.)

S219. One way to keep track of the distances between points is to set up a difference table for the labels of the vertices introduced in the hint. Suppose we choose vertices A, B, C, D, E, where we assume that they are listed in increasing order. Then the table below records the distances between adjacent vertices in the first row, the distances between vertices separated by a single vertex in the second row, the distances between vertices separated by two vertices in the third row, and so forth.

Vertices	A		B		C		D		E
Adjacent differences		a		b		c		d	
Every other differences			e		f		g		
Every third differences				h		i			
Every fourth differences					j				

For example, if we select points 0, 1, 3, 8, and 17, the difference table is

0		1		3		8		17
	1		2		5		9	
		3		7		7		
			8		5			
				4				

Note that the difference between 17 and 3 is 14, but 14 is the same as -7 modulo 21, so the distance between vertex 3 and vertex 17 is 7. Similarly, the distance between vertices 17 and 1 is 5, and the distance between vertices 17 and 0 is 4.

One organized approach that will eventually lead to a solution is to start with vertices at vertex 0 and use backtracking, increasing the vertex numbers systematically until there is a repeat in the difference table, then backtracking to the next option.

One set of vertices that have all ten distances represented is 0, 2, 7, 8, and 11.

0		2		7		8		11
	2		5		1		3	
		7		6		4		
			8		9			
				10				

Q219. Can you, from 21 people, form 21 committees of 5 so that no pair of persons serves on two different committees?

S220. We continue with the solutions begun in the hint.

(a) Let C be any other island in the group.

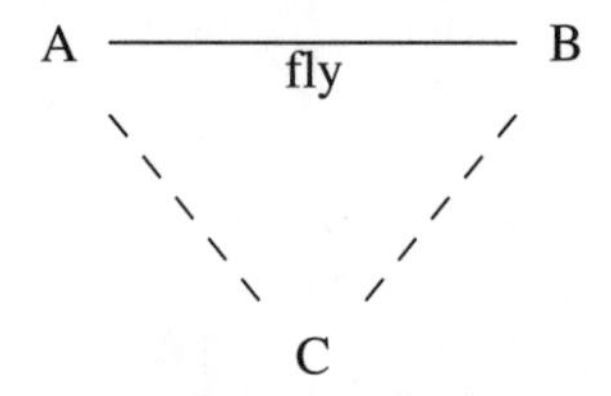

There cannot be a boat connexion between A and C, nor between B and C, because otherwise one could get from A to B by boat with at most one stop (at C), contrary to our assumption in the hint. Consequently, there must be a plane connexion between A and C, or between B and C. It follows that we can get from A to C *directly* by plane, or *indirectly* by plane with at most one stop (at B).

(b) By assumption, there is a direct flight from A to B, and a direct boat connexion from C to D.

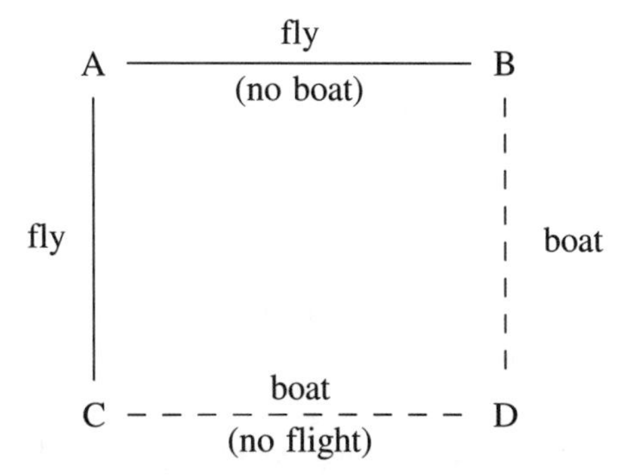

It is clear that both distances AC and BD cannot be traversed by the same mode of transportation. If, for example, both were by plane, then we could have flown from C to D with at most two stops (at A and B), contrary to our assumption. Similarly, if both were boat connexions, we could have gotten from A to B by boat with two stops (at C and D), contrary to our assumption. By symmetry, we may assume that the distance AC is covered by plane and BD by boat. Now if AD is a flight, we could have flown CAD, and if AD is by boat, then we could have boated ADB. Both alternatives contradict our assumption, and thus the claim of the problem is true.

S221. In answer to the question in the hint, S must be an even number, because each direct connexion between two islands, say A and B, is counted twice, once as an outgoing flight from A to B, and once as an outgoing flight from B to A (see the **Handshake lemma** in the Treasury). But if Wayside is not included, then S has the form $7+6x+8y$, which is odd; a contradiction.

S222. Suppose there are i roads leading into Lofton and o roads leading out of Lofton. Then there are $36\times 7+i$ roads leading into the various towns on the island, and $36\times 6+o$ roads leading out of the various towns on the island. Now every road that leads into one town is also a road that leads out of another town, and therefore, $36\times 7+i=36\times 6+o$, or equivalently, $36+i=o$. But there are at most 36 roads into or out of Lofton, so $o=36$, $i=0$, and there's no road leading into Lofton. Therefore the governor can't get back home on this system of roads.

S223. We continue in the same way as in the hint. Suppose we have numbered k towns from 1 to k in such a way that $1\to 2\to 3\to\cdots\to k$. Choose another town, call it Z. Again, there are three cases to consider: (i) $\mathrm{Z}\to 1$, (ii) $k\to \mathrm{Z}$, (iii) neither of the first two alternatives hold. (As in the hint, we note that the first two cases are not mutually exclusive.) Here's what we do in each of these cases:

(i) Let Z be given number 1 and increase the numbers of the k different towns by 1.

(ii) Let Z be given number $k+1$.

(iii) Because neither Z $\to$ 1 nor $k \to$ Z holds, we must have $1 \to$ Z and Z $\to k$. Among all the numbers $1, 2, \ldots, k-1$, choose the largest number m such that $m \to$ Z. There must be such a number because we know that $1 \to$ Z, and this relation doesn't hold for all k towns because, for instance, Z $\to k$ (that is, not $k \to$ Z).

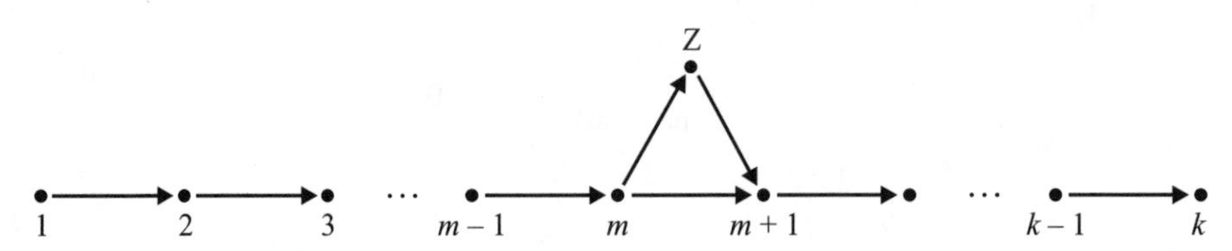

We have $m \to$ Z, and Z $\to m+1$. We now give Z the number $m+1$, and increase by one the number of those towns previously numbered $m+1, m+2, \ldots, k$. Repeat this argument until all towns are included in the ordering.

S224. The spiteful governor (and secretary) can assign road directions in the following way. Number the towns from 1 to 37, and assign the direction of the road between towns k and m so that it goes from k to m if $k < m$. It is clear that from a town, one can travel only to those towns with a larger number, one can never return to the town of origin.

S225. We continue with the solutions begun in the hint.

(a) If there is a road directly from A to B, we are done. Suppose this is not the case. Because $\mathcal{A}$ and $\mathcal{B}$ each consist of 18 towns, there must be a town D in both $\mathcal{A}$ and $\mathcal{B}$ (otherwise the island would have at least $18 + 18 + 2 = 38$ towns; the extra two towns for A and B). Consequently, we can travel from A to D, and then from D to B, that is, A $\to$ D $\to$ B.

(b) If there is a direct road from A to B we are done. Also, we're done if there is a town in both sets $\mathcal{A}$ and $\mathcal{B}$ (the same argument as in (a)). So suppose neither of these hold.

Because there are 37 towns on the island, the number of towns other than A and B and those in $\mathcal{A}$ or $\mathcal{B}$, is equal to $37 - 2 - 14 - 14 = 7$; we will denote the set of these seven towns by $\mathcal{C}$.

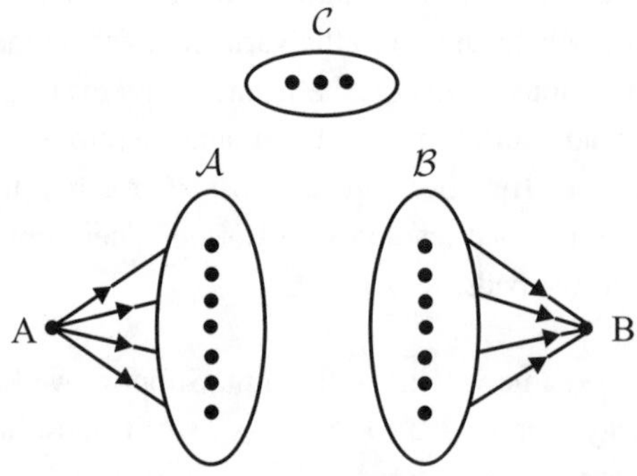

Each road that leads out of a town in $\mathcal{A}$ must go to a town in either $\mathcal{A}$, $\mathcal{C}$, or $\mathcal{B}$. We need to prove that at least one of them goes to $\mathcal{B}$. Because $\mathcal{A}$ contains 14 towns, there are $14 \times 14 = 196$ roads leading out of the towns in $\mathcal{A}$. At most $(14 \times 13)/2 = 91$ of these outgoing roads lead to another town in $\mathcal{A}$ (with 14 towns, there are only $(14 \times 13)/2$

different pairs of towns and each of these pairs account for at most one road between them). Furthermore, at most $14 \times 7 = 98$ roads lead from some town in $\mathcal{A}$ to some town in $\mathcal{C}$. Thus, at most $91 + 98 = 189$ roads lead from a town in $\mathcal{A}$ to a town in either $\mathcal{A}$ or $\mathcal{C}$. This leaves at least $196 - 189 = 7$ roads which lead from $\mathcal{A}$ to $\mathcal{B}$. This shows that the conditions of the problem can be satisfied.

S226. According to the hint, there is at least one person who did not hand out more than five business cards. Let us suppose that A handed out five cards, to B, C, D, E, and F, and that A did not exchange cards with G, H, J or K.

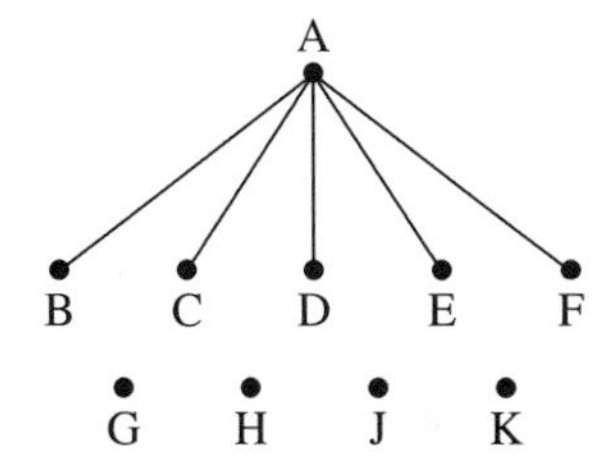

Each line segment in the figure represents one exchange. If we continue to draw a line between each pair of persons who exchanged business cards, we would have the largest number of lines if we draw a line segment from each person in the second row to each person in the third row. That contributes an additional $5 \times 4 = 20$ new lines, so there might have been as many as $2(20+5) = 50$ business cards exchanged. (If A had handed out fewer than five cards then it is easy to see that fewer than 50 cards would have been exchanged.)

Here's another way to think about it. The idea is that if two people exchange cards, we'll put them into two different camps. For example, if A and B exchange cards put them in different camps, similarly with A and C. Now B and C can be in the same camp because they don't exchange cards. In this way we build up a *bipartite* graph, with no people in the same camp exchanging cards. It's possible that everyone in one camp would exchange cards with everyone in the other camp. If there were 5 in each camp, there would be $2(5 \times 5) = 50$ exchanges.

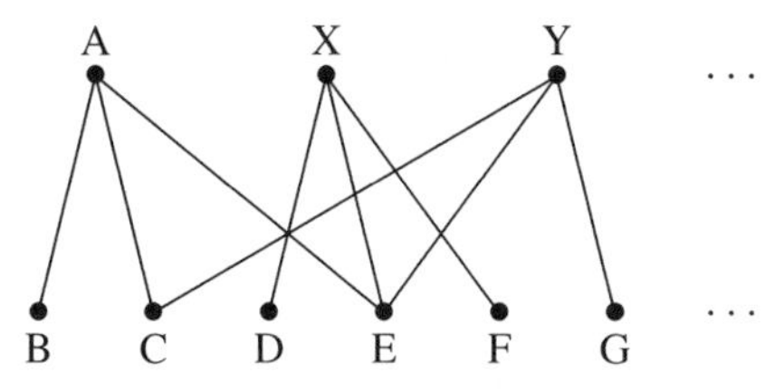

S227. (a) Yes, it is possible. For example, take the numbers 1, 2, 3, 5. The pairwise sums are 3, 4, 5, 6, 7, and 8.

(b) In the hint we showed that we may assume that the smallest number is 0. Then the next two must be consecutive, in order that the two smallest sums differ by one, say $a-1$ and a. Then, in order that the two biggest sums differ by one, the next number must be $a+1$, and call the largest number b. Then the sums of pairs are $a-1, a, \ldots, a+b, a+b+1$ and if these are consecutive, $(a-1)+9 = a+b+1$ so that $b = 7$. But the sum of all ten pairs of numbers is $4(0+(a-1)+a+(a+1)+b)$ because each number occurs in four pairs. Also, the sum of the arithmetic progression $a-1+a+\cdots+a+b+1$ is $5(a-1+a+b+1)$, so that $12a+4b=10a+5b$, so $2a=b=7$, which is impossible. (More succinctly, the sum of the sums is a multiple of 4 because each number occurs in just four sums. But, the sum of ten consecutive numbers is odd!)

A near miss is given by the set $\{0,1,2,4,7\}$ with sums $\{1,2,3,4,5,6,7,8,9,11\}$. These numbers are the first few terms of the Conway-Guy sequence. For example, if you take the set $\{7-0,7-1,7-2,7-4\}=\{3,5,6,7\}$, then the 2^4 subset sums 0, 3, 5, 6, 7, 8, 9, 10, 11, 12, 13, 14, 15, 16, 18, 21 are all distinct. The next few terms are 13, 24, 44. Try it out for 44: that is, write down the 2^7 subset sums of the set $\{44-0,44-1,44-2,44-4,44-7,44-13,44-24\}=\{20,31,37,40,42,43,44\}$. There are several unsolved problems in this area of mathematics; for more information, see Richard K. Guy, "Sets of integers whose subsets have distinct sums," *North-Holland Math. Stud.*, **60** (1982) pp. 141–154.

S228. After the second round, the remaining advisors are numbered $1,3,5,\ldots,33$, and the questioning in the next round begins with advisor 3. Again, add one to each number and divide by 2, to give each person a new number $\frac{1+1}{2},\frac{3+1}{2},\frac{5+1}{2},\ldots,\frac{33+1}{2}$, that is, $1,2,3,\ldots,17$. Those that are taken away on this round are numbered $2,4,6,\ldots,16$ and those that survive are numbered $1,3,5,\ldots,17$. It is now the top-ranked advisor (number 1) who is first on the next round of questioning. To obtain the new set of numbers for this round, we subtract one and divide by 2. The new numbers for the remaining advisors are $0,1,2,\ldots,8$, and those that are taken away are numbered 0 (the top advisor), 2, 4, 6, and 8. We are left with four advisors with numbers 1, 3, 5, 7, and the questioning starts with 3. From here, we see that 3, 7, and 5, will be taken away, in that order. Therefore Bob has number 1 in this final numbering.

We now work backwards to find Bob's original number. To do this, we must undo the renumbering operations that led to his final number 1. His number previous to 1 must have been 3 because $\frac{3-1}{2}=1$, and before that, it must have been 5, because $\frac{5+1}{2}=3$, and before that, 9, because $\frac{9+1}{2}=5$. Thus, Bob was originally ninth in the pecking order.

Put another way, in the first round, all the even numbers (that is, those congruent to 0 mod 2) are eliminated, including the last number, 68. On the next round, 1 is passed over and numbers congruent to 3 mod 4 are eliminated, including the last number, 67. On the 3rd round, 1 is again passed over, and numbers congruent to 5 mod 8 are removed, including 61, so that 65 is still there and is passed over, with the 4th round eliminating numbers congruent to 1 mod 16, namely, 1, 17, 33, 49, 65 and leaving 9, 25, 41, 57. So the numbers 25, 57, congruent to 25 mod 32 are next removed, and finally 41, leaving 9.

A shorter solution to the problem goes as follows. If there had only been four advisors in the circle, numbered $1,2,3,4$, those with numbers $2,4,3$ would be eliminated in that order. If there had been eight advisors, those with numbers $2,4,6,3,7,5$ would be eliminated in that order. In both cases, the top advisor (number 1) remains. The same thing happens with 16, 32, or 64 advisors in the circle. Because we started with 68 advisors, there were 64 left after those numbered 2, 4, 6, 8 had been taken away. We can think of the questioning continuing from this point with 64 advisors. Since advisor 9 is in the first position of what follows, it will be that advisor who remains after all others have been eliminated (just as we have previously seen).

This problem is a special case of the Josephus problem, and there is an enormous literature related to generalizations of it. For more information see W. W. Ball and H. S. M. Coxeter, *Mathematical Recreations and Essays* , 12th edition, University of Toronto Press, 1974, pp. 32–36.

S229. From the hint, add the tens and units in the multiplication.

$$10\mathrm{I} + \mathrm{I}(10\mathrm{S} + \mathrm{I}) \le 98$$

$$10\mathrm{I}(\mathrm{S} + 1) \le 98 - \mathrm{I}^2$$

If $\mathrm{I} \ge 3$, then we would have $\mathrm{S} + 1 \le (98 - 3^2)/30 < 3$ and $\mathrm{S} = 1 = \mathrm{M}$, a contradiction. This gives $\mathrm{I} = 2$ and $\mathrm{S} + 1 \le 94/20 = 4.7$, so that $\mathrm{S} = 3$, $12 \times 32 = 384$, $\mathrm{O} = 8$, $\mathrm{L} = 4$.

$$(40 + \mathrm{A})(10\mathrm{D} + 8) = (10\mathrm{R} + \mathrm{E})(10\mathrm{F} + \mathrm{A})$$

$$8\mathrm{A} \equiv \mathrm{EA} \pmod{10}$$

E is not 8 or 3, so $8 \equiv \mathrm{E} \pmod 2$ and $\mathrm{A} = 0$ or 5. If $\mathrm{E} = 0$, $\mathrm{A} = 5$, the right side is divisible by 50, and the left side is not divisible by 5^2, a contradiction. So $\mathrm{E} = 6$. If $\mathrm{A} = 5$, then $45(5\mathrm{D} + 4) = (5\mathrm{R} + 3)(10\mathrm{F} + 5)$ and the only way to make the right side divisible by 9 is $\mathrm{F} = 7$ (and $\mathrm{R} = 0$ or 9), but then the right side is divisible by 5^2 and the left side is divisible only by 5. So $\mathrm{A} = 0$, $4(5\mathrm{D} + 4) = \mathrm{F}(5\mathrm{R} + 3)$, the left side is divisible by 4 and not 8, so that $\mathrm{R} = 5$, $\mathrm{F} = 7$, and $\mathrm{D} = 9$. With the key deciphered, the encrypted message 98 56 12 70 384 40 32 98 translates to DO RE MI FA SOL LA SI DO.

S230. It may happen that none of the 20 speakers have attended each other's talks. To see this how this could happen, suppose the speakers stand in a circle and that each speaker has heard precisely those talks given by the first nine persons to his/her immediate left. Then it is easy to see that no two mathematicians heard each other speak.

> **Q230.** What if each speaker had gone to exactly ten invited talks? Can we be sure that at least two invited speakers had heard each other's talk?

S231. Let k denote the number of players in the tournament. There were $k(k-1)/2$ games between the players, so they share $k(k-1)/2$ points. The top four players have accumlated $12\frac{1}{2}$ points. For $k(k-1)/2 \ge 12\frac{1}{2}$, k must be at least 6. The other $k - 4$ players score at most $1\frac{1}{2}$ points each, so

$$\frac{k(k-1)}{2} \le 12\frac{1}{2} + (k-4)\frac{3}{2}$$

$$k^2 - k \le 25 + 3k - 12$$

$$k^2 - 4k \le 13$$

$$k^2 - 4k + 4 \le 17$$

$$(k-2)^2 \le 17$$

$$k - 2 \le \sqrt{17}$$

so k cannot be more than 6. We conclude that $k = 6$.

Because $k = 6$, there were 15 points scored in the tournament, so the other two players shared $2\frac{1}{2}$ points. Therefore, the player in fifth place must have scored $1\frac{1}{2}$ points, and the sixth place player scored 1 point.

Here is one possible tournament outcome with these scores.

	1	2	3	4	5	6	Score
1	—	$\frac{1}{2}$	1	1	1	1	$4\frac{1}{2}$
2	$\frac{1}{2}$	—	$\frac{1}{2}$	$\frac{1}{2}$	1	1	$3\frac{1}{2}$
3	0	$\frac{1}{2}$	—	$\frac{1}{2}$	1	1	3
4	0	$\frac{1}{2}$	$\frac{1}{2}$	—	$\frac{1}{2}$	0	$1\frac{1}{2}$
5	0	0	0	$\frac{1}{2}$	—	1	$1\frac{1}{2}$
6	0	0	0	1	0	—	1

S232. (a) After the conclusions established in the hint, it suffices to show that we can increase the number of coins by $k+1$ in one of the specified bowls and simultaneously increase the number of coins in the other seven bowls by only k coins. If we could do this, we could increase the number of coins in a bowl with the fewest coins by one more than the others would increase. By repeating this procedure sufficiently often we could eventually get the same number of coins in each bowl.

To determine how to carry out such a procedure, suppose that all the bowls are empty, and that we want $k+1$ coins in the bowl number 1, and k coins in each of the other seven bowls (assume the bowls are numbered in order from 1 to 8). The total number of coins required for such a procedure is $(k+1)+7k = 8k+1$. Because we distribute five coins each time, it has to be the case that $8k+1 = 5t$, where we make t distributions of five coins. The simplest solution is $k = 3$, $t = 5$, so we go 1 2 3 4 5; 6 7 8 1 2; 3 4 5 6 7; 8 1 2 3 4; 5 6 7 8 1 and the first bowl will have $3+1$ coins, with 3 in each of the others.

Q232a. Suppose we have b bowls, and we add a coin to c consecutively placed bowls. Can you get the same number of coins in each bowl?

(b) Number the bowls from 1 to 7.

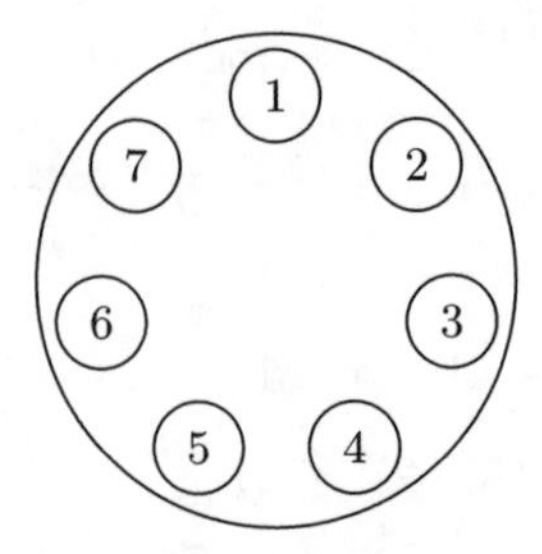

Consider bowls numbered 3, 6, and 7. We cannot put coins in two of these in the same step, so they require $3+6+7 = 16$ steps, by which time $16 \times 2 = 32$ coins have been distributed. But we wish to distribute only $1+2+\cdots+7=28$ coins, so 4 must be removed, taking 2 more steps, 18 in all.

This minimum can be achieved: Put coins in 1 and 3, 3 and 5 twice, 5 and 7 three times, 7 and 2 four times, 4 and 6 six times, and take away coins from 2 and 4 twice.

Q232b. What other distributions can be achieved? What happens when you have a different number of bowls?

S233. (a) The sum of all three columns is $1+2+\cdots+9=45$, so the tens and hundreds columns add to 30, and these must therefore be 8 and 22, respectively. The partitions of 8 into three distinct parts are 1 2 5 and 1 3 4, and those of 22 are 6 7 9 and 5 8 9, so we have

$$\begin{aligned}(\text{hundreds, tens, units}) &= \big(\{6,7,9\},\{1,2,5\},\{3,4,8\}\big)\\ \text{or} &\ \big(\{6,7,9\},\{1,3,4\},\{2,5,8\}\big)\\ \text{or} &\ \big(\{5,8,9\},\{1,3,4\},\{2,6,7\}\big)\end{aligned}$$

with $3\times 6\times 6\times 6=648$ ways of filling in the stars.

(b) From the hint we have established that $c\geq 7$, and that d, e are 7, 9 in some order.

$$\begin{array}{rrrrrr} & & & 3 & * & e\\ & & \times & d & * & 2\\ \hline 2 & c & * & 3 & & \\ & 1 & * & * & * & \\ & & & * & * & *\\ \hline 3 & * & * & * & * & * \end{array}$$

If $d=9$, then the hundreds product is $2763=9\times 307$, $2853=9\times 317$, or $2943=9\times 327$, and the units product is respectively, $2\times 307=614$, $2\times 317=634$, or $2\times 327=654$. The first two contain illegal ones and threes, whereas in the last case the tens product must be $4\times 327=1308$, $5\times 327=1635$, or $6\times 327=1962$. Again, the first two contain illegal threes, and the last gives $295300+19620+654=315574$, containing an illegal 1.

So $d=7$ and $e=9$. The hundreds product is smaller than $7\times 400=2800$, so $c=7$, and the possible multiples of 7 are $2793=7\times 399$ or $2723=7\times 389$. The latter would only give a carry of 1 from the thousands column, so the multiplicand is 399 and we need $5\times 399=1995$ to give the necessary carries of 2 from both the hundreds and the thousands columns. So there is just one solution.

$$\begin{array}{rrrrrr} & & & 3 & 9 & 9\\ & & \times & 7 & 5 & 2\\ \hline 2 & 7 & 9 & 3 & & \\ & 1 & 9 & 9 & 5 & \\ & & & 7 & 9 & 8\\ \hline 3 & 0 & 0 & 0 & 4 & 8 \end{array}$$

S234. Let s be the side of the square, and a, b, c, d be the distances of P from A, B, C, D, respectively.

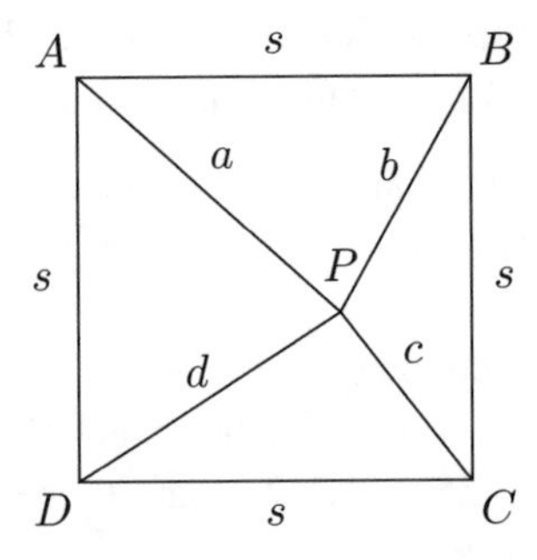

Peter can certainly find the perimeter of triangle PCD, because if we add together $s+a+d = 4.7$ and $s+b+c = 4.3$ and subtract $s+a+b = 5.7$, we see that $s+b+c = 3.3$.

But he's wrong in saying that the point P is inside the square! As $s + c + d = 3.3$, s must be less than 1.65 by the **Triangle inequality**. But since $s + a + b = 5.7$, we must have $a + b > 4.05$. But if P were inside the square, then $a + b$ would be less than $AC + BC = s\sqrt{2} + s < (1 + \sqrt{2})(1.65) < 2.42 \times 1.65 < 4 < 4.05$, a contradiction.

One cannot help but wonder: Is there such a point P with these particular perimeters, and if so, where is it? If such a point exists, how many such points might there be?

Q234. Where, if anywhere, did Peter make his point?

S235. Suppose the radius of the black disk is more than 2 cm. One of the green disks (of radius 1 cm) must cover the center of the black disk, and therefore cannot cover any of the large disk's perimeter. Because the sides of an inscribed regular hexagon are more than 2 cm in length, no single small disk can cover one of these sides by itself; that is to say, a small disk will cover less than one-sixth of the perimeter of the large disk. Thus, seven small disks of radius 1 cannot entirely cover the large disk.

It is possible to cover a black disk of radius 2. Place six of the small disks so their centers are at the midpoints of the sides of the inscribed regular hexagon ABCDEF, and place the seventh disk so that its center coincides with the center of the large disk.

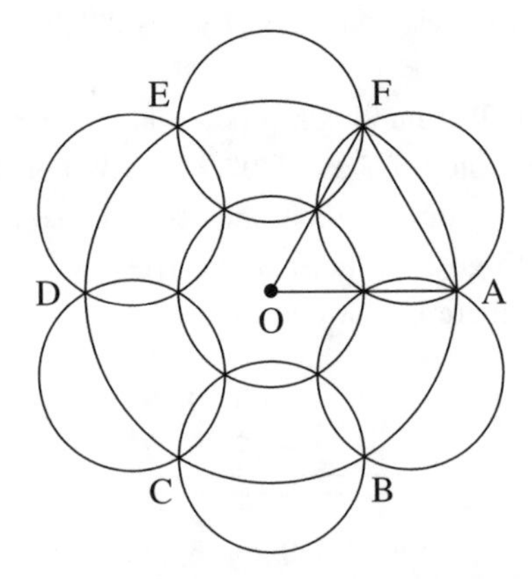

Because the radius of the large disk and the side length of the regular hexagon is 2 cm, the entire black disk will be covered by the seven smaller green disks.

S236. We work in rods throughout. Divide the square into an 8×1 rectangle and two 4×7 rectangles. Because $65 = 8^2 + 1^2 = 4^2 + 7^2$, the diagonals of each rectangle are length $\sqrt{65}$ and we may place the sentries at their centers so that no point of the square is more than $\sqrt{65}/2$ from a sentry.

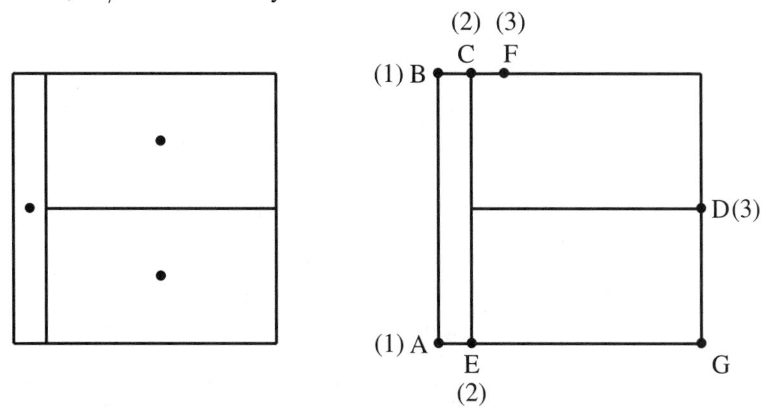

Suppose it were possible to arrange for a shorter distance. Of the four corners, one sentry must guard two, say A and B are guarded by (1). Then C must be guarded by a different sentry, say (2), because AC $= \sqrt{65}$, and D must be guarded by a third sentry (3) because BD $>$ CD $= \sqrt{65}$. Then E must be guarded by (2), because BE $=$ DE $= \sqrt{65}$, and F by (3) because AF $>$ EF $= \sqrt{65}$. But now G is unguarded, being more than $\sqrt{65}$ from B, C, F. So it is not possible to reduce the distance below $\sqrt{65}/2$.

> **Q236.** Because times are good, the City has employed five sentries to guard the square. However, in an effort to cut taxes, the mayor decides to cut the security force on the square from five to four, claiming that four sentries are just as effective as five, in the sense that no matter how the five sentries are placed, the maximum distance between a trouble spot and a sentry is no less than the maximum distance that could be attained by using four sentries. Is the mayor right about this?

S237. (a) If one chooses numbers A and B, they are replaced by $A + B$. Then later, if one chooses $A + B$ and C, they are replaced by $(A + B) + C$. This is the same result one would get if one first chose A and C, and then $A + C$ and B; that is, $(A + C) + B$. Thus, the order in which one chooses the numbers does not affect the final value, which after fourteen operations is equal to $1 + 2 + 3 + \cdots + 15 = 120$.

(b) Observe that $A + B + AB = (A + 1)(B + 1) - 1$. The pair of numbers A and B is replaced by $X = (A+1)(B+1) - 1$. If you choose this number and another number C, they are replaced by $(X+1)(C+1) - 1 = (A+1)(B+1)(C+1) - 1$. This is the same as if you had first chosen numbers A and C and then $(A + 1)(C + 1) - 1$ and B. The final number, therefore, does not depend on the order in which the numbers are chosen. The operation is both associative and commutative, and after fourteen such operations, the final result is $(1+1)(2+1)(3+1)\cdots(15+1) - 1 = 16! - 1 = 3041 \times 6880233439$, which is certainly larger than a billion.

S238. Obviously, we can tile the 10×10 grid with 25 square tetrominoes. We shall show that it is not possible to tile it with tetrominoes of type T, Z, L, or I.

T-tetromino: Color the grid in the usual checkerboard manner. Each tetromino covers an odd number of black squares, so that 25 of them will cover an odd number, and therefore will not cover 50.

Z-tetromino: Color the grid diagonally with four colors:

1	2	3	4	1	2	3	4	1	2
4	1	2	3	4	1	2	3	4	1
3	4	1	2	3	4	1	2	3	4
2	3	4	1	2	3	4	1	2	3
1	2	3	4	1	2	3	4	1	2
4	1	2	3	4	1	2	3	4	1
3	4	1	2	3	4	1	2	3	4
2	3	4	1	2	3	4	1	2	3
1	2	3	4	1	2	3	4	1	2
4	1	2	3	4	1	2	3	4	1

The Z-tetromino uses either all four colors, or two squares each of two adjacent colors. But the numbers of squares colored 1, 2, 3, 4 are 26, 25, 24, 25, which gives a contradiction, as we shall see. For example, suppose there are a tetrominoes which use all four colors and b that use 1 and 2, c which use 2 and 3, d which use 3 and 4, and e which use 4 and 1. Then

$$26 = a + 2b + 2e \quad \text{and} \quad 25 = a + 2b + 2c,$$

which implies that a is both even and odd.

L-tetromino: Color the grid so that every other column is black:

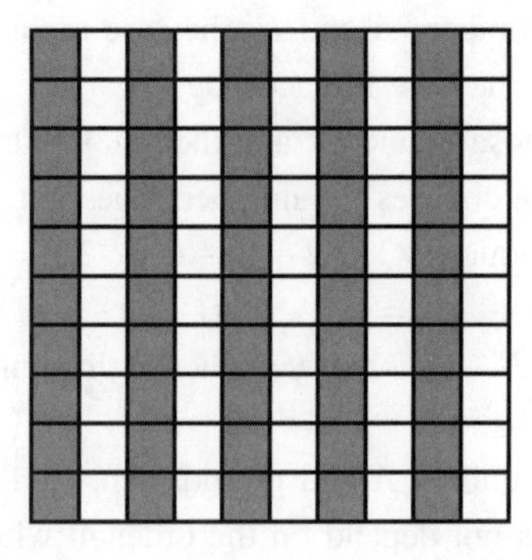

The L-tetromino covers either one or three black squares in the grid. Twenty-five such tetrominoes will therefore cover an odd number of black squares, whereas there are 50 black squares in the grid.

I-tetromino: Color the grid as follows:

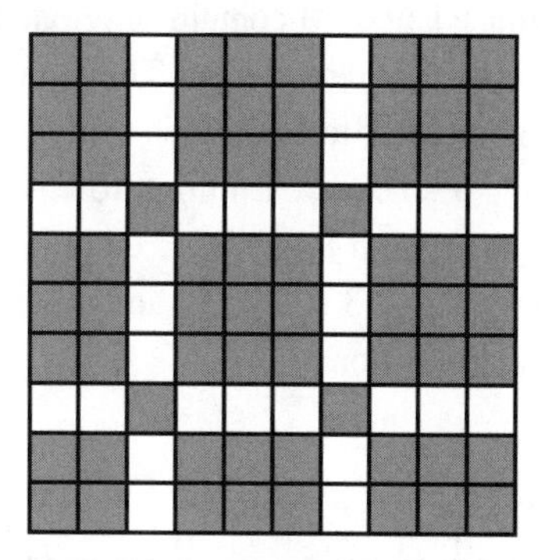

Each I-tetromino covers an odd number of black squares, so 25 such tetrominoes won't cover 50 black squares.

S239. (a) Suppose that A has k digits and let $a_1a_2a_3\dots a_k$ be the sequence of digits in A. Then B is the number $a_1a_2\dots a_ka_1a_2\dots a_k$ and we can write

$$B = a_1a_2\dots a_k\underbrace{00\dots0}_{k}+a_1a_2\dots a_k$$

so that

$$B = (a_1a_2\dots a_k)(1\underbrace{00\dots0}_{k}+1) = (a_1a_2\dots a_k)(1\underbrace{00\dots0}_{k-1}1)$$

and, because k is an odd number, $10^k + 1$, is divisible by $10 + 1 = 11$, and so also is B. See **Geometric series** in the Treasury.

Another approach is to use the rule for divisibility by 11 (see **Divisibility tests** in the Treasury): a decimal number $b_1b_2b_3\dots b_n$ is divisible by 11 if and only if the alternating sum $b_1 - b_2 + b_3 - b_4 + \cdots$ is divisible by 11. Because k is odd, the alternating sum

$$a_1 - a_2 + a_3 - \cdots + a_k - a_1 + a_2 - a_3 + \cdots - a_k = 0,$$

so, by the rule, B is divisible by 11.

(b) Suppose, again, that A has k digits and let $a_1a_2\dots a_{k-1}a_k$be its decimal representation. Then the digits in B are $a_1a_2\dots a_{k-1}a_ka_ka_{k-1}\dots a_2a_1$, and we can write

$$B = a_1(10^{2k-1} + 1) + a_2(10^{2k-3} + 1) \times 10 + a_3(10^{2k-5} + 1) \times 10^2 + \cdots$$
$$+ a_{k-1}(10^3 + 1) \times 10^{k-2} + a_k(10 + 1) \times 10^{k-1}.$$

Again, within each parentheses we find a number of the form $10^m + 1$, where m is odd, each of which is divisible by 11. Thus B is divisible by 11.

Alternatively, according as k is odd or even,

$$a_1 - a_2 + a_3 - \cdots + a_k - a_k + a_{k-1} - \cdots - a_1 = 0,$$

or

$$a_1 - a_2 + a_3 - \cdots - a_k + a_k - a_{k-1} + \cdots - a_1 = 0,$$

and therefore B is divisible by 11.

S240. (a) From the hint, row 5 and column 5 are all black, and squares 4 5 6 7 in column 2 are black. As columns 1 and 10 contain only one black square, squares 2 to 9 of row 6 are black. Column 6 is now determined. This completes row 1, so that column 9 is squares 2 to 10. Row 7 is 2 and 5 to 9; column 7 is then 2 3 5 6 7 and row 2 is 4 5 6 7 9. Row 3 intersects column 9 so square 2 in that row is white and row 3 and column 2 are fixed. Row 8 implies that column 8 is 3 to 7; column 3 implies that row 4 is 2 3 5 8 9; row 8 implies that column 4 is 2 3 5 6 9 10, and column 7 implies that rows 9 and 10 are both 2 to 6 and 9, and we're done.

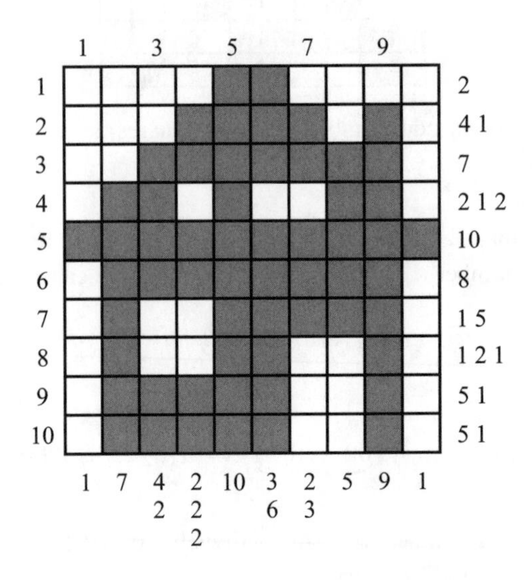

(b) Columns 3 and 5 are both 1 2 and 4 to 11; rows 1 and 2 are both 3 4 5; column 4 is 1 to 8 and rows 5 to 7 imply that columns 1 and 7 are 4 5 6. Row 4 is 1 to 7. Rows 1 and 6 are now complete and rows 7 to 10 imply that columns 2 and 6 are both 4 and 11, making columns 1 to 8 complete. We are left with the following grid, occupying the last five rows and 9 columns.

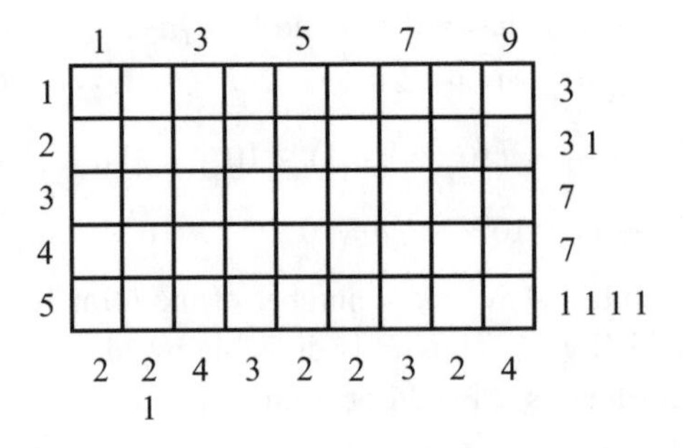

In this, the last column must intersect rows 3 and 4, which are therefore both 3 to 9. Column 1 is 1 2; column 2 is 1 2 5; row 1 now implies that column 3 is 1 to 4, and that column 9 is 2 to 5. Rows 1 to 4 are now complete and row 5 is fixed by the column numbers as 2 4 7 9.

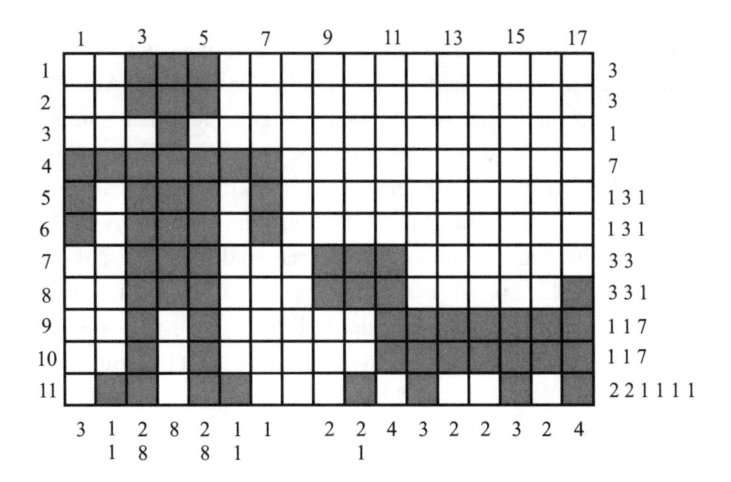

S241. From the hint, row 18 is 1 to 4, 6 to 10, 12 to 14, 16, 17, 19 to 22. Columns 1 to 4 are now determined and then the first blocks in rows 8 to 16 are 4 to 12, 3 to 13, 2 to 10, 1 to 10 (twice), 1 to 9 (thrice) and 1 to 8. Columns 5 to 10 are 7 to 16, 6 to 16 and 18, 5 to 16 and 18, 4 to 16 and 18, 5 to 15 and 18, and 6 to 12, 17, 18. By watching the columns we see that row 17 is 1 to 4, 10 to 12, 14, 17, 19 to 21. Columns 14, 17, 19, 20 end in 16 to 18, 10 to 18, 15 to 18, and 16 to 18. Columns 21, 22, and row 15 are complete. Row 7 and column 11 intersect in a black square. Column 16 intersects row 9, which is then complete. Row 6 intersects columns 16 to 18, the last of which is then complete. Column 12 contains 12 to 14, so row 13 is complete; row 12 is completed with 15 to 17, and row 11 with 12 to 17. Column 16 is completed with 2 to 12. Row 8 contains 15, 19 and 20; column 20 is then complete and column 19 is completed with 6 and 7. Row 5 doesn't intersect with column 17, so column 17 is completed with 6 to 8. Column 15 is 2 to 4, 6 to 8, 10 to 12, and 16. Row 6 is completed with 13, row 14 with 11, and the picture is completed with 12 and 13 in row 4 and 13 in row 5.

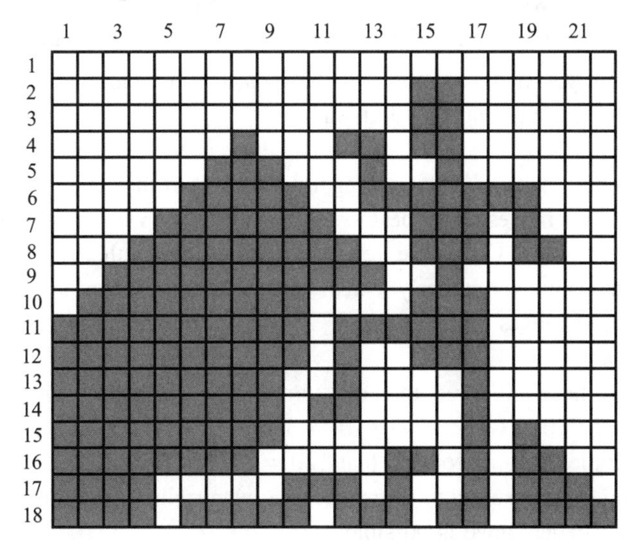

S242. Each number you add to the list at the end of each move is less than 13. Therefore, at all times, at least one of the numbers on the paper is less than 13. At the end, when only two numbers remain on the paper, one of which is 102, the other number must be less than 13. Should you now take these two numbers and replace them with the remainder when their sum is divided by 13, the final number on the paper would be the same as if the remainder obtained when the sum of all 500 numbers is divided by 13. Because $500 = 38 \times 13 + 6$, there are 38 copies of the remainders $1, 2, \ldots, 12$ and 0, and six extra remainders 1, 2, 3, 4, 5, 6. All except the last six cancel each other out, 1 with 12, 2 with 11, and so on, and the final remainder will be $1 + 2 + 3 + 4 + 5 + 6 - 13 = 8$. The remainder of 102 divided by 13 is 11 and therefore the other number must be 10.

S243. In the hint we showed how to place the animals in the cages in a compatible way, by disregarding the restriction of only two animals per cage. We must now describe how to move the animals between cages so that no cage will contain more than two animals.

Suppose, first, that one of the cages, for example $\mathcal{A}$, has more than three animals in it. Then there must be some cage, $\mathcal{B}$, with at most one animal, and this animal cannot be incompatible with all the animals in $\mathcal{A}$. Consequently, we can move that animal from $\mathcal{A}$ to $\mathcal{B}$.

Thus, we may assume that no cage has more than three animals in it. What can we do if we find two cages $\mathcal{A}$ and $\mathcal{B}$ with three animals apiece? Then, again, there must be a cage with at most one animal, and one can certainly move one of these animals (from $\mathcal{A}$ or $\mathcal{B}$).

It remains to consider what happens if there is exactly one cage with three animals, two cages with two animals, and one cage with one animal. If the animal in the last cage (by itself in that cage) is compatible with some animal in the first cage, then the procedure is clear. If not, then the animal in the last cage is compatible with the animals in cages 2 and 3. Furthermore, each animal in cage 1 is incompatible with at most two of the four animals in cages 2 and 3. Therefore we can move an arbitrary animal from cage 1, let us say animal A, to either cage 2 or 3, whichever cage A finds an animal in that it is compatible with, and then move the other animal in that cage (2 or 3) to cage 4. For a different treatment see **Dirac's theorem** in **Hamilton circuits** in the Treasury.

> **Q243.** Can you make a connexion between this problem and the problem of finding a Hamilton circuit on a graph?

S244. We continue with the solution begun in the hint. If the "chain reaction" procedure described in the hint moves each guest to their proper place we can continue the argument as in the next paragraph. Otherwise, repeat the procedure another time, starting with one of the guests who is not yet properly seated. Once again, the sum of the "distances" moved in this sequence of moves is a multiple of 16. Repeat this procedure until everyone is in the proper place.

We already have observed that the sum of the "distances" for all the guests is necessarily equal to $0 + 1 + 2 + \cdots + 15 = 120$. On the other hand, we have also seen that it is a multiple of 16. It follows that $120 = 16n$ for some number n. However, this equation has no integer solution. Consequently, our original assumption is not possible, so such a simultaneous move is possible.

The proposition need not hold if the number of people is 15 instead of 16. An example of this is shown below where the numbers from 0 to 14 on the inside of the circle denote the placecards and the numbers on the outside denote the placement of the guests around the table. The numbers within the parentheses denote each person's distance from their proper place (measured in the number of places they must move to their right). Observe that different guests have different distances, so that regardless of how they rotate as a group, only one person will be properly placed.

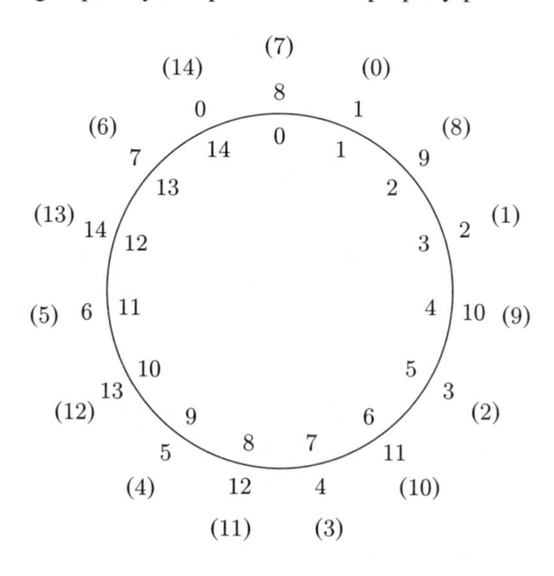

> **Q244.** Suppose that n guests, numbered $0, 1, 2, \ldots, n-1$ sit down in a random manner and that *none* of the guests is properly placed in front of their placecard. Is it possible for each of the guests to move the same number of places to their right so that at least two persons are in their proper place?

S245. (a) Of course one can mark off the cakes Bruce tastes by brute force checking, but we would like to find a more general approach. Note that the cakes Bruce tastes are numbered 0, 1, 3, 6, 10, 15, and so forth, the triangular numbers $n(n+1)/2$. All we need to do is take this sequence modulo 25: 0, 1, 3, 6, 10, 15, 21, $28 - 25 = 3$ (a repetition), $3+8 = 11$, $11+9 = 20$, $20+10-25 = 5$, $5+11 = 16$, $16+12 = 3$ (another repetition), $3 + 13 = 16$, and now Bruce is visiting the cakes in reverse order (forward 14 from 16 is the same as backwards 11 from 16, which brings Bruce back to 5; then, forwards 15 from 5 is backwards 10 from 5, which brings Bruce to 20, and so forth). Therefore, Bruce tastes less than half of the different cakes (eleven of them: 0, 1, 3, 5, 6, 10, 11, 15, 16, 20, 21) and he doesn't taste cakes numbered 2, 4, 7, 8, 9, 12, 13, 14, 17, 18, 19, 22, 23, or 24.

(b) Bruce tastes the cakes numbered $0, 1, 4, 9, 16, \ldots$, which we recognize as the squares, and we will get the appropriate cake number by taking this sequence modulo 25. So, the cakes visited are $0, 1, 4, 9, 16, 25 = 0$ (a repetition), $6^2 = 36 \equiv 11$, $7^2 = 49 \equiv 24$, $8^2 = 64 \equiv 14$, $9^2 = 81 \equiv 6$, $10^2 \equiv 0$ (a repetition), $11^2 = 121 \equiv 21$, $12^2 = 144 \equiv 19$. and then in reverse $13^2 = (-12)^2 \equiv 19$, $14^2 = (-11)^2 \equiv 21$, and so forth. In this case as well, Bruce will taste only 11 different cakes: 0, 1, 4, 6, 9, 11, 14, 16, 19, 21, 24. (For more discussion of this problem see **Triangular numbers** in the Treasury.)

S246. (a) Let T_k denote the number of different ways of tiling a $2 \times k$ rectangle with 1×2 tiles. Then, it is easy to check that $T_1 = 1$, $T_2 = 2$, and $T_3 = 3$.

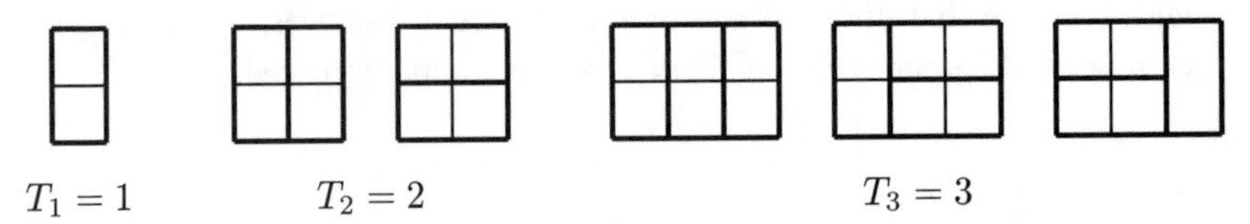

$T_1 = 1$ $T_2 = 2$ $T_3 = 3$

Now consider a $2 \times k$ rectangle, where $k \geq 3$, and focus attention on the two squares along the left edge. There are two ways these squares can be covered: (i) by a vertical tile, or (ii) by two horizonal tiles.

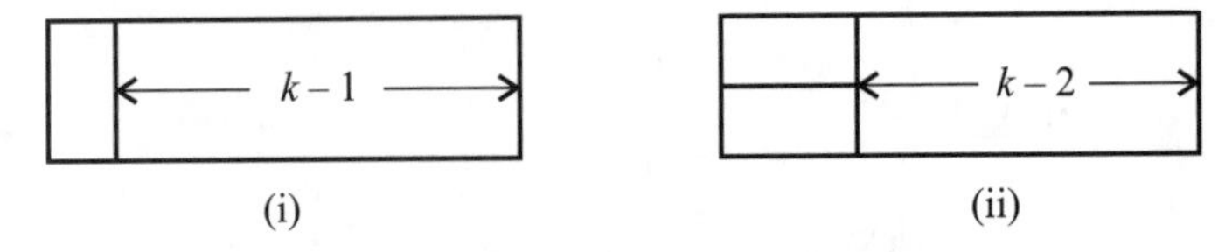

(i) (ii)

After the vertical tile is placed in the first case, there remains a $2 \times (k-1)$ rectangle left to tile, and, with our notation, that can be done in T_{k-1} ways. Similarly, after the two horizontal tiles are placed in the second case, there remains a $2 \times (k-2)$ rectangle left to tile, and that can be done in T_{k-2} ways. It follows that $T_k = T_{k-1} + T_{k-2}$.

Because $T_1 = 1$ and $T_2 = 2$, we can now compute, $T_3 = T_2 + T_1 = 3$, $T_4 = T_3 + T_2 = 5$, $T_5 = T_4 + T_3 = 8$, and so forth, until we find $T_{14} = T_{13} + T_{12} = 610$. The number of different ways to tile the 2×14 rectangle with 1×2 tiles is therefore 610.

This is the sequence of **Fibonacci numbers** (see the Treasury), but they are usually numbered $u_0 = 0$, $u_1 = 1$, $u_2 = 1$, $u_3 = 2$, $u_4 = 3$, $u_5 = 5, \ldots$ so that our T_{14} is u_{15}.

Another approach is to count the number of horizontal pairs that can occur in a tiling. The following table makes this explicit.

All dominoes vertical	$\binom{14}{0} =$	1
One horizonal pair	$\binom{13}{1} =$	14
Two horizonal pairs	$\binom{12}{2} =$	66
Three horizonal pairs	$\binom{11}{3} =$	165
Four horizonal pairs	$\binom{10}{4} =$	210
Five horizonal pairs	$\binom{9}{5} =$	126
Six horizonal pairs	$\binom{8}{6} =$	28
Seven horizonal pairs	$\binom{7}{7} =$	1

So the total is

$$\binom{14}{0} + \binom{13}{1} + \binom{12}{2} + \binom{11}{3} + \binom{10}{4} + \binom{9}{5} + \binom{8}{6} + \binom{7}{7} = 610.$$

This approach shows an interesting connexion between the Fibonacci numbers and the binomial coefficients. (See **Pascal's triangle** in the Treasury.)

(b) The configuration of twelve squares shown below can be covered in exactly seven different ways with six dominoes.

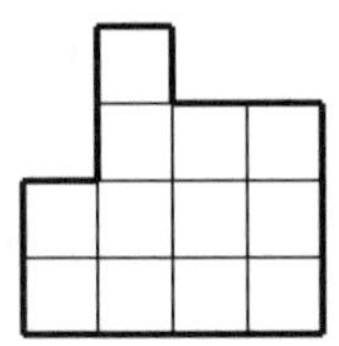

Q246a. Is there a configuration of twelve 1×1 squares that can be tiled by dominoes in exactly k ways, for each of the cases $k = 0, 1, 2, \ldots, 13$?

Q246b. Are there more ways of tiling a 2×8 rectangle with dominoes than there are a 4×4 square?

S247. (a) Divide the numbers $1, 11, 111, \ldots$ by 2010 and consider the sequence of their remainders (see the hint). Because the remainders are all less than 2010, there will be remainders in this list that are equal. Suppose the numbers

$$A = \underbrace{11\ldots1}_{k} \quad \text{and} \quad B = \underbrace{11\ldots1}_{m}$$

give the same remainder when they are divided by 2010, where $A < B$ (equivalently, $k < m$). Then

$$A = 2010k_1 + r, \quad \text{and} \quad B = 2010m_1 + r,$$

and therefore, $B - A = (m_1 - k_1)2010$, which is divisible by 2010.

(b) Using the same reasoning as in part a, we know there is a number of the form $C = \underbrace{11\ldots1}_{n}\underbrace{00\ldots0}_{k}$, which is divisible by 2011. But $C = \underbrace{11\ldots1}_{n} \times 10^k$ and 2011 is a prime number. Because 2011 does not divide 10^k, it must divide $\underbrace{11\ldots1}_{n}$.

For another solution, see **Fermat's little theorem** in the Treasury.

S248. One approach is brute-force backtracking. It's clear that zero is at the top, since if it's anywhere else, then two of the numbers would be equal. The first number of the row determines the entire row, so let's begin by supposing that the second row begins with 1. It continues with 9. Then try for the third row, in order: 2 9 no! 3 8 1 no! 4 7 2 Yes!, so on to the fourth row: 3 1 no! 5 9 no! 6 8 9 no! 8 6 1 no! so try again for the third row: 5 6 3 and then the fourth row: 2 3 4 1 no! 7 8 8 no! 8 7 9 no!, so back to the third row: 6 5 4 and fourth row 2 4 no! 3 3 no! 7 9 no! 8 8 no! Back to the third row again: 7 4 5 and fourth row 2 5 no! 3 4 no! 6 1 no! 8 9 no! Finally, the third row: 8 3 6 and fourth row 2 6 no! 4 4 no! 5 3 no! 7 1 no! and we've failed.

This backtracking is tedious, time-consuming and fraught with error. Much better to program your computer to do it, especially if you want to try out larger numbers.

Before we go on to the second row, 2 8 let's see if we can save ourselves some effort. You've probably noticed that if you do find a solution then another is found by left-right reflexion, so at least we don't need to try 9 1 for the second row.

Here's another shortcut. If we take a complete set of residues modulo 10 (see **Residues** in the Treasury), namely

$$0 \ 1 \ 2 \ 3 \ 4 \ 5 \ 6 \ 7 \ 8 \ 9$$

and multiply them by any number relatively prime to 10 (that is, a number with no common factor with 10) (for example, 3, 7, 9) we get each of the following respectively:

$$\begin{array}{cccccccccc} 0 & 3 & 6 & 9 & 2 & 5 & 8 & 1 & 4 & 7 \\ 0 & 7 & 4 & 1 & 8 & 5 & 2 & 9 & 6 & 3 \\ 0 & 9 & 8 & 7 & 6 & 5 & 4 & 3 & 2 & 1 \end{array}$$

the same numbers again, though in a different order. So any solution we find will give four solutions, by multiplying the numbers in the triangle by 1, 3, 7, 9 modulo 10. So not only do we not need to try 9 1 as a second row, but neither do we need to try 3 7 nor 7 3. That leaves 2 8 and its multiples by 3, 7, 9: 6 4 or 4 6 or 8 2, and this covers all cases.

There are only three possiblities for row three:

```
  0          0          0
 2 8        2 8        2 8
5 7 1      7 5 3      9 3 5
```

and it doesn't take long to see that there is no fourth row that fits the first and last figures. Either 1 6 9 4 or 6 1 4 9 will fit the middle one. So if we don't count reflexions as different, there are just four solutions (the second in each pair is obtained by multiplying the first by 7):

```
   0          0          0          0
  2 8        4 6        2 8        4 6
 7 5 3      9 5 1      7 5 3      9 5 1
1 6 9 4    7 2 3 8    6 1 4 9    2 7 8 3
```

Q248. Is there a modular triangle for $t = 15$?

S249. In the hint we showed that there is no such n of shape $3k+1$. Nor then is there one of shape $3k+2$, since subtracting each digit, including leading zeros, from 9, gives a **One-to-one correspondence** (see Treasury) between the two possibilities:

$$27 - (3k+2) = 3(8-k) + 1$$

or

$$27 - (3k+1) = 3(8-k) + 2.$$

So it remains to consider $n = 3k$ for $k = 0, 1, 2, 3, 4$, since any such solutions will find their counterparts for $9-k$. $k = 0$ does not give a solution, and if $k = 2$, 3 or 4, there are six numbers with digits $k-1$, k, $k+1$, and six with digits $k-2$, k, $k+2$. But for $k = 1$, $n = 3$, whose partitions are 3, 21, 111 we have three permutations of 003, six of 012 and one of 111, just ten numbers

$$3, \ 30, \ 300, \ 12, \ 21, \ 102, \ 120, \ 201, \ 210, \ 111$$

with digital sum 3. Correspondingly, subtracting each digit from 9, we find the only other solution, $n = 24$:

$$996,\ 969,\ 699,\ 987,\ 978,\ 897,\ 879,\ 798,\ 789,\ 888$$

S250. (a) Choose a player A who has won the most games. Suppose that A has k wins against players $C_1, C_2, \dots, C_k$, and that A lost to B. B could not have won all the games against $C_1, C_2, \dots, C_k$ because, with the win over A, B would have had more wins than A, contrary to our choice. Consequently, A satisfies the conditions necessary to win a gold medal.

(b) We continue with the solution started in the hint, which we summarize in the following diagram.

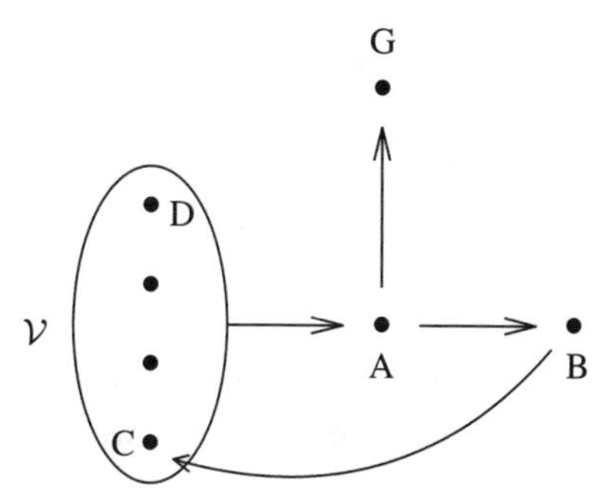

Here, A and B are gold medal winners, and A has defeated B; $\mathcal{V}$ is the set of all players V such that V $\to$ A (there must be at least one player in this set because B is a gold medal winner who lost to A), and G denotes an arbitrary player, different from A, who is not in the set $\mathcal{V}$.

If we think of the tournament as restricted to just the players in $\mathcal{V}$, then among these there will be a "group-gold-medal winner", a person D who for every player E in $\mathcal{V}$, D $\to$ E, or D $\to$ F $\to$ E, for some player F in $\mathcal{V}$. Because A did not lose to G, we must have A $\to$ G. Consequently, D $\to$ A $\to$ G. This means that D not only won a "group-gold-medal" in group $\mathcal{V}$, but even fulfilled the conditions for a gold medal in the entire tournament. Thus, if there are two gold medal winners, there must be at least three; that is, there can never be exactly two gold medal winners.

> **Q250.** Obviously, in a tournament between just two players, only one of them will get a gold medal. In a tournament with three players, however, it is possible for each of them to get a gold medal (for example, if A $\to$ B $\to$ C $\to$ A. Is it possible, in a tournament between four players, that each them could get a gold medal? What about a tournament with five players?

S251. In the hint, we've seen that K must be even. Now we can interpret the problem as the same as the problem of scheduling a round-robin tournament for K players. Think of the streets as the players in the tournament. There's a house numbered i at the intersection of each pair of streets (players) who play each other in the ith round. The procedure for scheduling a round-robin tournament is described in the Treasury (see **Round-robin tournaments**). Here are the house number assignments for 8 streets.

Players (Streets); Rounds (House Number)

Rounds (House Number)	1	2	3	4	5	6	7	8
1	7	6	5	8	3	2	1	4
2	6	5	4	3	2	1	8	7
3	5	4	8	2	1	7	6	3
4	4	3	2	1	7	8	5	6
5	3	8	1	7	6	5	4	2
6	2	1	7	6	8	4	3	5
7	8	7	6	5	4	3	2	1

$\Longrightarrow$

Street	1	2	3	4	5	6	7	8
1	–	6	5	4	3	2	1	7
2	6	–	4	3	2	1	7	5
3	5	4	–	2	1	7	6	3
4	4	3	2	–	7	6	5	1
5	3	2	1	7	–	5	4	6
6	2	1	7	6	5	–	3	4
7	1	7	6	5	4	3	–	2
8	7	5	3	1	6	4	2	–

S252. In the hint we saw that $a_1 + a_2 + \cdots + a_k \geq (k-1)k/2$. Because the same inequality holds for the b_i and c_i, it follows that

$$\begin{aligned}3(k-1)k/2 &\leq (a_1 + a_2 + \cdots + a_k) + (b_1 + b_2 + \cdots + b_k) + (c_1 + c_2 + \cdots + c_k)\\ &= (a_1 + b_1 + c_1) + (a_2 + b_2 + c_2) + \cdots + (a_k + b_k + c_k)\\ &= 10 + 10 + \cdots + 10 = 10k.\end{aligned}$$

Divide this inequality by k and multiply by 2 and we have $3(k-1) \leq 20$, so $k \leq 7$. So it is not possible to place eight coins on the grid, but we can place 7, as shown below.

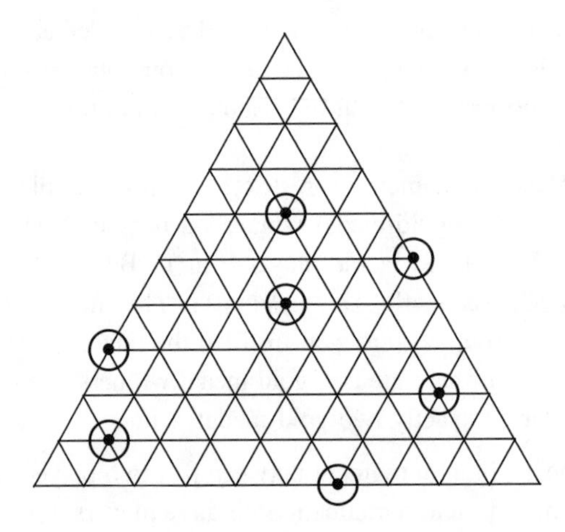

Q252. How many coins can you put on other sizes of triangular grid with no two on a line?

S253. We continue with the solution begun in the hint. The third question can be the following.

Question 3. "Is the number between 1 and 10?" By symmetry, we can consider only the answer "No." You can then draw a line under the numbers from 1 to 10, and the result is the following row.

$$\underline{\underline{5}}\ \underline{\underline{6}}\ \underline{\underline{7}}\ \underline{\underline{8}}\ \underline{9}\ \underline{10}\ 11\ 12\ \underline{13}\ \underline{14}\ \underline{15}\ \underline{16}$$

Because the chosen number cannot be a number with two lines under it, we can reduce the row to the following:

$$\underline{9}\ \ \underline{10}\ \ 11\ \ 12\ \ \underline{13}\ \ \underline{14}\ \ \underline{15}\ \ \underline{16}.$$

Question 4. "Is your chosen number either 11 or 12?" If the answer is "No" we have

$$\underline{9}\ \ \underline{10}\ \ \underline{11}\ \ \underline{12}\ \ \underline{13}\ \ \underline{14}\ \ \underline{15}\ \ \underline{16}$$

whereas, if the answer is "Yes" we have

$$\underline{\underline{9}}\ \ \underline{\underline{10}}\ \ 11\ \ 12\ \ \underline{\underline{13}}\ \ \underline{\underline{14}}\ \ \underline{\underline{15}}\ \ \underline{\underline{16}}$$

which reduces to

$$11\ 12.$$

In the last three questions you have to separately consider both answers as in Question 4. If the last answer is "No," you can eliminate half the remaining numbers with each of the next three questions. For example, you can first ask about numbers 9, 10, 11, 12; if you get a "Yes," you will have two lines under numbers 13, 14, 15, 16, so they can be eliminated. With a "No" answer, you can eliminate numbers 9, 10, 11, 12. After the last such question there will only be one number left.

If the answer to question 4 is "Yes," you can ask three times if one of the numbers, for example, 11, is the chosen number. If the answer to two of these questions is "Yes" then 11 is the chosen number; if the answer to two of these questions is "No" then 12 is the chosen number.

A very interesting solution to this problem uses the Hamming code; read about it in the Treasury under **Error-correcting codes**.

S254. You can find my five numbers with only two questions! Your first question can be, for example, "What is the sum $a_1b_1 + a_2b_2 + a_3b_3 + a_4b_4 + a_5b_5$ if $b_1 = b_2 = b_3 = b_4 = b_5 = 1$?" My answer, of course, will be the sum A of my chosen numbers. Suppose that A is a number with k digits; that is, $10^{k-1} \leq A < 10^k$. Then you know that none of my numbers has more than k digits. Your next question should use the numbers $b_1 = 1, b_2 = 10^k, b_3 = 10^{2k}, b_4 = 10^{3k}$, and $b_5 = 10^{4k}$. If you now partition the answer you receive from me from the right into five groups of k digits each, you will find my five numbers.

$$\underbrace{\ldots a_5}_{k\,\text{digits}}\ \underbrace{\ldots a_4}_{k\,\text{digits}}\ \underbrace{\ldots a_3}_{k\,\text{digits}}\ \underbrace{\ldots a_2}_{k\,\text{digits}}\ \underbrace{\ldots a_1}_{k\,\text{digits}}$$

For example, suppose I have chosen numbers 12, 135, 85, 7, and 299. My answer to your first question is $A = 12 + 135 + 85 + 7 + 299 = 538$. You know then that none of my numbers has more than three digits ($k = 3$), and my answer to your second question is 299007085135012.

The same reasoning would work had I chosen any set of numbers, not necessarily five. You could always determine my numbers by asking just two questions of this sort.

S255. There are many ways to solve this problem. This solution is by Mark Krusemeyer.

Dan is right. In fact, his computer can also generate the number 1 (which he uses for division), starting with any nonzero number a, $a * a = 0$, and then $0 * a = 1$.

First we take care of some special cases. If $a = 0$ or $b = 0$, or both, define $a \times b$ to be $1 * 1$. If $a = 0$ and b is arbitrary, define $a - b$ to be $(b * (-1)) * 1$ (which is $1 - (1 - b/(-1))$). If $a = 0$ and b is arbitrary, define $a + b$ to be $(b * 1) * 1$ (which is $1 - (1 - b)$).

Now suppose that neither a nor b is zero. One way to proceed is to observe that

$$1 * b = 1 - \frac{1}{b}, \qquad \text{and} \qquad a * 1 = 1 - a,$$

so that

$$(1 * b) * 1 = 1 - \left(1 - \frac{1}{b}\right) = \frac{1}{b}.$$

In other words, we can invert any nonzero number. Then to multiply a by b, just divide a by $1/b$.

Also, note that

$$(a * b) * a = 1 - \frac{1 - a/b}{a} = 1 - \frac{1}{a} + \frac{1}{b},$$

so that

$$\big((a * b) * a\big) * 1 = \frac{1}{a} - \frac{1}{b}.$$

To subtract, apply this expression to $1/a$ and $1/b$ (which can be computed from a and b, as we've seen).

Finally, to add, first compute $b * 1 = 1 - b$, then subtract 1 from this (which we can do) to get $-b$, and then subtract this from a.

Here are explicit formulas for nonzero a and b.

$$a/b = (a * b) * 1$$

$$\begin{aligned} a \times b &= \frac{a}{1/b} \\ &= \big(a * (1/b)\big) * 1 \\ &= \Big(a * \big((1 * b) * 1\big)\Big) * 1 \end{aligned}$$

$$\begin{aligned} a - b &= \frac{1}{1/a} - \frac{1}{1/b} \\ &= \Big(\big((1/a) * (1/b)\big) * (1/a)\Big) * 1 \\ &= \bigg(\Big(\big((1 * a) * 1\big) * \big((1 * b) * 1\big)\Big) * \big((1 * a) * 1\big)\bigg) * 1 \end{aligned}$$

$$\begin{aligned} a + b &= a - \big(1 - (1 - b)\big) \\ &= a - \big(1 - (b * 1)\big) \\ &= a - \big((b * 1) * 1\big) \\ &= \Bigg(\bigg(\big((1 * a) * 1\big) * \Big(\Big(1 * \big((b * 1) * 1\big)\Big) * 1\Big)\bigg) * \big((1 * a) * 1\big)\Bigg) * 1 \end{aligned}$$

S256. Presumably there are many solutions, but here is one given by Cedric Smith based on Herbert Wright's idea.

From the hint we know that Hal noticed that Bishop Hog had written on top of the 13 coins the letters

D E F R O C K I N G H A L

He also noticed that he had asked his assistant to do the following weighings

	Coins put on the	
	Left-hand pan	Right-hand pan
First weighing	C O I N	F A K E D
Second weighing	F O R	A C H I L D
Third weighing	——	F I N D E R H O G

Now, since the fake coin weighs almost the same as a genuine one, it is obvious that whichever coin is faked, the balancing weights must be put on the left-hand pan at each weighing. Also, if the weight of a good coin is g grams, then the balancing weights, say w_1, w_2, w_3 respectively, in the three weighings must be approximately g, $3g$, and $9g$ so that it is immediately obvious which of the reported weights applies to the first weighing, which to the second weighing, and which to the third.

Now let us number off the 13 coins as c_1 = D, c_2 = E, c_3 = F, up to c_{13} = L. Now, for any coin c_r we can define a 3-element vector $V(r)$ as follows: $V(r) = (v_1(r), v_2(r), v_3(r))$ where $v_1(r) = 1$, -1, or 0 according as in the first weighing the coin c_r is placed on the left pan, the right pan, or left off the scales respectively, and similarly for $v_2(r)$, $v_3(r)$. Thus we get the following table of values.

Coin # r	Coin	Vector $V(r)$
1	D	$(-1\ -1\ -1)$
2	E	$(-1\ 0\ -1)$
3	F	$(-1\ 1\ -1)$
4	R	$(0\ 1\ -1)$
5	O	$(1\ 1\ -1)$
6	C	$(1\ -1\ 0)$
7	K	$(-1\ 0\ 0)$
8	I	$(1\ -1\ -1)$
9	N	$(1\ 0\ -1)$
10	G	$(0\ 0\ -1)$
11	H	$(0\ -1\ -1)$
12	A	$(-1\ -1\ 0)$
13	L	$(0\ -1\ 0)$

Also, let W be the vector (w_1, w_2, w_3) and M the vector $(1, 3, 9)$. If, as above, g is the weight of a good coin and $g + f$ the weight of the faked coin, (where f may be positive or negative and is small compared with g), and c_r is the faked coin, then the result of the weighings can be expressed in the form

$$W = gM - fV(r) \tag{1}$$

We now bring in the idea of the scalar product of vectors: If $A = (a_1, a_2, a_3)$ and $B = (b_1, b_2, b_3)$, then the scalar product of A and B is defined to be $A \cdot B = a_1b_1 + a_2b_2 + a_3b_3$. If A and B are nonzero vectors such that $A \cdot B = 0$, we say that A and B are orthogonal to each other. It is known that if A and B are nonzero vectors and their coordinates are not proportional to one another (that is, neither is a multiple of the other) there is a nonzero vector E that is orthogonal to both A and B, that is, $A \cdot E = 0$ and $B \cdot E = 0$; furthermore, any such vector has the form kE, with $k \neq 0$.

Thus, for each r there is a nonzero vector $U(r)$ orthogonal to both M and $V(r)$ as follows.

Coin # r	Coin	Vector $U(r)$
1	D	$(3\ -4\ 1)$
2	E	$(3\ 8\ -3)$
3	F	$(3\ 2\ -1)$
4	R	$(12\ -1\ -1)$
5	O	$(6\ -5\ 1)$
6	C	$(9\ 9\ -4)$
7	K	$(0\ 3\ -1)$
8	I	$(3\ 5\ -2)$
9	N	$(3\ -10\ 3)$
10	G	$(3\ -1\ 0)$
11	H	$(6\ 1\ -1)$
12	A	$(9\ -9\ 2)$
13	L	$(9\ 0\ -1)$

Note that no $U(r)$ is a multiple of a different $U(s)$. From (1) it follows that, if c_r is the fake, $W \cdot U(r) = 0$. We cannot have $W \cdot U(s) = 0$ for $s \neq r$, for otherwise both $U(r)$ and $U(s)$ would be orthogonal to both W and M which is impossible. So to locate the fake we simply calculate $W \cdot U(r)$ for different r, and the r satisfying $W \cdot U(r) = 0$ is the fake c_r.

Having found the fake, equation (1) gives 3 equations for g and f, which can be solved to find g and f, and hence $g + f$.

Responses to Queries

R6. It is helpful to keep track of not only the intervals of time we can reach, but also the states we can be in at those times. Let the pair (a, b) indicate how much time *remains* in the 9-minute glass (the first coordinate) and the 13-minute glass (the second coordinate).

When we find ourselves in a certain state at a certain time, we have four alternatives, $TT, T\overline{T}, \overline{T}T, \overline{TT}$, which mean, respectively, turn both glasses over, turn the 9-glass but not the 13-glass, leave the 9-glass alone and turn the 13-glass, and leave both glasses as they are.

The following figure describes the options and consequences, given that we are in state $(0, 0)$ at time t, which we symbolize by writing $(0, 0)_t$.

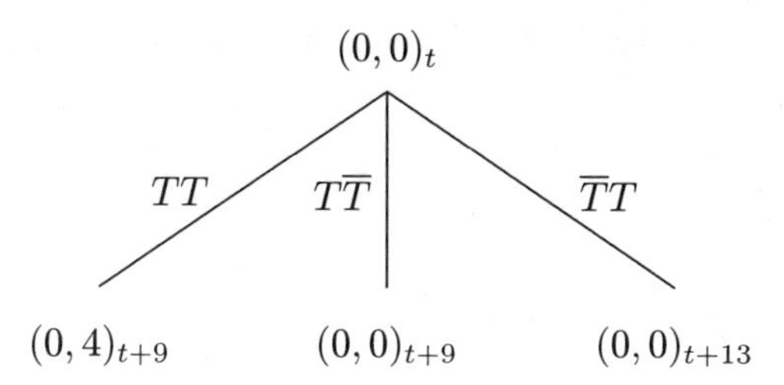

Using this information, we can begin to construct a table that compares measurable times with the attainable states at those times. For example, from $(0, 0)_9$ we can reach $(0, 0)_{22}$ by applying $\overline{T}T$, and from $(0, 0)_{13}$ we can reach $(0, 4)_{22}$ by applying TT. By repeated use, we can reach the following states.

Times	States
0	$(0,0)$
9	$(0,0)$, $(0,4)$
13	$(0,0)$
18	$(0,0)$, $(0,4)$
22	$(0,0)$, $(0,4)$
26	$(0,0)$
27	$(0,0)$, $(0,4)$
31	$(0,0)$, $(0,4)$

Now consider the options and consequences starting from the state $(0,4)$.

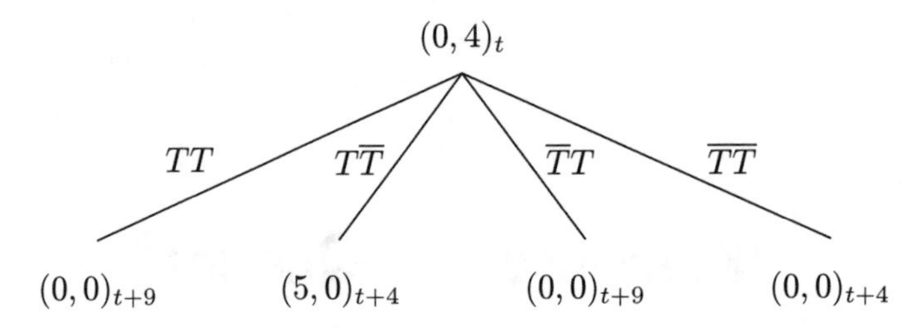

Applying this to the $(0,4)$ states in the preceding table, we can augment the list of attainable times and states.

Times	States
0	$(0,0)$
9	$(0,0),\ (0,4)$
13	$(0,0),\ (5,0)$
18	$(0,0),\ (0,4)$
22	$(0,0),\ (0,4),\ (5,0)$
26	$(0,0),\ (5,0)$
27	$(0,0),\ (0,4)$
31	$(0,0),\ (0,4),\ (5,0)$

Continuing in this way, by examining what happens from state $(5,0)$, and then from the states it generates, and so forth, we can extend the table in length and breadth. This analysis yields a few new entries in this range, namely, 17, 21, and 25, as shown by the following procedures.

$$(5,0)_{13} \xrightarrow{TT} (0,9)_{17}, \qquad (0,9)_{17} \xrightarrow{TT} (5,0)_{21}, \qquad (5,0)_{21} \xrightarrow{TT} (0,9)_{25}.$$

But more importantly, we discover the sequences

$$(5,0)_{13} \xrightarrow{\overline{T}T} (0,8)_{18} \xrightarrow{T\overline{T}} (1,0)_{26} \xrightarrow{\overline{T}T} (0,12)_{27}$$

$$(0,12)_t \xrightarrow{TT} (8,0)_{t+1} \qquad \text{and} \qquad (8,0)_t \xrightarrow{TT} (0,12)_{t+1}$$

The first of these shows that we can reach state $(0,12)$ in 27 seconds, and repeated application of the last pair of procedures shows that we can measure all time intervals beyond 27.

In summary, the measurable time intervals under the stated conditions are: 9, 13, 17, 18, 21, 22, 23, and 25 minutes onwards. If you were allowed to start the glasses ahead of time, you could measure any number of minutes.

For Rikki-Tikki-Tavi. (Under this heading we will occasionally suggest problems for the reader to continue to investigate.)

Replace the nine-minute and thirteen-minute glasses with m-minute and n-minute glasses, where m and n are relatively prime positive integers. What is the largest integer time interval one cannot measure with these glasses?

R8. Replace the parrot's "O" and "K" with the mathematician's 0 and 1, and denote the kth digit of the word by $d(k)$. From the definition, we find that $d(1) = 0$, $d(2) = 1$, $d(4) = 0$, $d(8) = 1$, and $d(2^k) \equiv k \pmod 2$. Also, from the definition, if n is not a power of 2, $d(n) = 1 - d(n - 2^t)$, where 2^t is the largest power of 2 which is less than n. From this, we can compute

$$\begin{aligned} d(2020) &= 1 - d(2020 - 1024) = 1 - d(996) \\ &= 1 - \big(1 - d(996 - 512)\big) = d(484) \\ &= 1 - d(484 - 256) = 1 - d(228) \\ &= 1 - \big(1 - d(228 - 128)\big) = d(100) \\ &= 1 - d(100 - 64) = 1 - d(36) \\ &= 1 - \big(1 - d(36 - 32)\big) = d(4) \\ &= 0. \end{aligned}$$

The 2020th letter (which doesn't exist until the twelfth day) is "O."

This sequence is known as the Thue-Morse sequence and it has many interesting properties. For example, it does not contain any substring of the form WWW where W is any string of zeros and ones. The Thue-Morse sequence is usually written with the terms starting at the zeroth term: $x(0) = 0, x(1) = 1, x(2) = 1, \ldots$, and it is not difficult to show that $x(2n) = x(n)$, $x(2n + 1) = 1 - x(2n)$, and that $x(n)$ is the modulo 2 sum of the digits in the base 2 representation of n. For example, $d(2020) = x(2019) = x(11111100011_2) = 8 \pmod 2 = 0$. For more, see Manfred R. Schroeder, *Fractals, Chaos, and Power Laws: Minutes from an Infinite Paradise*, W. H. Freeman, New York, 1991.

R9. The possible cuts correspond to the set of partitions of 12 into the sum of four parts. These are easy to enumerate:

1. 9111	6. 6321	11. 5322
2. 8211	7. 6222	12. 4431
3. 7311	8. 5511	13. 4422
4. 7221	9. 5421	14. 4332
5. 6411	10. 5331	15. 3333

It is necessary, but easy, to check that each of these divisions accomplishes the construction.

R11. Whatever the length of a strip of paper, it is always possible, by folding, to find a piece of half the length. By folding again, one can find any number of quarters; then again, eighths, and so on. So, if we start with $2/3$ meter, we can find any length $\frac{d}{3\times 2^k}$ meters for $k \geq 0$ and d odd and less than 2^{k+1}.

R17. Each of the solutions previously found for sums of consecutive positive integers can be extended to a different solution with consecutive integers by adjoining consecutive integers downward to zero plus the negatives of these numbers (to cancel them out). This yields the following additional solutions.

$$\begin{aligned}105 &= \big((-51)+(-50)+\cdots+(-1)+0+1+\cdots+51\big)+\big(52+53\big)\\ &= \big((-33)+(-32)+\cdots+32+33\big)+\big(34+35+36\big)\\ &= \big((-18)+(-17)+\cdots+17+18\big)+\big(19+\cdots+23\big)\\ &= \big((-14)+(-13)+\cdots+13+14\big)+\big(15+\cdots+20\big)\\ &= \big((-11)+(-10)+\cdots+10+11\big)+\big(12+\cdots+18\big)\\ &= \big((-5)+(-4)+\cdots+4+5\big)+\big(6+\cdots+15\big)\\ &= 0+\big(1+2+\cdots+13+14\big)\end{aligned}$$

There is one additional sum (which corresponds to the sum of one consecutive integer, $105 = 105$), namely,

$$105 = \big((-104)+(-103)+\cdots+103+104\big)+105.$$

The cancellation of positive and negative parts makes it clear there can be no further solutions. Conclusion: there are 15 ways of writing 105 as a sum of two or more consecutive integers.

Here's the general result (replacing 105 by n): Each odd divisor, d, of n gives rise to exactly two representations of n as a sum of consecutive integers, namely d consecutive integers of which the middle one is n/d:

$$\frac{n}{d}-\frac{d-1}{2},\ldots,\frac{n}{d},\ldots,\frac{n}{d}+\frac{d-1}{2}$$

and $2n/d$ consecutive integers, of which the middle pair is $(d-1)/2, (d+1)/2$:

$$\frac{d+1}{2}-\frac{n}{d},\ldots,\frac{d-1}{2},\frac{d+1}{2},\ldots,\frac{d-1}{2}+\frac{n}{d}$$

The first of these is nonnegative if $n/d-(d-1)/2 \ge 0$, or equivalently, if $8n+1 \ge (2d-1)^2$, and the second is nonnegative if $(d+1)/2 - n/d \ge 0$, or equivalently, if $8n+1 \le (2d+1)^2$.

Note that $d = 1$ leads to the representation n itself (one integer), or as the sum of $2n$ consecutive integers

$$-(n-1)-(n-2)-\cdots-1+0+1+2+\cdots+(n-1)+n$$

Every representation of n as a sum of consecutive integers has either an odd or an even number of terms, so it must be one of the above types. Note that $105 = 3\times5\times7$ has 8 odd divisors, so that there are 16 representations of 105 as a sum of consecutive integers; 15 of them have more than one term, as we have seen. In general, if $n = 2^a p_1^{a_1} p_2^{a_2}\cdots$ is the prime factorization of n, where p_1, p_2, ... are distinct odd primes, then the number of representations is $2(a_1+1)(a_2+1)\cdots$; that is, twice the number of odd divisors of n. See **Divisors** in the Treasury.

R18. Just one Varmlander is needed. To see this, first consider the case where all 17 Venusians are men. The graph below shows how the dance can be done (each Venusian man is holding hands with three other people and the Varmlander is holding hands with seven others).

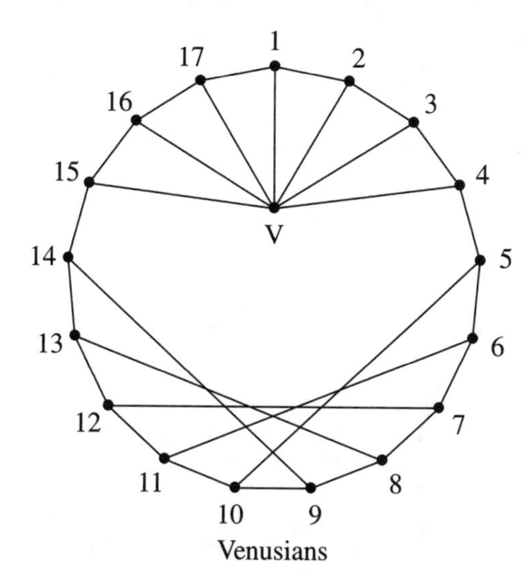

A dance for any combination of 17 Venusian men and women can be arranged as follows. We may replace 1 by a woman by asking 9 and 10 to stop holding hands, and each to hold 1 instead. Similarly 10 and 11 can leave go and each hold 2 who is then pentavalent and can take the place of a woman, and so on. Some of them may need rather long arms!

R21. There are only four ways to partition 14 into three distinct parts not greater than seven: 761, 752, 743, and 653. Three of these have to fit together on the "outside" of the three lines. Only three of the four possible sets of three can fit together as required: 761, 752, 653; 761, 743, 653; 752, 743, 653. The figures below show how these can fit.

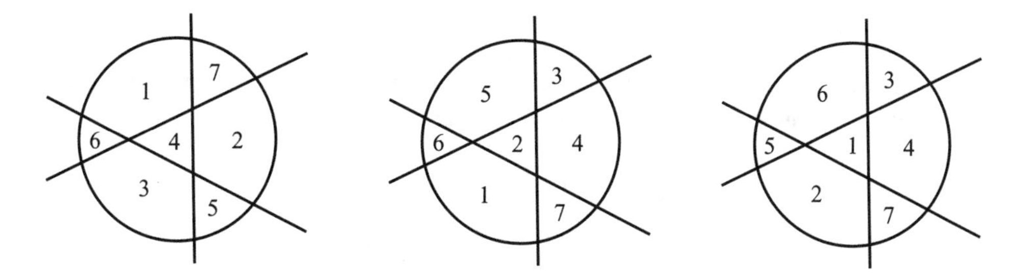

If you read about **nim-addition** in the Treasury, you'll notice that the nim-sum of the numbers on either side of any of the straight lines is zero, and that these three pictures are manifestations of the **Fano Configuration**!

R22. We want to find all the members of the Farey sequence of order 20 that lie between $\frac{3}{5}$ and $\frac{13}{20}$. We'll show you how to do this under **Farey sequence** in the Treasury. They are

$$\frac{3}{5}, \frac{11}{18}, \frac{8}{13}, \frac{5}{8}, \frac{12}{19}, \frac{7}{11}, \frac{9}{14}, \frac{11}{17}, \frac{13}{20}.$$

Thus, the second fewest number of employees possible is 11 (corresponding to $\frac{7}{11}$), then 13 (from $\frac{8}{13}$), then, in order, 14, 16 ($= 8 \times 2$), 17, 18, 19, and any number from 21 onwards ($21 = 8 + 3$, $22 = 11 \times 2$, $23 = 5 + 18$, $24 = 8 \times 3$, $25 = 11 + 14, \ldots$).

R32a. One approach is judiciously to connect the four small circuits shown.

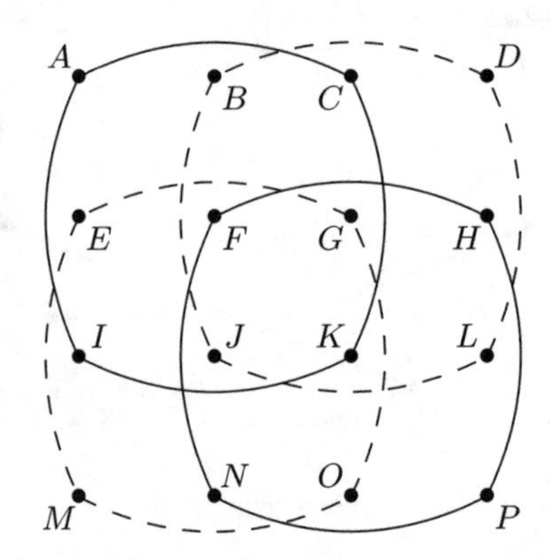

For this, delete AI and DL, then add AD and IL (this connects the top two loops). Next, delete EM and HP, then add EH and MP (this connects the bottom two loops). Finally, delete AD and MP, then add AM and DP (this connects the final two loops). The resulting tour has both vertical and horizontal symmetry.

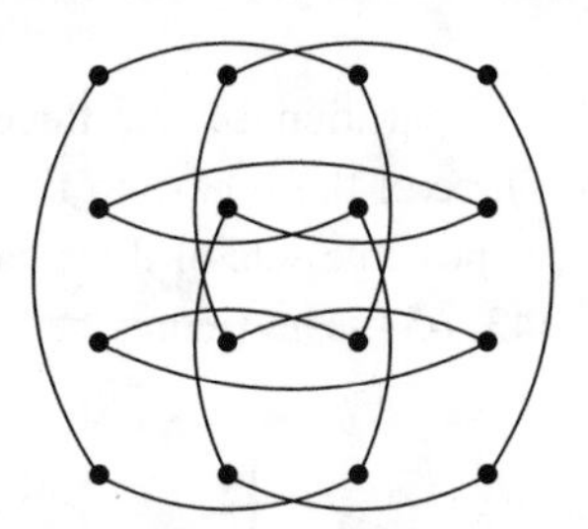

Thus we have two solutions to the problem. Are there others? To answer this, recall that solutions correspond to Hamilton circuits on the following graph.

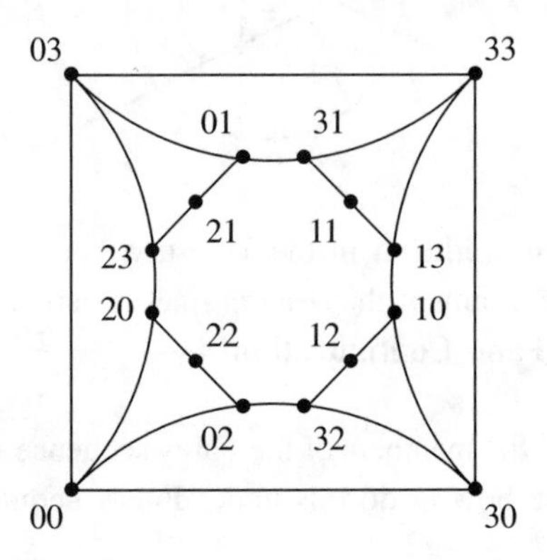

We can find *all* Hamilton circuits on this graph by noting that vertices 11, 12, 21, 22 are each 2-valent, so that we must use the eight edges at 45°. We must join the inner octagon with the outside square, so by symmetry, we may as well use the edge 03-23. The vertices 01, 02, 10, 13, 20, 23, 31, 32 are each 3-valent, so we must omit an edge at each of them. We will not use 03–01, since it would form a 4-circuit, nor will we use 23–20. So we use 20–00, and not 00–02, which would form another 4-circuit. So we use 02–32 and not 32–30. We use 01–31 and not 31–33.

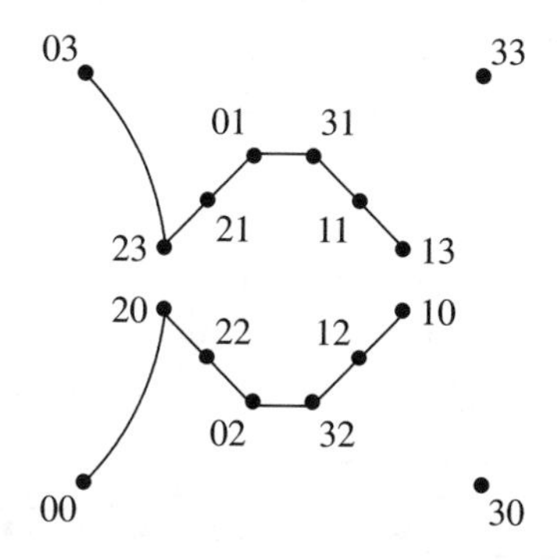

There are two ways of finishing off, depending on whether you use 10–13 or not. The rest is forced. The first figure uses 03–33–30–00 and has symmetry about a horizontal axis, the latter uses 10–30–33–13 and 00–03 and has symmetry about both a horizontal axis and a vertical axis.

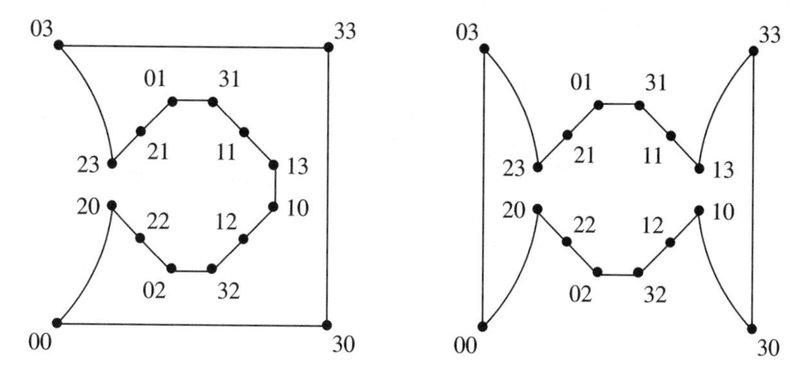

Of course, these diagrams are schematic; these correspond to the two tours previously constructed.

R32b. Yes, there is a solution on a 5×5 checkboard. One way to construct it is to make five small circuits like the one below, one in each row.

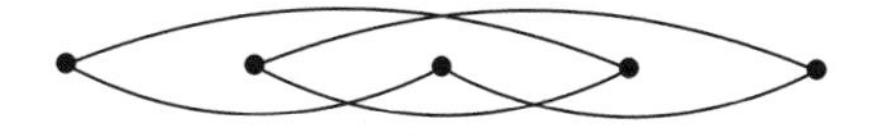

Connect these loops to complete the circuit (which is symmetric about the center of the board).

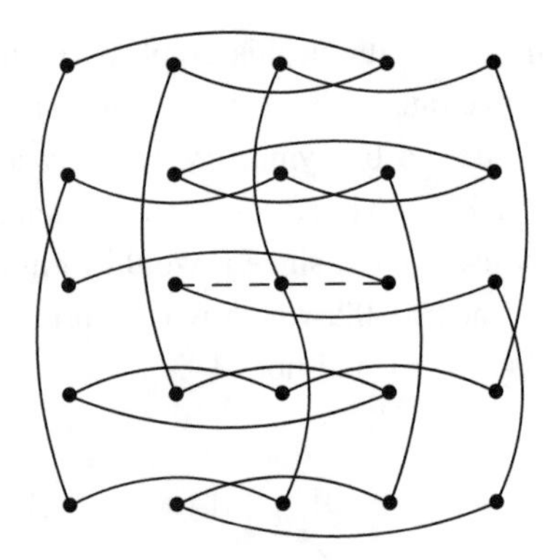

The procedure for producing this tour can be used to make symmetric tours on larger boards; we'll illustrate this by giving two examples, one that generalizes to any odd by odd sized board, and the other to any even by even board.

For the odd case, begin by producing a loop for a single row of the board, as shown.

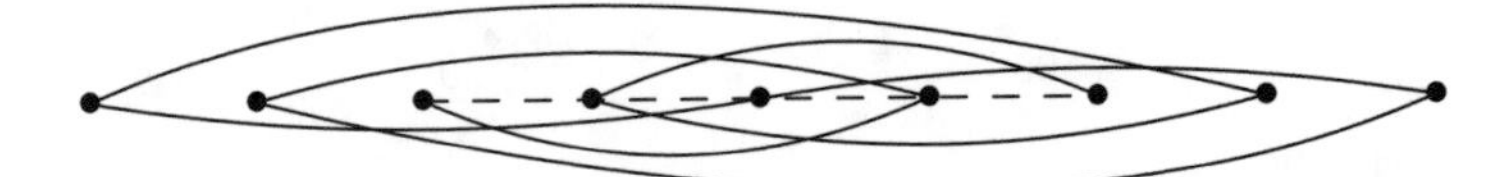

Now reproduce this loop in each of the rows, and connect the loops (one way to do this is to make the row connexions in the same order in which the vertices in a single row are connected by the row loop). This gives a tour such as that shown below, which has symmetry about the center of the board.

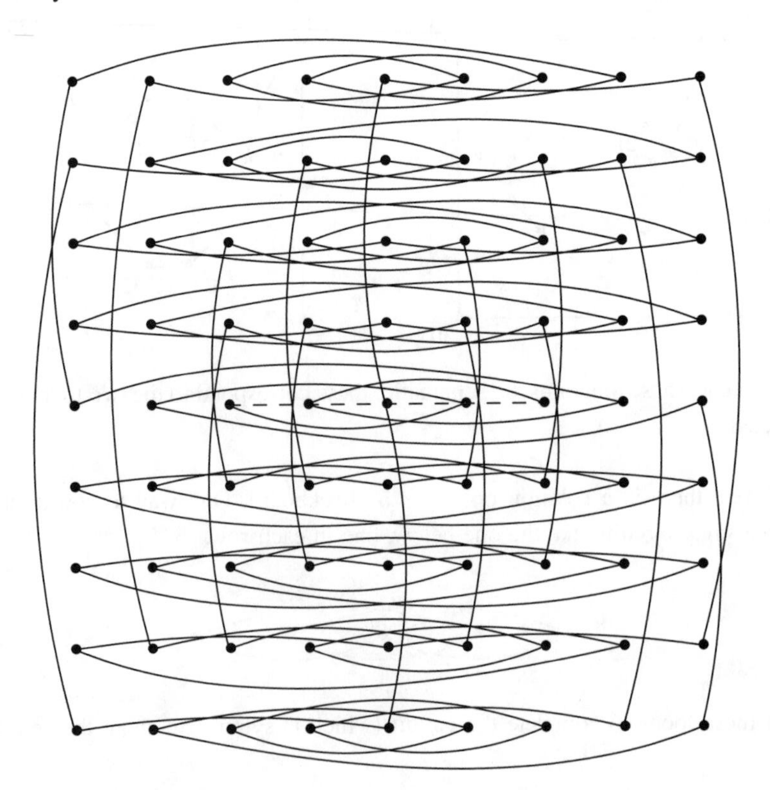

For the even by even board, begin with a loop on a single row, such as the one below.

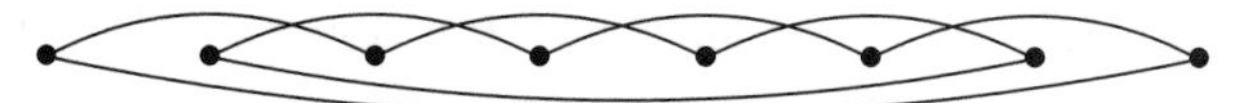

Now replicate this loop on each row and then connect the loops (in the same order as the vertices are connected in the row circuit). Here's the result.

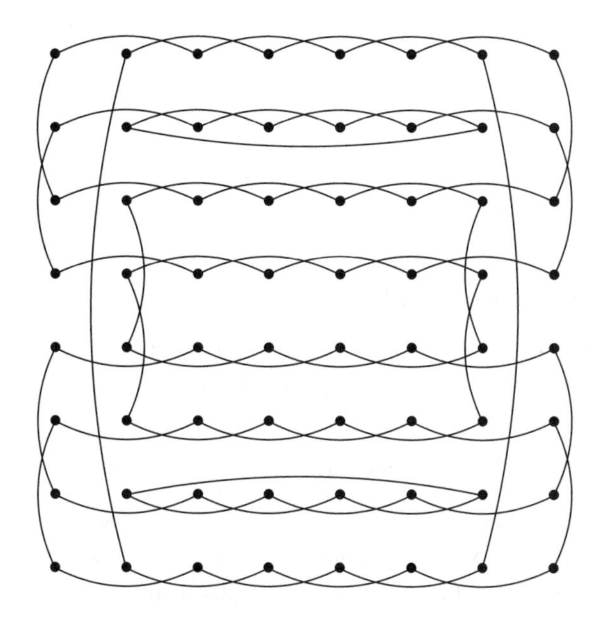

R34. (a) $\{9,5,4,2\} \to \{4,2\} \to \{2\} \to \emptyset$ (3 moves)

This example suggests that the best move at each stage is to empty as many jars as possible. In our first move, for example, we reduced the number of empty jars by 2; no other move does as well in this regard. This example makes a case for the following algorithm:

> **#1. Empty the Most Jars Algorithm (EMJA):** Empty as many jars as you can. If there is more than one way of doing this, choose the move that makes the resulting numbers as small as possible in their **lexicographic order,** after common factors are taken out (because $\{12,8,4\}$ can be treated in essentially the same way as $\{3,2,1\}$). For example, if the cookie numbers are $\{12,9,5,1\}$, there is no way to make just two nonempty jars, but there are many ways to reduce it to three: $\{12,8,4\} = \{3,2,1\}$ and eighteen others, listed under **lexicographic order** in the Treasury. The algorithm will choose $\{3,2,1\}$, because it is smallest in the lexicographic ordering.

(b) $\{17,16,11,4,2,1\} \to \{6,5,4,2,1\} \to \{2,1\} \to \{1\} \to \emptyset$ (4 moves)

Had we applied the Empty the Most Jars Algorithm, the first move would have been to position $\{11,4,2,1\}$. However, this requires four more moves to empty, so it is not optimal. The better approach in this example is to take as many cookies as possible at each stage. Here, we first removed 33 cookies, then 12, then 2, then 1. So another procedure that evidently works in some cases is the following algorithm:

#2. Take the Most Cookies Algorithm (MCA): At each move, take as many cookies as possible. In the case of a tie, make the move that makes the resulting cookie numbers as small as possible in their lexicographic order, after common factors are taken out.

(c) $\{36, 33, 5, 1\} \to \{5, 4, 1\} \to \{1\} \to \emptyset$ (3 moves)

The Most Cookies Algorithm and the Empty the Most Jars Algorithm lead to $\{5, 3, 1\}$ on the first move. However, this doesn't lead to the optimal solution. In this case, the best approach is to take 32 from two of the jars, so here's another idea.

#3. Binary Algorithm (BA): For k as large as possible, take 2^k cookies from all jars that contain at least 2^k cookies.

(d) $\{11, 5, 4, 2\} \to \{6, 4, 2\} \to \{2\} \to \emptyset$ (3 moves)

None of these algorithms is optimal in all cases, even though each of them is superior to the other two in certain situations.

Cookie Numbers	EMJA	MCA	BA	Optimal
$\{9, 5, 4, 2\}$	3 moves	4 moves	4 moves	3 moves
$\{17, 16, 11, 4, 2, 1\}$	5 moves	4 moves	5 moves	4 moves
$\{36, 33, 5, 1\}$	4 moves	4 moves	3 moves	3 moves
$\{11, 5, 4, 2\}$	4 moves	4 moves	4 moves	3 moves

For Rikki-Tikki-Tavi. Suppose you are given n jars with $k_1 > \cdots > k_n$ cookies, respectively. You are permitted to choose any subset of these jars and take the same number of cookies from these jars. Is there a procedure that will always lead to the optimal solution (for emptying the jars in the fewest moves)?

R37. There is a unique solution when $n = 2$: namely 2, 1, 1, 2.

By looking at the possible parity patterns (uniquely determined by the parities of the first two terms), we find that for $n = 3$, there are five essentially different solutions:

1, 3, 2, 1, 3, 2
3, 1, 2, 3, 1, 2
3, 1, 2, 1, 3, 2
1, 2, 3, 1, 2, 3
3, 2, 1, 1, 2, 3

Four other solutions can be found by writing the first four sequences in reverse order.

There are no solutions for $n = 4$ or $n \geq 6$ (by parity considerations; each subsequence of three consecutive terms contains two odds and one even, and in these cases, there are not enough odd terms, and there are too many even terms, to fill out the sequence as required).

R38. There are many solutions. First off, there are 15 ways to partition the special gifts so that each party receives two. For each of these, there are several ways to partition the remaining nine gifts into sets of three of the same total value. For example, if the special gifts are divided $(10, 11)$, $(12, 13)$, $(14, 15)$, then there are nine ways to partition the gifts into five parts of equal value:

$$\begin{array}{lll}
(10,11,3,7,9) & (12,13,4,5,6) & (14,15,1,2,8)\\
(10,11,4,6,9) & (12,13,3,5,7) & (14,15,1,2,8)\\
(10,11,2,8,9) & (12,13,4,5,6) & (14,15,1,3,7)\\
(10,11,4,6,9) & (12,13,2,5,8) & (14,15,1,3,7)\\
(10,11,5,6,8) & (12,13,2,4,9) & (14,15,1,3,7)\\
(10,11,2,8,9) & (12,13,3,5,7) & (14,15,1,4,6)\\
(10,11,3,7,9) & (12,13,2,5,8) & (14,15,1,4,6)\\
(10,11,4,7,8) & (12,13,1,5,9) & (14,15,2,3,6)\\
(10,11,3,7,9) & (12,13,1,6,8) & (14,15,2,4,5)
\end{array}$$

Perhaps the fairest apportionment is to partition the special gifts so that each receives two of the same total value; this can be done in only one way: $(10,15)$, $(11,14)$, and $(12,13)$. The other gifts can then be partitioned into $(9,5,1)$, $(8,4,3)$, $(7,6,2)$, or $(9,4,2)$, $(8,6,1)$, $(7,5,3)$. Each of these sets of three can be matched with the sets of two in six different ways, giving twelve solutions in which each person receives two special gifts of the same total value and three regular gifts of the same total value.

R41. In the first part, we want to find integers x and y so that

$$7x + 11y = 1.$$

One solution is $x = -3$, $y = 2$. Other solutions are

$$x = -3 + 11k,$$

$$y = 2 - 7k,$$

where $k = 0, \pm 1, \pm 2, \ldots$, and it can be shown that this gives all possible solutions. When $k = 1$ we get $x = 8$, $y = -5$, which translates into "Fill the 7-bucket 8 times, each time transferring as much as possible to the 11-bucket; empty the 11-bucket (when full) 5 times."

In the second part, we want integers x and y so that

$$7x + 11y = 2.$$

From the first part, $7(-3) + 11(2) = 1$, so multiplying by 2, we get $7(-6) + 11(4) = 2$. Thus, one solution is $x = -6$, $y = 4$; that is, fill the 11-bucket 4 times, empty as much as possible into the 7-bucket, and empty this 6 times. Other solutions are

$$x = -6 + 11k,$$

$$y = 4 - 7k$$

for $k = 0, \pm 1, \pm 2, \ldots$. When $k = 1$ we get $x = 5$, $y = -3$, which takes fewer steps.

R45. The multiplier is 258; the three products end in 4, 5, 6; the final product is $16uv46$; the multiplicand is $6x7$, where x is 3 or 8 and 8 is too big. Therefore the answer is $637 \times 258 = 164346$.

In the second problem, the second digit of the multiplier is 2; the hundreds product is $5y6286$, where y is 0 or 1; the multiplicand is $25z143$ where z is 3 or 8. The first digit of the multiplier is 3; $z = 3$, and the third digit of the multiplier is also 3: $253143 \times 3235 = 818917605$.

R46. Here are figures to show the we can have 4, 5, 6, or 7 triangular regions with six lines.

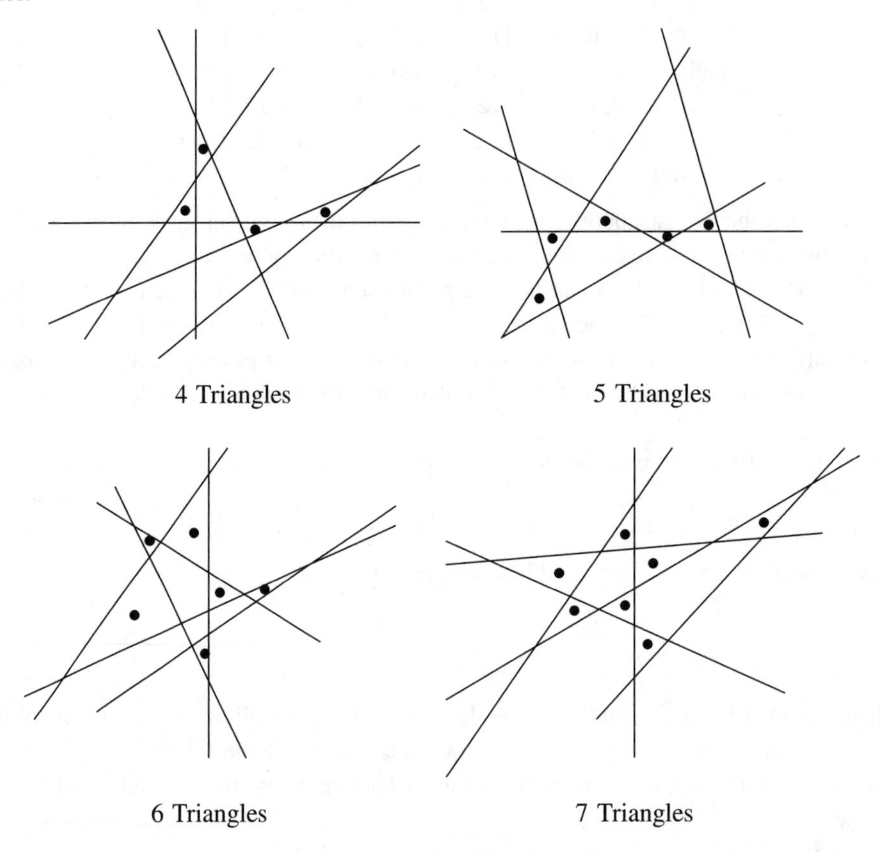

4 Triangles 5 Triangles

6 Triangles 7 Triangles

Rikki-Tikki-Tavi will want to know how many triangular regions can be made with an arrangement of n lines. An upper bound is $n(n-2)/3$, because there are at most $n - 2$ finite segments cut off on each line and each segment can be the edge of at most one triangular region. Notice that we're talking about regions, not triangles in general, which may contain several polygonal regions. Assume that all lines are long enough to meet any others, unless they are parallel.

R47. As in the solution, label the digits as shown.

$$\begin{array}{cccc} & a & b & c \\ & d & e & f \\ \hline g & h & i & j \end{array}$$

In the solution, we may assume that that $a < d$, $b < e$, $c < f$. Also, if the summands are x and y, and the total z, then z (and $x + y$) are multiples of 9.

It's clear that $g = 1$ and that there is a carry from the hundreds to the thousands. According as there are (1) no other carries, (2) a carry from the tens to the hundreds,

(3) a carry from the units to the tens, (4) both of these last, we have the equations

$$\begin{array}{llll}
(1) & a+d = h+10 & b+e = i & c+f=j \\
(2) & a+d+1 = h+10 & b+e = i+10 & c+f=j \\
(3) & a+d = h+10 & b+e+1 = i & c+f=j+10 \\
(4) & a+d+1 = h+10 & b+e+1 = i+10 & c+f=j+10
\end{array}$$

The total sum of the digits $S = a+b+c+d+e+f$ and $T = h+i+j$ is 44, so S and T have the same parity. If we add the sets of equations, we see that cases (2) and (3) cannot arise. Moreover, (1) gives $S = T+10$ so that $S = 27$, $T = 17$, while (4) gives $S = T+28$, with $S = 36$, $T = 8$.

In case (1), 0 can only occur where the carry is; that is, $h = 0$. Now $T = 17$ implies that i and j are 8 and 9 in some order.

In case (4), $T = 8$ requires that hij are 026 or 035 in some order.

The equations in (1) and (4) are all of the form $r+s = t+u$, where $u = 10$, 0, 0 in (1), and $u = 9$, 9, 10 in (4). The columns of the sum are distinct triples (r, s, t) selected from the following table of possibilities:

$u=0$	$t =$	0	2	3	5	6	8	9
	r, s						2,6	2,7
							3,5	3,6
								4,5
$u=9$	r, s	2,7	3,8					
		3,6	4,7	4,8				
		4,5	5,6	5,7	6,8	7,8		
$u=10$	r, s	2,8	3,9	4,9				
		3,7	4,8	5,8	6,9			
		4,6	5,7	6,7	7,8	7,9		

In case (1), the only ways of selecting two distinct triples (r, s, t) from $u = 0$ are 268 and 459 or 358 and 279 and in each case the remaining triple, 370 or 460, appears in $u = 10$, yielding the solutions

$$\begin{array}{r}
\ \ \leftrightarrow\ \\
3\ 4\ 2 \\
7\ 5\ 6 \\
\hline
1\ 0\ 9\ 8
\end{array}
\qquad\qquad
\begin{array}{r}
\ \ \leftrightarrow\ \\
4\ 3\ 2 \\
6\ 5\ 7 \\
\hline
1\ 0\ 8\ 9
\end{array}$$

where the arrows indicate that we can swap the units and tens columns.

In case (4) the only pairs of distinct triples from $u = 9$ are

270	270	360	450	450	472	562
483	685	472	382	786	685	483

and in four cases the missing triple

695	493		796	392		

is in $u = 10$ (in fact 8 and 9 do not occur in the total in case (4) and so must be chosen from $u = 9$ and $u = 10$ respectively) and we have the solutions

$$\begin{array}{r} \leftrightarrow \\ 2\ 4\ 6 \\ 7\ 8\ 9 \\ \hline 1\ 0\ 3\ 5 \end{array} \qquad \begin{array}{r} \leftrightarrow \\ 4\ 3\ 7 \\ 5\ 8\ 9 \\ \hline 1\ 0\ 2\ 6 \end{array} \qquad \begin{array}{r} \leftrightarrow \\ 2\ 6\ 4 \\ 7\ 8\ 9 \\ \hline 1\ 0\ 5\ 3 \end{array} \qquad \begin{array}{r} \leftrightarrow \\ 4\ 7\ 3 \\ 5\ 8\ 9 \\ \hline 1\ 0\ 6\ 2 \end{array}$$

where now we may swap the tens and hundreds columns.

The total number of solutions is $(2+4) \times 2 \times 8 = 96$.

R48a. If one of the houses is inside the triangle formed by the other three, the gazebo should be at the "middle" house. This can be verified as in the original problem by applying the triangle inequality to show that if the gazebo is anywhere else, then the amount of walking is greater.

R48b. If there are only three houses, and if the triangle has an angle $\geq 120°$, build the gazebo at that vertex. Otherwise, the gazebo should be at the famous Fermat-Torricelli point, the point from which the angles subtended by the sides are all equal (to $120°$) (it is not difficult to to show, using high-school geometry, that any other point will increase the sum of the distances).

Here's a heuristic demonstration based on physical evidence. Draw a map of the houses on a table and drill three holes, one at each house. Suspend a one-kilogram weight on a string through each hole and tie the three strings together in a single knot. Let go and the knot will come to rest. Because the weights minimize their total potential energy, they will be as low as possible, with a minimum amount of string on the table (the walking distance we're trying to minimize). Because the weights are equal, the tensions in the three strings will be equal (to one kilogram in each). Because the weights are in equilibuium, these equal tensions must make equal angles (of $120°$) with each other. (However, if one of the angles in the triangle is equal to or greater than $120°$, the knot will come to rest at that vertex, so in this case it's best to build the gazebo there.)

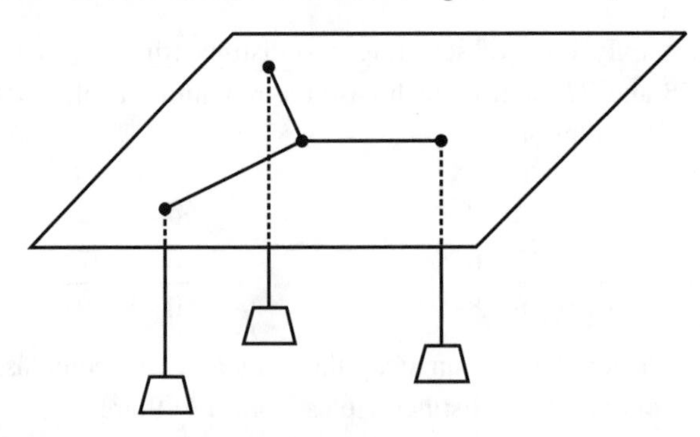

This method can be modified, using different weights proportional to the numbers of people from each house who wish to visit the gazebo.

R48c. Suppose there is no gazebo at all, but you wish to construct a paved trail that will connect the four houses. What is the minimal total length of such a trail?

The following is a good attempt at a solution, but is not best possible, if an angle of the quadrangle is more than 120 degrees.

Choose the pair of opposite edges of smaller total length ($AC+BD$ in the following figure). Draw equilateral triangles externally on these two edges, together with their circumcircles. If the centers of these circles are P and Q, let the line PQ cut the circles in X and Y. Then the minimal length path comprises the five segments AX, CX, XY, YB, YD.

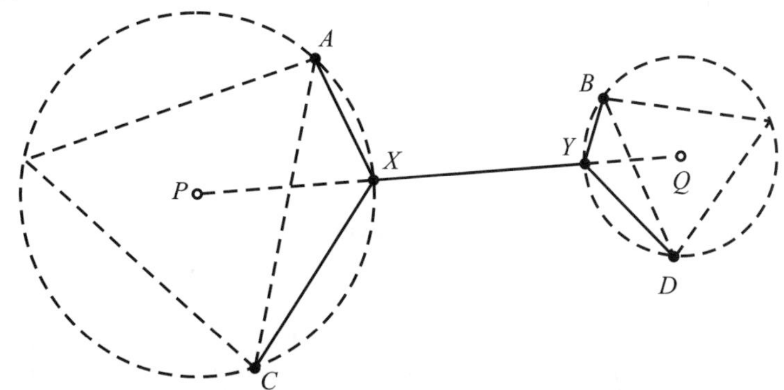

As an example, consider the case of four points located at the vertices of a square of side length 2 centered at the origin. In the first figure, the total path length is $4 \times \sqrt{2} = 5.656854248\ldots$. In the second case, the total length is $2 + 2\sqrt{3} = 5.464101615\ldots$

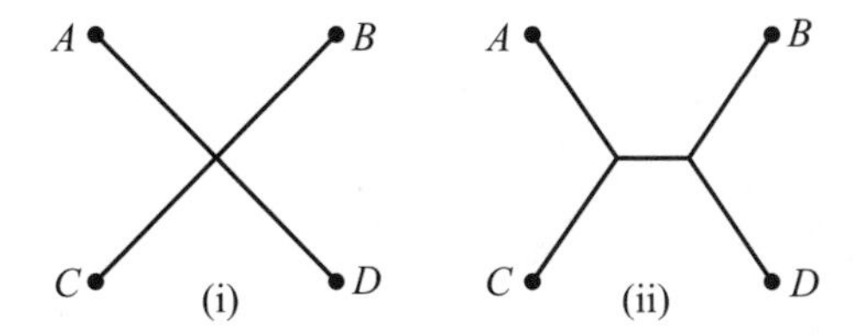

Problems such as this are physically solvable by using the surface tension in soap films (see R. Courant, "Soap film experiments with minimal surfaces," *American Mathematical Monthly*, **47** (1940), pp. 167–174).

Two parallel glass plates P and Q are joined by three or more perpenpendicular bars (here, we have four bars at A, B, C, D). If we immerse this object in a soap solution and withdraw it, the film solves the problem by forming a system of vertical planes joining the fixed bars which make angles of $120°$ at X and Y as in the three-house solution.

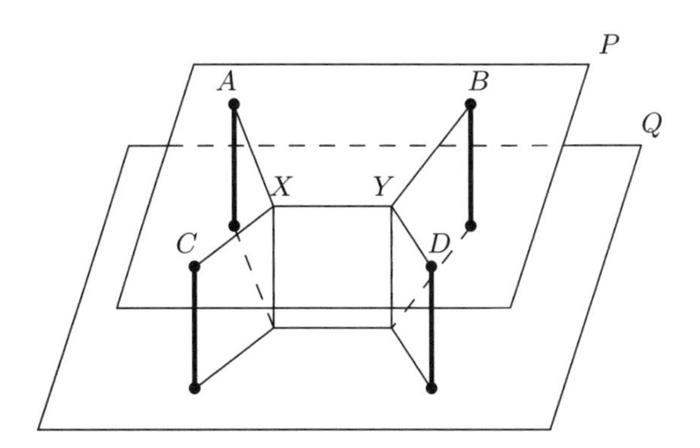

R54. Any number of the form $55-2x$, with $2 \leq x \leq 27$. Each such x can be partitioned into distinct parts of sizes 2 to 10.

R57. There are three statements to consider:

C : It is not true that both of these have green hair. $\equiv$ A is truthful or B is truthful.

D : Either A has green hair or B has blue hair. $\equiv$ A is a liar or B is truthful.

E : I am a liar or (C and D are both liars or both truthful).
$\equiv$ I am truthful, and C and D do not have the same colored hair.

The table below has an "x" where these statements are consistent. There are just two cases (indicated by arrows) where all three statements are consistent. In each of them B is truthful, so I should go through the mountains.

	A	B	C	D	E	C: A is true, or B is true.	D: A is a liar, or B is true.	E: I am a liar, or $C \equiv D$.
→	T	T	T	T	T	x	x	x
	T	T	T	T	L	x	x	
	T	T	T	L	T	x	x	
	T	T	T	L	L	x		
	T	T	L	T	T		x	
	T	T	L	T	L		x	
	T	T	L	L	T			
	T	T	L	L	L			
	T	L	T	T	T	x		x
	T	L	T	T	L	x	x	
	T	L	T	L	T	x	x	
	T	L	T	L	L	x	x	
	T	L	L	T	T			
	T	L	L	T	L		x	
	T	L	L	L	T		x	x
	T	L	L	L	L		x	
→	L	T	T	T	T	x	x	x
	L	T	T	T	L	x	x	
	L	T	T	L	T	x		
	L	T	T	L	L	x		
	L	T	L	T	T		x	
	L	T	L	T	L		x	
	L	T	L	L	T			x
	L	T	L	L	L			
	L	L	T	T	T		x	x
	L	L	T	T	L		x	
	L	L	T	L	T			
	L	L	T	L	L			
	L	L	L	T	T	x	x	
	L	L	L	T	L	x	x	
	L	L	L	L	T	x		x
	L	L	L	L	L	x		

R59. The following moves verify the first claim.

$$\underline{01} \leftrightarrow \underline{10}0 \leftrightarrow 11\underline{10} \leftrightarrow 1\underline{111}1 \leftrightarrow 11\underline{01} \leftrightarrow \underline{11}100 \leftrightarrow 1\underline{000} \leftrightarrow \underline{111} \leftrightarrow \underline{10}$$

Because of what we've just shown, we can replace the substring 01 with 10. Also, the string of equivalences includes the claims of (ii) and (iii).

To show (iv), we begin by showing that we can generate any string (of length greater than one) with just one 1. The chain of moves

$$\underline{10} \to \underline{01} \to \underline{10}0 \to \underline{01}0 \to \underline{10}00 \to \underline{01}00 \to \underline{10}000 \to \cdots$$

shows how to get a string with 1 followed by an arbitrary number of trailing zeros. To get the desired string with only one 1 in it, make the string $1000\ldots0$ of the proper length, and then move the 1 into position by repeated applications of (i).

We can generate a string of three ones by applying $10 \to 01 \to 111$. Now consider an arbitrary string with r ones and s zeros, where r is odd and greater than 1 and $s \geq 0$. Begin by generating a string of r consecutive ones by repeated application of (iii). Next add s zeros to the end of this string by repeated application of (ii). Finally, use repeated application of (i) to move the zeros and ones into the required positions.

For the final part, the string $\underline{11}0 \to 0000$ shows we can get a string of four zeros. Also, repeated use of

$$\underline{000}0 \to 1\underline{10} \to 1\underline{01} \to \underline{11}00 \to 00000$$

shows that we can make strings of zeros of arbitrary length greater than or equal to 4.

The string $1\underline{10} \to 1111$ shows we can get a string of four ones. Repeated use of $1\underline{10} \to 1\underline{01} \to 1100$ shows that we can generate strings with two initial ones followed by an arbitrary positive number of zeros. In particular, we can generate 11 followed by $3n$ zeros. Each of the n substrings, 000, can be replaced by 11, which shows we can get strings of ones of arbitrary even length.

Now consider an arbitrary string with r ones and s zeros, where r is even, $r \geq 4$, and $s > 0$. First generate a string of r ones, then add the appropriate number of zeros using repeated application of

$$11\underline{11} \to 1\underline{10}00 \to 1\underline{01}00 \to 11\underline{000}0 \to 11110$$

In the final step, use repeated application of (i) to properly position the zeros and ones.

R65. Yes, there's another solution.

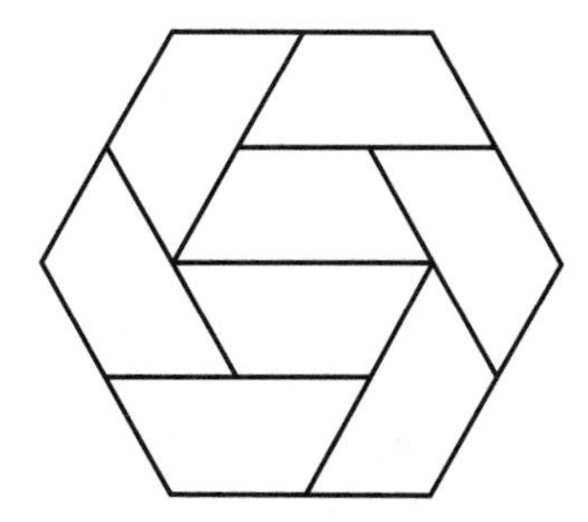

R70. In order to count the number of different solutions, we have to be very clear about what we mean by different. We also need to know something about the symmetries of the cube. We will count two solutions as being the same if we can rotate one into the other. If you want to count those as different, then you must multiply the numbers here by 24, which is the number of rotational symmetries of the cube. These are illustrated in figures (i), (ii), (iii). You can (i) rotate the cube through $90°$, $180°$ or $270°$ about an axis through the center of opposite faces (there are three such axes), or (ii) through $\pm 120°$ about a diagonal axis (there are four of these), or (iii) through $180°$ about an axis through midpoints of opposite edges (there are six of these), or (iv) you can leave it alone. This gives a total of $(3 \times 3) + (2 \times 4) + (1 \times 6) + 1 = 24$ rotational symmetries.

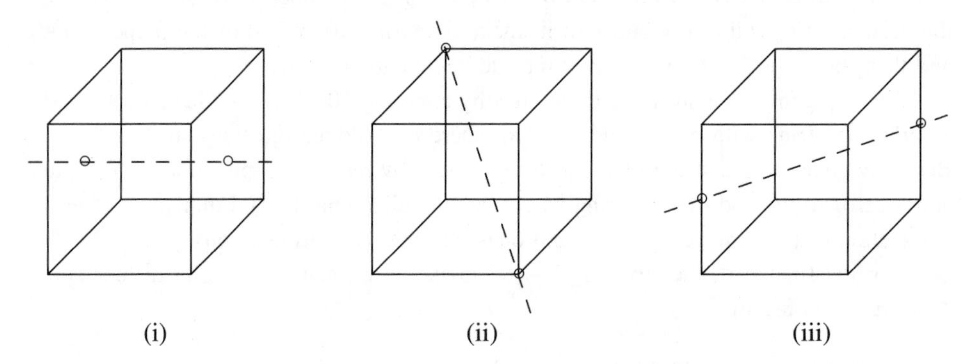

(i) (ii) (iii)

There are eight corner cells on the cube. It takes eight blocks to occupy these cells, and none of these eight blocks will occupy the *Den*, the middle cell of the cube. The piece that occupies the Den is the *Denizen*.

The Denizen has either an arm or an elbow in the Den. As we are not counting rotations as different, we can always rotate a solution so that the Denizen is either in the Throne position or in the Gallows position.

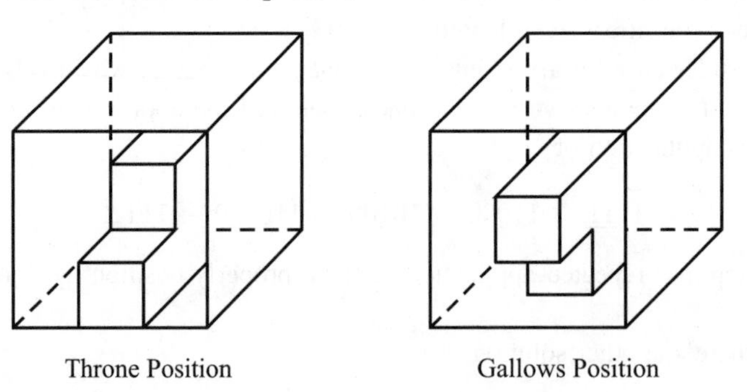

Throne Position Gallows Position

We can dispose of the solutions with the Denizen in the Gallows position by absorbing them among those with the Denizen in the Throne position. The piece under the Gallows is an *elbow* piece and occupies a corner; there are four possible positions for this piece. In each of these cases the Denizen and the elbow piece form a *triad* of three positions. For example, if we add the elbow as in figure (i), this gallows position can be rearranged as in (ii) or (iii). Figure (ii) already has the Denizen in the Throne position, and figure (iii) can be rotated into figure (iv), which is also a Throne position. That is to say, each triad gives three solutions, with types Throne – Gallows – Throne.

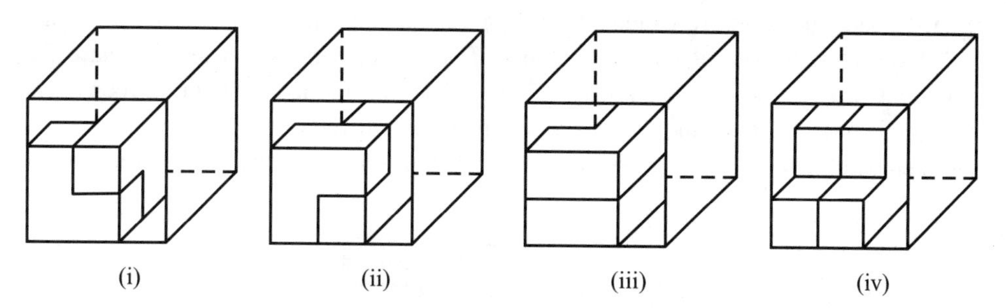

(i) (ii) (iii) (iv)

We shall count the solutions by the number of triads they contain. Omitting the extensive details, it turns out that there are 18 solutions with four triads, 70 solutions with three triads, 30 solutions with two triads, 72 solutions with one triad, and 32 solutions with no triads. This makes 222 solutions in all. Of these, 74 are Gallows solutions and 148 are Throne solutions. If we don't count reflexions as different, there are 111 different solutions, 37 Gallows and 74 Throne.

Details can be found in Richard K. Guy & Marc Paulhus, Nine Cubits or Simple Soma, *Coll. Math. J.*, **33**(2002) pp. 188–195.

R72. Here's one solution.

1	9	28	3	22	30	93
23	19	5	26	12	8	93
18	17	16	15	14	13	93
7	11	25	6	20	24	93
29	21	4	27	10	2	93
78	77	78	77	78	77	

R79. If the side of the square picture is s and the borders have widths a, a, a, b, then $s^2 + 652 = (s + 2a)(s + a + b)$ where the units are $\frac{1}{4}$-th of an inch.

The number 652 is 4 times 163, where -163 is a famous mathematical number, the discriminant of Euler's $n^2 + n + 41$ formula whose values have no prime factors smaller than 41. So, when we add squares to 652, we don't find many factorizations of the result, only a few that allow us to cut out the square, and only one (at the top of right half of the following table) whose borders agree with the conditions of the question.

s	s^2+652	a,a,a,b	s	s^2+652	a,a,a,b
2	$656 = 4 \times 164$	$1,1,1,161$	37	$2021 = 43 \times 47$	$3,3,3,7$
2	$656 = 8 \times 82$	$3,3,3,77$	39	$2173 = 41 \times 53$	$1,1,1,13$
2	$656 = 16 \times 41$	$7,7,7,32$	43	$2501 = 41 \times 61$	—
6	$688 = 8 \times 86$	$1,1,1,79$	49	$3053 = 43 \times 71$	—
6	$688 = 16 \times 43$	$5,5,5,32$	—	—	—
10	$752 = 16 \times 47$	$3,3,3,34$	80	$7052 = 82 \times 86$	$1,1,1,5$
14	$848 = 16 \times 53$	$1,1,1,38$	162	$26896 = 164 \times 164$	$1,1,1,1$

So the picture is a $9\frac{1}{4}$-inch square and the frame is $10\frac{3}{4} \times 11\frac{3}{4}$ inches.

R81. Each "lights-off" room must be visited an odd number of times and each "lights-on" room must be visited an even number of times (perhaps not at all). Because the final pattern is that of a checkerboard, Clicker must turn the lights off in 32 rooms and this requires at least 64 moves (one move to enter the room and one to leave). The checkerboard pattern can be achieved in 64 moves as shown below, so this is a shortest path.

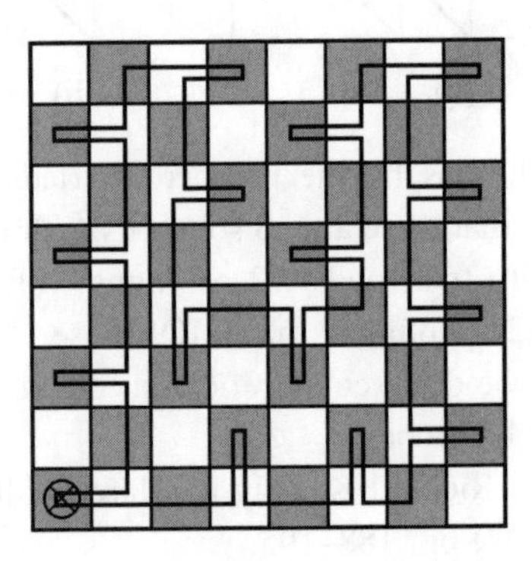

R84. We want to find solutions of $(10a+b)(100c+10d+e) =$ a 4-digit number, or a 5-digit one, whose digits f, g, h, i (and j) are all different. As the digits add to 45, a multiple of 9, we can work modulo 9 to get

$$(a+b)(c+d+e) \equiv f+g+h+i(+j) \equiv -a-b-c-d-e \pmod 9$$

$$(a+b+1)(c+d+e+1) \equiv 1 \pmod 9)$$

where a, b, c, e are never 0 and b is not 1. So, modulo 9, the pair $a+b$, $c+d+e$ must be $0,0$ or $1,4$ or $3,6$, or $7,7$ in some order, and the product $(a+b)(c+d+e)$ must be 0 or 4. Therefore the digits are to be chosen from one of the six rows

ab or *ba*		*c*, *d*, *e* (in some order)
18 27 36 45	with	018 027 036 045 126 135 189 234 279 369 459 468 567
19 28 37 46	with	013 049 058 067 139 148 157 238 247 256 346 589 679
12 39 48 56	with	015 024 069 078 123 159 168 249 258 267 348 357 456 789
13 49 58 67	with	019 028 037 046 127 136 145 235 289 379 469 478 568
15 24 69 78	with	012 039 048 057 129 138 147 156 237 246 345 489 579 678
16 25 34 79	with	016 025 034 079 124 169 178 259 268 349 358 367 457

Most cases are ruled out by coincidences. A complete search yields seven solutions for the first part and nine for the second:

$18 \times 297 = 5346$	$12 \times 483 = 5796$	$42 \times 138 = 5796$
$27 \times 198 = 5346$	$39 \times 186 = 7254$	
$28 \times 157 = 4396$	$48 \times 159 = 7632$	

$27 \times 594 = 16038$	$45 \times 396 = 17820$	$39 \times 402 = 15678$
$36 \times 495 = 17820$	$54 \times 297 = 16038$	$78 \times 345 = 26910$
$63 \times 927 = 58401$	$46 \times 715 = 32890$	$52 \times 367 = 19084$

R86. We'll count the essentially different ways of doing it. That is, if we can get one worm from another by rotating or reflecting the grid, or by swapping a worm's head with its tail, we'll say they are essentially the same placement.

If its tail is next to its head, we'll call it a *ringworm* and count it as many times as there are different ways of taking the two neighboring squares.

There are just two types of ringworm:

k	j	i	h
l	a	b	g
m	p	c	f
n	o	d	e

o	p	c	d
n	a	b	e
m	j	i	f
l	k	h	g

The first has left-right symmetry, so we count it nine times, for head-tail in positions ab, bc, cd, de, ef, fg, gh, hi, ij and the second has the symmetries of the rectangle and we count it only five times: ab, bc, cd, de, ef.

There are 23 different worms which are not ringworms (can you find them all?) giving a total of 37 worms altogether.

Rikki-Tikki-Tavi will want to find out how many ways an n^2 numberworm can be put into an $n \times n$ square.

And on page 21 of Martin Gardner, *The Incredible Dr. Matrix*, Charles Scribner's Sons, New York, 1967, 1976, there is a picture of a numberworm forming an **antimagic square**:

1	2	3
8	9	4
7	6	5

all the row sums, 6, 21, 18, column sums, 16, 17, 12, and diagonals 15, 19, are different. This is not a **pandiagonal** antimagic square, since the other (wrap-around) 'diagonals' 8—2—5, 6—4—1, 8—6—3, 7—2—4 give some duplicate totals. Rikki-Tikki-Tavi will want to find a pandiagonal antimagic square; and to find a magic ringworm—how small a one can there be?

R88. By filling in the path on those squares with only two outlets, beginning in the corners, we arrive at the path on the top. There are three options for the portion of the path through square b. If the path turns a corner in square b (there are two ways this can happen) then the path through a is forced (in each case) to connect with c, making a loop that doesn't pass through all the squares of the board. Therefore, the path through b is horizontal (enters and exits on the two vertical sides). Similarly, the path through square e is horizontal. Knowing this, the remainder of the path is uniquely determined using the principle that the path through a square with only two outlets is completely determined. The unique solution is shown on the bottom.

For Rikki-Tikki-Tavi. For each m and n, find the least number of blocked squares (guideposts) you need in a rectangle with m rows and n columns to force a unique rook circuit.

The best known results are due to Marc Paulhus who has kindly given us permission to reproduce them here.

It's not too hard, when you know the answers, to verify exact results which have a dimension less than 5, or one that is a multiple of 4.

Rectangles of size $1 \times n$ have no rook tours, and those of size $2 \times n$ have a unique tour.

For rectangles of size $3 \times 3n + j$ with $n > 0$ and $j = 0$ or ± 1 it is necessary and sufficient to use $n + j$ guideposts to enforce a unique tour. You can check this by noting that consecutive posts have to be on squares of opposite colors. One way of placing them is

$$\begin{array}{llllll} \text{for} \quad j = 0, & \text{put} \quad n & \text{guideposts at} & (3r+1, 1) & \text{for} & 0 \le r < n \\ \phantom{\text{for}} \quad j = 1, & \phantom{\text{put}} \quad n+1 & & (3r+1, 1) & & 0 \le r \le n \\ \phantom{\text{for}} \quad j = -1, & \phantom{\text{put}} \quad n-1 & & (3r, 1) & & 0 < r < n. \end{array}$$

Find the unique tours in the following 3×8, 3×9, 3×10 rectangles, where we've put ⊗ for a guidepost on a white square and ⊕ for one on a black square. Of course, there must be the same number of each kind, unless both dimensions are odd, when there is one more of the former.

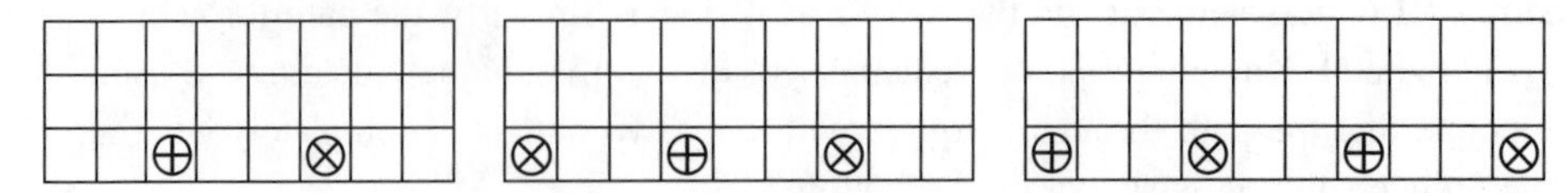

Rectangles of size $4 \times n$ only need 2 guideposts, if you place them at (1,1) and at $(n-1, 2)$ or $(n-1, 3)$ according as n is even or odd. Check these 4×10 and 4×11 rectangles.

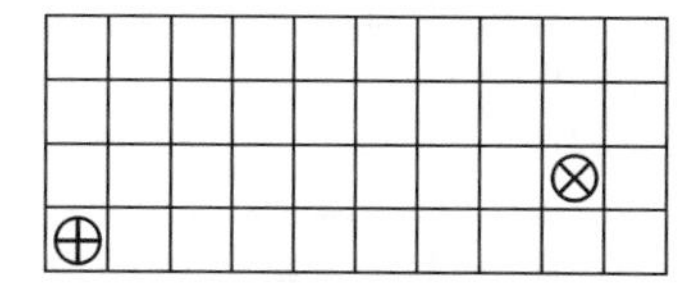

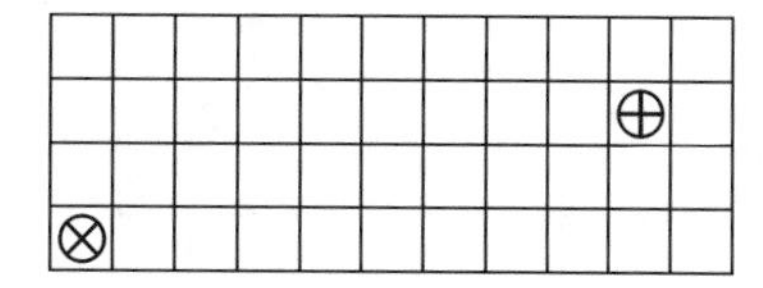

If one dimension of the rectangle is a multiple of 4, say $4m$, then $2m$ guideposts are seen to suffice, if you stack m copies of the $4 \times n$ examples that we've just seen. Check that the standard 8×8 chessboard has a unique rook circuit if 4 guideposts are placed as follows:

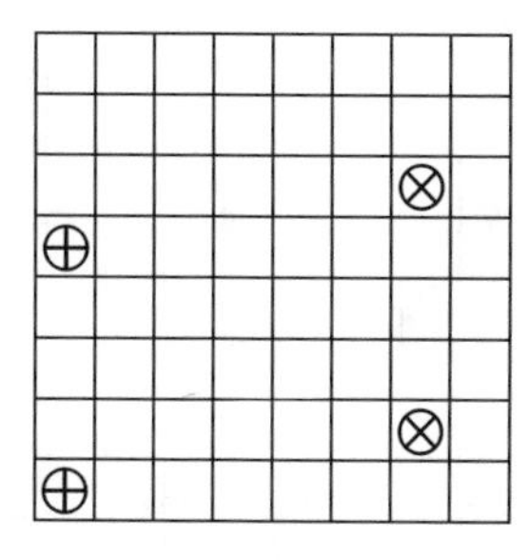

We are less certain if the dimensions of the rectangle are $4m + i$ by $4n + j$, with $m, n > 0$ and $i, j = 1$, 2 or 3. The least numbers of guideposts are believed to be

$$\begin{array}{lll} 2(m+n)-1 & \text{if} \quad ij = 1 & \text{first diagram} \\ 2(m+n) & \text{if} \quad ij \text{ is even} & \text{next three} \\ 2(m+n)+1 & \text{if} \quad ij = 3 \text{ or } 9 & \text{last row} \end{array}$$

The fact that these numbers suffice is illustrated by the following six diagrams, for $m = n = 2$ and $(i, j) = (1, 1)$, $(1, 2)$ (first row); $(2, 2)$, $(2, 3)$ (second row); $(3, 1)$, $(3, 3)$ (last row). For the cases $(i, j) = (2, 1)$, $(3, 2)$, or $(1, 3)$ reflect the appropriate diagram about its rising diagonal.

In each diagram the framed $(4m + i) \times 4$ and $4 \times (4n + j)$ rectangles contain 2 guideposts, one of each color. These may be deleted (in order to reduce n or m to 1) or duplicated (to cover cases where n or m is greater than 2) producing rectangles of different width or height. Enjoy verifying the uniqueness of the rook circuit in each diagram.

$(i, j) = (1, 1)$

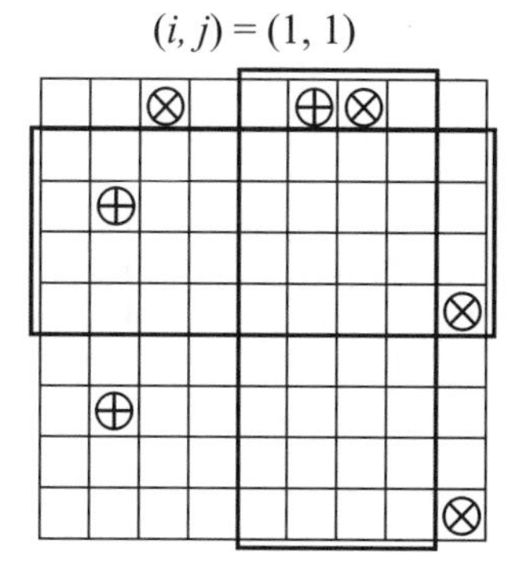

$(i, j) = (1, 2)$

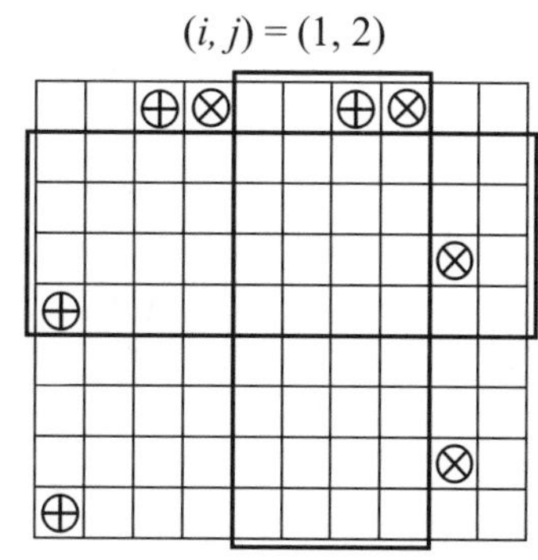

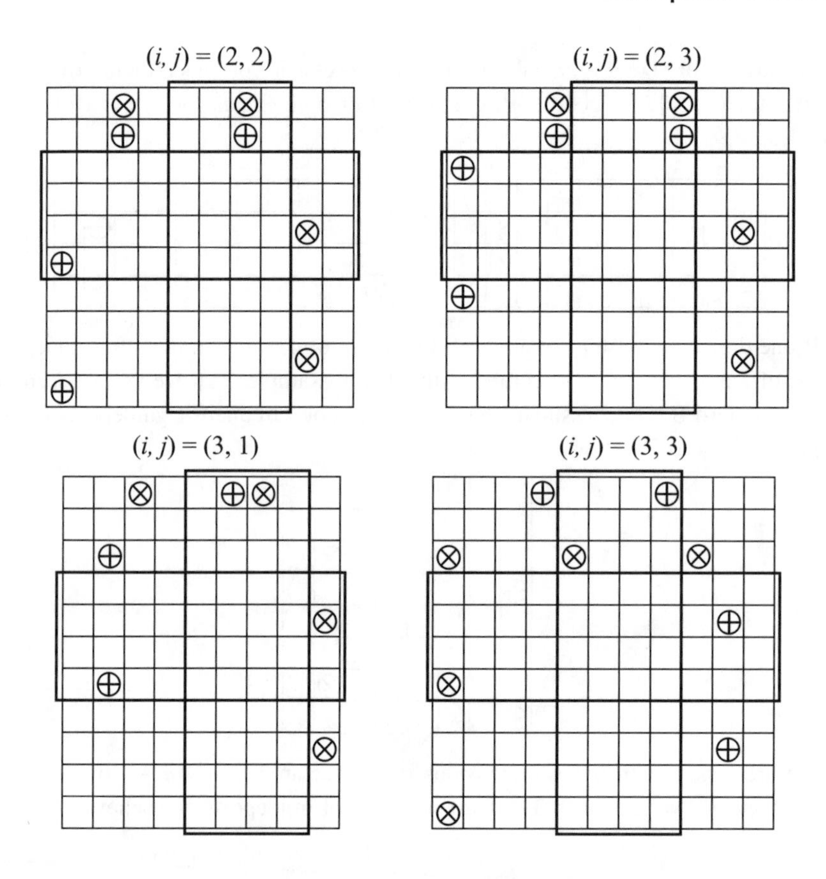

R89. We shall use the term "rook circuit" to mean a path of horizontal and vertical segments that passes through each square of the chessboard exactly once and returns to the starting square. We have seen that the number of horizontal segments in the circuit is equal to the number of vertical segments (because they alternate).

If we disregard rotations and reflexions, there are exactly 12 different *minimal* rook circuits (circuits with the minimal number of segments) on the 8×8 board. To construct these, it is helpful to know that in a minimal circuit, either every row contains a horizontal segment, or every column contains a vertical segment. For otherwise some row, say, would fail to contain a horizontal segment and then each square in this row must be visited by a vertical segment of the circuit.

It follows that a minimal circuit on the $2n \times 2n$ board will consist of $2n$ vertical segments and $2n$ horizontal segments. Here are the 12 minimal circuits for the 8×8 board—the shaded area is the inside of the region described by the circuit.

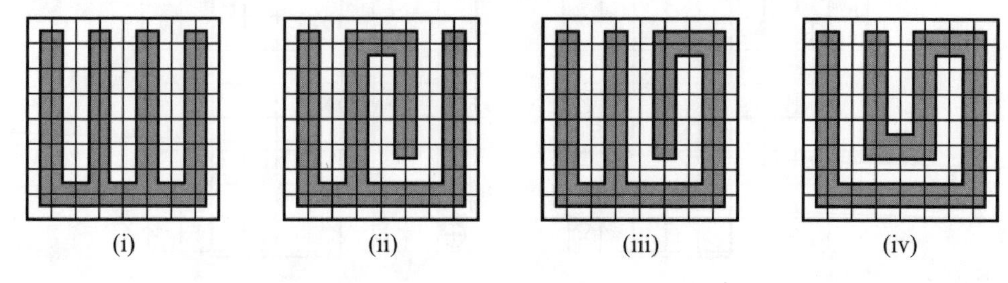

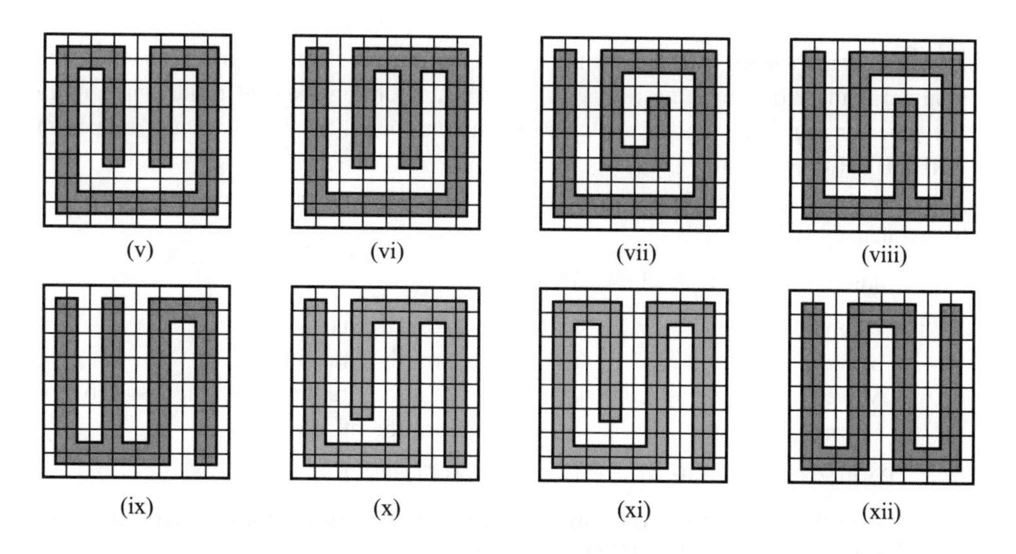

(v) (vi) (vii) (viii)

(ix) (x) (xi) (xii)

Rikki-Tikki-Tavi wants to know how many essentially different minimal rook circuits there are on a $2n \times 2n$ chessboard?

R90. Let 1 and 0 represent heads and tails, so that the arrangement HHTHHTH translates to 1101101. Flipping over the first five coins corresponds to **nim-addition** (see the Treasury): $1101101 \oplus 1111100 = 0010001$, (TTHTTTH).

The vertices in the following graph represent particular arrangements for the seven coins. For example, one can go from position 1110000 (3 heads) to position 1001110 (four heads), by nim-adding 0111110 (by turning the middle five coins).

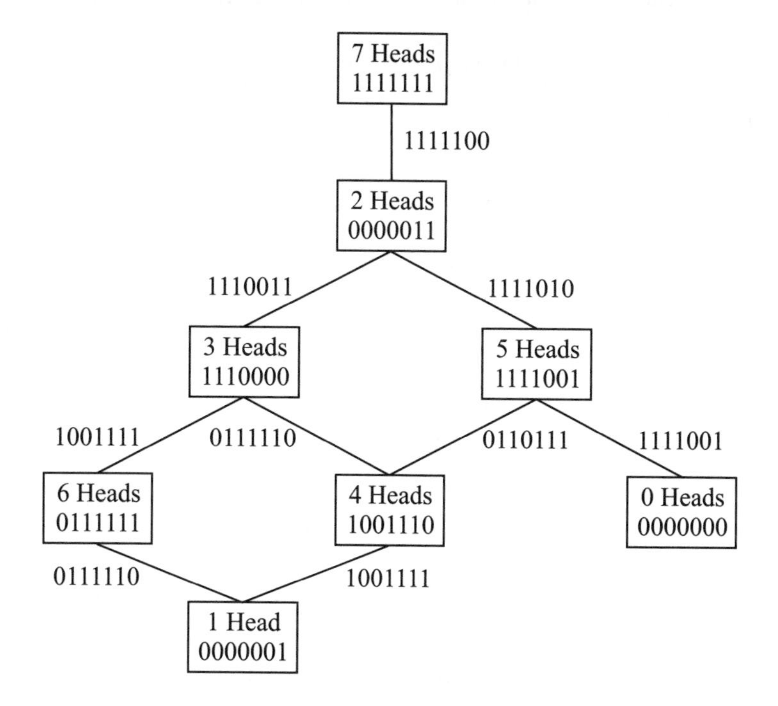

This graph can be used to find the solution to our problem, and more generally, the minimal number of moves necessary to go from any arrangement of heads and tails to any other arrangement. For example, what (minimal number of moves) will transform HHHHHHH to HTHTHTH?

The graph shows us how to get from seven heads to a particular instance of four heads, namely HTTHHHT.

Column #:		1	2	3	4	5	6	7	
Position		1	1	1	1	1	1	1	HHHHHHH
Move	$\oplus$	1	1	1	1	1	0	0	Turn the first five
Move	$\oplus$	1	1	1	0	0	1	1	Turn the first three and the last two
Move	$\oplus$	0	1	1	1	1	1	0	Turn the middle five
Position		1	0	0	1	1	1	0	HTTHHHT

Now exchange columns three and four, and six and seven, and the final sum gives the desired arrangement HTHTHTH:

Column #:		1	2	4	3	5	7	6	
Position		1	1	1	1	1	1	1	HHHHHHH
Move	$\oplus$	1	1	1	1	1	0	0	Turn the first five
Move	$\oplus$	1	1	0	1	0	1	1	Turn the first two, the last two, and the middle
Move	$\oplus$	0	1	1	1	1	0	1	Turn the second, third, fourth, fifth, and seventh
Position		1	0	1	0	1	0	1	HTHTHTH

How can we get (in a minimal number of moves) from, say, HHTHHTH to THHHHHH?

Let x represent the *nim-sum* of the moves needed to reach 0111111 from 1101101. Then

$$1101101 \oplus x = 0111111.$$

Adding 1101101 to each side gives

$$\begin{aligned} x &= 1101101 \oplus 0111111 \\ &= 1010010 \end{aligned}$$

It follows that $1111111 \oplus x = 1111111 \oplus 1010010 = 0101101$. This calculation shows that the same moves that transform 1101101 to 0111111 will transform 1111111 to 0101101. But, from previous work, we know how to find the fewest moves that will take HHHHHHH to THTHHTH (seven heads to four heads). The steps are shown below (the moves on the left come from the graph; the moves on the right are obtained by exchanging the first two columns and the last two columns):

Column #:		1	2	3	4	5	6	7
Position		1	1	1	1	1	1	1
Move	$\oplus$	1	1	1	1	1	0	0
Move	$\oplus$	1	1	1	0	0	1	1
Move	$\oplus$	0	1	1	1	1	1	0
Position		1	0	0	1	1	1	0

Column #:		2	1	3	4	5	7	6
Position		1	1	1	1	1	1	1
Move	$\oplus$	1	1	1	1	1	0	0
Move	$\oplus$	1	1	1	0	0	1	1
Move	$\oplus$	1	0	1	1	1	0	1
Position		0	1	0	1	1	0	1

The moves that constitute x are encoded in the calculations on the right: $x = 1111100 \oplus 1110011 \oplus 1011101 = 1010010$. Thus, the three moves that solve the problem are (in any order): (i) turn over the first five coins, (ii) turn over the first three and the last two, (iii) turn over the first, third, fourth, fifth, and seventh:

$$\text{HHTHHTH} \to \text{TTHTTTH} \to \text{HHTTTHT} \to \text{THHHHHH}.$$

R96. From the hint, $45 + (C + E + G)$ is a multiple of 4, so that $C + E + G = 7$, 11, 15, 19, or 23, and S is respectively, 13, 14, 15, 16, or 17. The smallest of these is 13, and the partitions of 13 into three distinct parts of size at most 9 are: 931, 841, 832, 751, 742, 652, and 643. Now $C + E + G = 7$, so C, E, G are 4, 2, 1 in some order. Only two of the partitions above, 841 and 742, have two parts in common with 4, 2, 1, and this forces exactly one solution (apart from rearrangements): $9+3+1 = 1+8+4 = 4+7+2 = 2+5+6$ ($E = 4$).

A keen reader will want to work out the other cases. For $S = 14$, E can be 1, 2, 3, 7 or 8; there are no solutions for $S = 15$; for $S = 16$, E can be 2, 3, 7, 8, or 9.

R99. Yes. Number the men from 1 to 16 according to the size of their galoshes, and let $a \to b$ mean that a takes b's galoshes, so that $a \leq b$. Suppose that $10 \to 16$, $9 \to 15$, $8 \to 14$, $7 \to 13$, $6 \to 12$, $5 \to 11$, $4 \to 10$, $3 \to 3$, $2 \to 2$, $1 \to 1$. Then 1 to 3 leave with their own galoshes, 4 to 10 with other people's galoshes, and 11 to 16 without. Galoshes 4 to 9 are left behind.

R102. No, knowing that every subset of eleven coins from her backpack necessarily contains at least three coins from the same country is not sufficient to conclude that 21 coins are from the same country. For example, one might have 20, 20, 20, 20, and 16 coins from five countries.

R104a. We can obtain any square where the sum of the numbers in the black squares is three more than the sum of the numbers in the white squares. For consider the square as shown below, where $a + c + e + g + k = b + d + f + h + 3$.

a	b	c
d	e	f
g	h	k

Now,

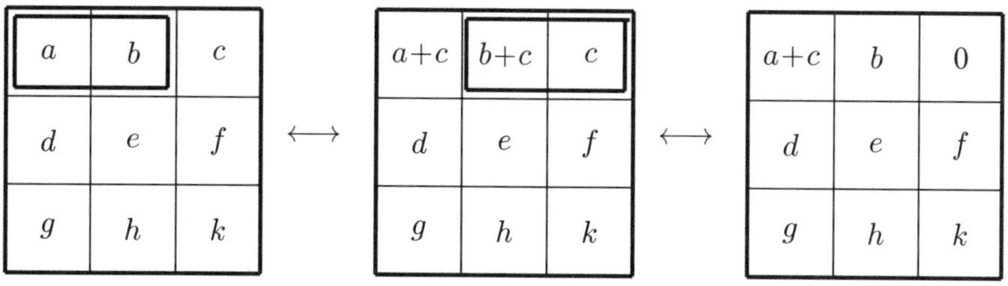

a	b	c
d	e	f
g	h	k

$\longleftrightarrow$

$a+c$	$b+c$	c
d	e	f
g	h	k

$\longleftrightarrow$

$a+c$	b	0
d	e	f
g	h	k

These steps demonstrate how we can "move" c to the top-left corner, and leave 0 in its place. With this same approach, we can move e, g, and k to the top-left corner, leaving 0 in their places; that is, we can get $a+c+e+g+k$ into the upper left square, and 0 in every other black square. In a similar way, we can get $b+d+f+h$ into the upper-middle square, and 0 in every other white square. From this position we can get 3 into the top-left square and zero everywhere else.

Thus, the original grid can be transformed to a grid with 3 in the top-left position and zero elsewhere. But, because the steps in the general procedure are reversible, we can get to any grid in which $a+c+e+g+k = b+d+f+h+3$.

R104b. Regardless of the order of the moves on the way to a minimal grid, we must reach a stage where the last entry in the middle row is 0,

a	b	0
c	d	0
x	0	y

where $x = 0, 1, 2, 3, 4$, or 5, and $y = 1, 2$, or 3.

We'll show that each of these possibilities for x and y can be achieved as part of a minimal square.

Case 1. Suppose $x+y > 3$. Then $b+c > a+d$ because $b+c = c+d+(x+y)-3$. Thus, we can reduce the square to the form

0	z	0
0	0	0
x	0	y

where $z = x+y-3$. There are 12 minimal squares of this form ($x = 1, 2, 3, 4, 5$, $y = 3$; $x = 2, 3, 4, 5$, $y = 2$; $x = 3, 4, 5$, $y = 1$).

Case 2. Suppose $x+y = 3$. Then $a+d = b+c$, and we can reduce the grid to

w	0	0
0	z	0
x	0	y

There are three grids of this form ($x = 0$, $y = 3$; $x = 1$, $y = 2$; $x = 2$, $y = 1$).

Case 3. Suppose $x+y < 3$. The $a+d > b+c$ and we can reduce the grid to the form

w	0	0
0	z	0
x	0	y

where $w + x + y + z = 3$, w, x, y, z are each 0, 1, or 2. There are 16 grids of this form (4 with exactly one of w, x, y, z equal to 0 and the others equal to 1, and 12 with one variable equal to 2 and another equal to 1).

Altogether, there are 31 minimal grids.

R108. As in the second solution, the millionth number will be the equivalent of 999999_{10}, which is 11333310_7, which is 11444410 when translated into our symbols.

R111. Yes, with an infinite collection of circles arranged in a honeycomb pattern. But, No! Such a construction can't be carried out with a *finite* collection, for, suppose that it can be done with c circles. Then, following the ideas of the original problem, there are $V = 6c/2 = 3c$ vertices and $E = 6c$ edges. Suppose that there are $F = c + t$ regions. Then, by **Euler's formula** (see the Treasury), $3c - 6c + c + t = 2$, so that $t = 2c + 2$. These t non-circular regions each have at least 3 edges, so $6c + 6 = 3t \leq E = 6c$, a contradiction.

Euler's formula can be used to investigate more general configurations. Call a set of circles in the plane an *n-system* if each circle touches just n others, either internally or externally. Apply Euler's formula to the "map" formed by such an n-system. Suppose there are c circles in the system. Each has n points of contact, dividing it into n arcs, a total of $E = nc$ edges. Each point of contact belongs to exactly 2 circles, so that there are $V = cn/2$ vertices. By Euler's formula, the number of regions is $F = E + 2 - V = cn/2 + 2$. These regions are either circles, c of them, or "star-shaped" regions between them, say s of them.

$$\frac{cn}{2} + 2 = c + s.$$

If the stars each have 3 "points," then the number of edges is $cn = 3s$,

$$\frac{cn}{2} + 2 = c + \frac{cn}{3}$$
$$n = 6 - \frac{12}{c}.$$

For example, $c = 3$, 4, 6, or 12 corresponds to $n = 2$, 3, 4, or 5 as shown below.

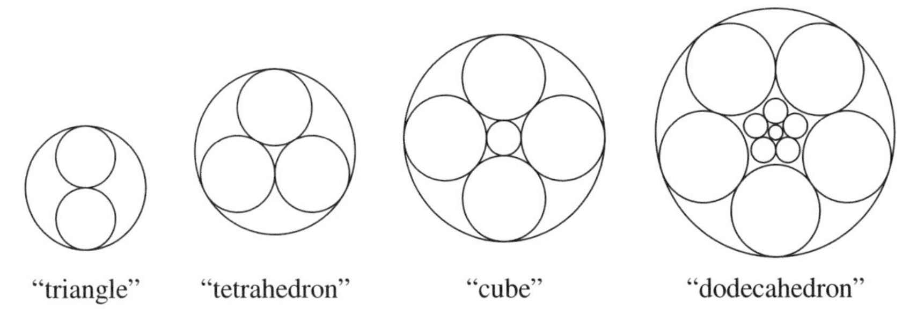

"triangle" "tetrahedron" "cube" "dodecahedron"

If the stars each have 4 points, the number of edges, $cn = 4s$ and $n = 4 - 8/c$; for example, $c = 4$ or 8 corresponds to $n = 2$ or 3 as shown.

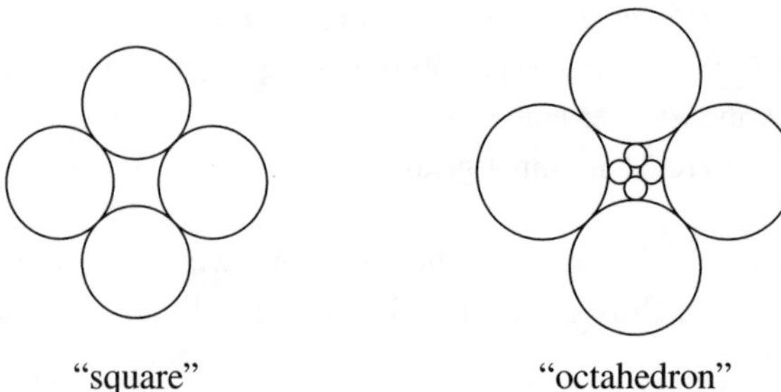

"square" "octahedron"

If the stars each have 5 points, $n = 10/3 - 20/3c$ (for example, $c = 5$ or 20 corresponds to $n = 2$ or 3, the pentagon and the icosahedron), whereas if the stars each have 6 points, $n = 3 - 6/c$ (for example, if $c = 6$ then $n = 2$, the hexagon).

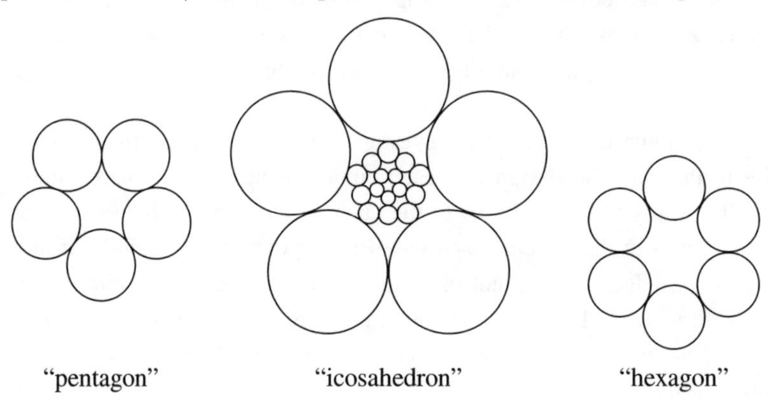

"pentagon" "icosahedron" "hexagon"

In the 3-point, 4-point and 6-point cases, we can also consider the case $c = \infty$, with $n = 6$, 4 and 3 respectively.

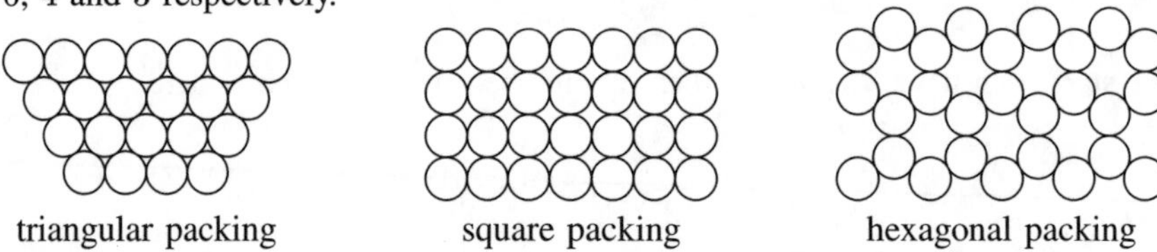

triangular packing square packing hexagonal packing

There is also the possibility of mixtures of stars; for example, with $n = 4$, $c = 8$, $s_4 = 2$, $s_3 = 8$, we get $E = 32, F = 18$ and $V = 16$ (the square antiprism), or with $n = 3$, $c = 12$, $s_3 = s_6 = 4$, we get $E = 36$, $V = 18$, and $F = 20$ (the truncated tetrahedron).

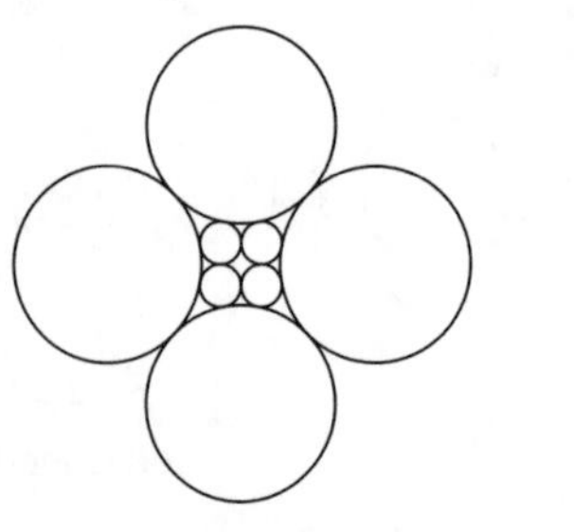

"square antiprism"
(2 squares, 8 triangles)

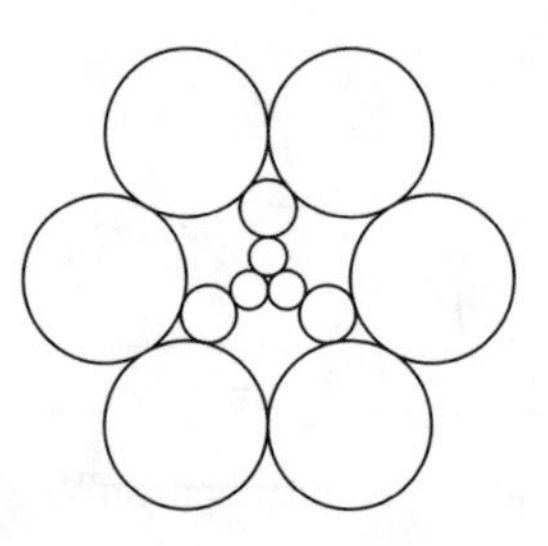

"truncated tetrahedron"
4 triangles, 4 hexagons

R113. For $n = 2$ we can, of course, use all the points, and for $n = 4$ we have seen that we can do so also. For $n \geq 6$ it is also possible. The examples in the following figure show how to increase n to $n + 4$ by drawing a border round a solution and connecting it by a bridge. For $n = 3$ it is only possible to use 7 out of the 9 points. For $n = 5$, one point has to be omitted; color the four corner dots and the midpoints of the sides black, and the rest, except for the middle, white. No two of the black dots can be connected to the same white dot without wasting a dot. So all edges meeting at a black dot are accounted for, and the edges at a midpoint must make an angle of 45° and not involve a boundary point, else it, or the adjacent corner, will be wasted.

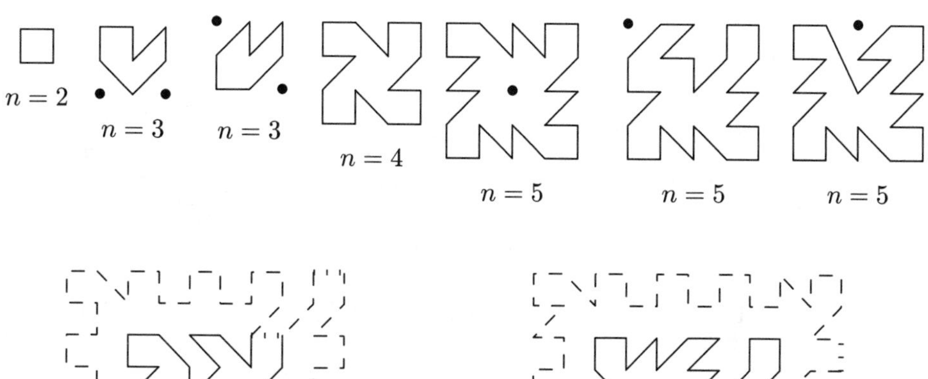

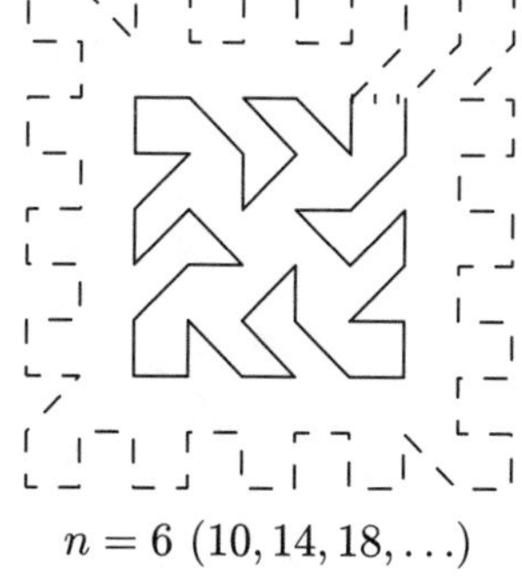

$n = 6\ (10, 14, 18, \ldots)$

$n = 7\ (11, 15, 19, \ldots)$

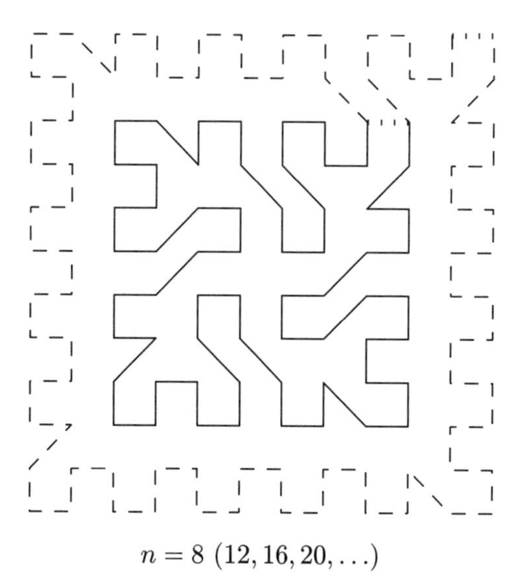

$n = 8\ (12, 16, 20, \ldots)$

$n = 9\ (13, 17, \ldots)$

R114. The same technique of bordering that we used in **S113** also works for triangular grids as shown in the following figures.

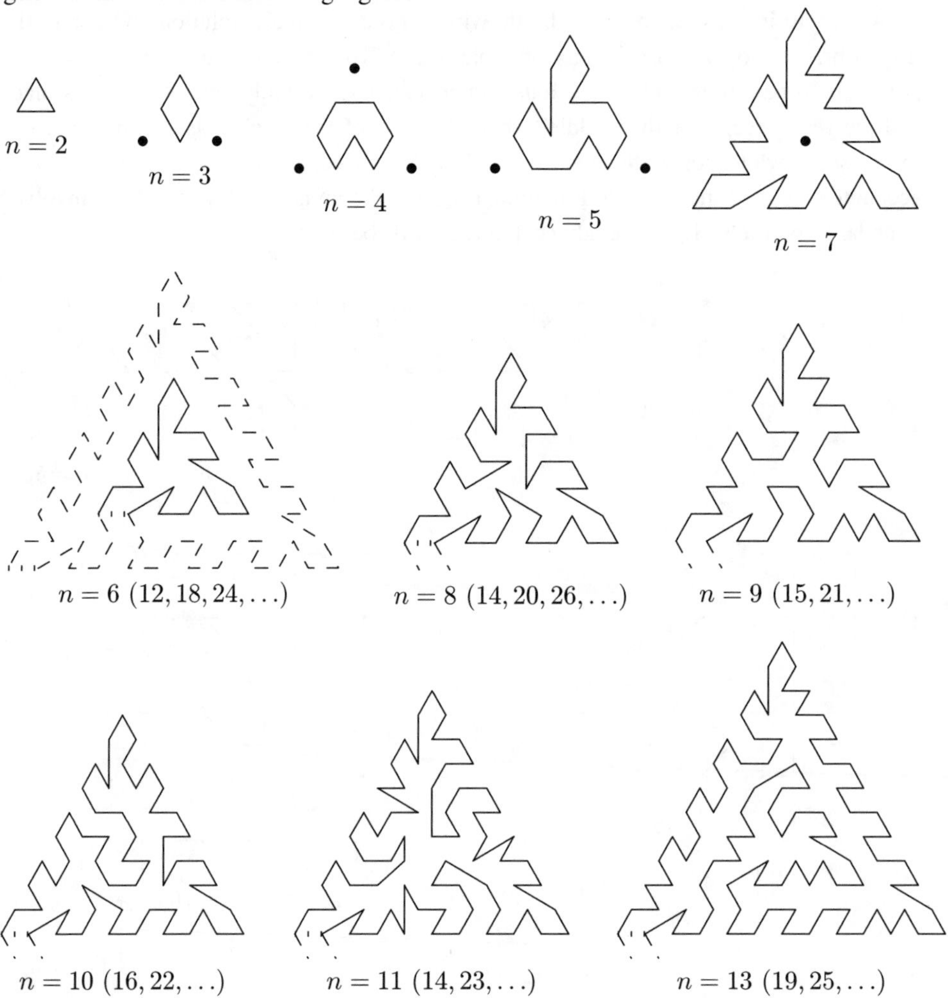

The hexagonal grid also can be handled by the bordering technique.

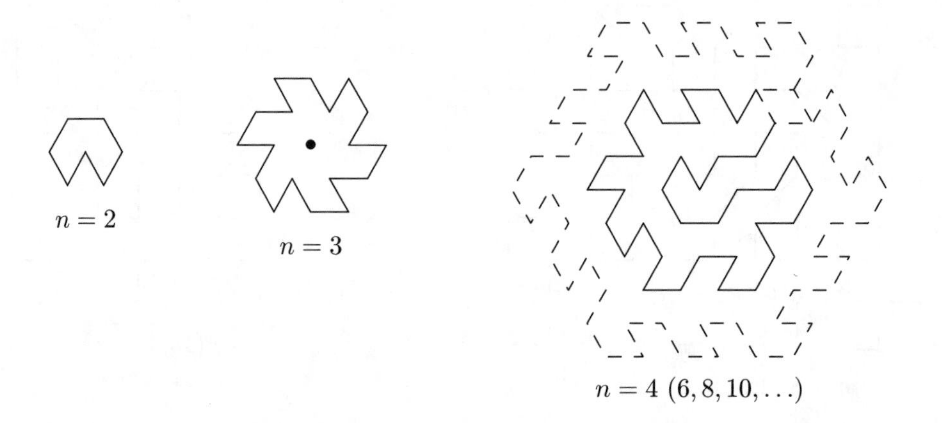

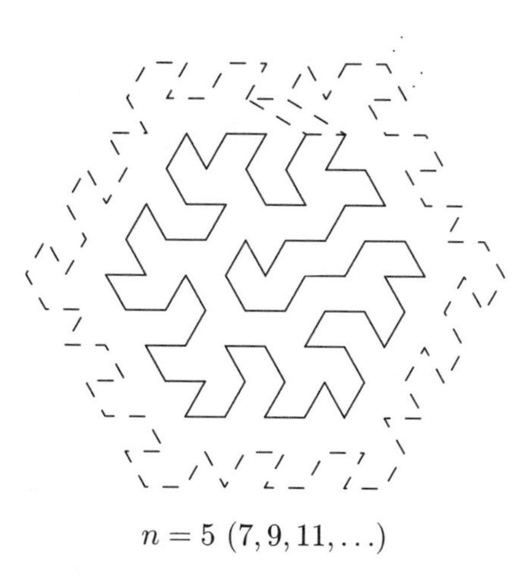

$n = 5\ (7, 9, 11, \ldots)$

The following table summarizes the investigations of **R113** and **R114**.

shape of grid	number of sides	adding a border increases n by	can use all points	but not for $n =$
triangular	3	6	for $n \geq 8$	$1, 3, 4, 5, 7$
square	4	4	for $n \geq 6$	$1, 3, 5$
hexagonal	6	2	for $n \geq 4$	$1, 3$

R117. First note that we can append any number of zeros to any solution that we find. Next, as the sums of the digits are equal, $2A \equiv A^2 \bmod 9$, so that $A \equiv 0$ or $2 \bmod 9$. We've already seen that $A = 0$, 2, and 9 are solutions. In addition, there are several families of solutions, for example, based on the number in the first column in the following table, we find the expression in the second column, having the values in the third column.

A	Formula	Other values
11	$10^n + 1,\ (n \geq 0)$	$101, 1001, 10001, \ldots$
18	$2 \cdot 10^n - 2,\ (n \geq 0)$	$198, 1998, 19998, \ldots$
29	$3 \cdot 10^n - 1,\ (n \geq 0)$	$299, 2999, 29999, \ldots$
38	$4 \cdot 10^n - 2,\ (n \geq 0)$	$398, 3998, 39998, \ldots$
45	$5 \cdot 10^n - 5,\ (n \geq 0)$	$495, 4995, 49995, \ldots$
47	$5 \cdot 10^n - 3,\ (n \geq 0)$	$497, 4997, 49997, \ldots$
99	$10^n - 1,\ (n \geq 0)$	$999, 9999, 99999, \ldots$
119	$12 \cdot 10^n - 1,\ (n \geq 0)$	$1199, 11999, 119999, \ldots$
146	$15 \cdot 10^n - 4,\ (n \geq 0)$	$1496, 14996, 149996, \ldots$
189	$2 \cdot 10^n - 11,\ (n \geq 0)$	$1989, 19989, 199989, \ldots$
245	$25 \cdot 10^n - 5,\ (n \geq 0)$	$2495, 24995, 249995, \ldots$
362	$38 \cdot 10^n - 18,\ (n \geq 0)$	$3782, 37982, 379982, \ldots$
452	$47 \cdot 10^n - 18,\ (n \geq 0)$	$4682, 46982, 469982, \ldots$
488	$5 \cdot 10^n - 12,\ (n > 0)$	$4988, 49988, 499988, \ldots$
1098	$11 \cdot 10^n - 2,\ (n \neq 1)$	$10998, 109998, 1099998, \ldots$
1145	$115 \cdot 10^n - 5,\ (n \geq 0)$	$11495, 114995, 1149995, \ldots$

There are also the sporadic solutions $144, 182, 335, 344, 351, 369, 459, 461, 468, 479, 639, 729, 794, 839, 848, 929, 954, \ldots$, which do not appear to belong to a family. These, and the numbers in the table, include all values of A less than 1000.

R119. Yes, there are 449 ways to tile the 4×10 board with two sets of tetrominoes. None of the solutions exhibit any symmetry (with respect to a rotation about the center, or with respect to a reflexion about a line through the center).

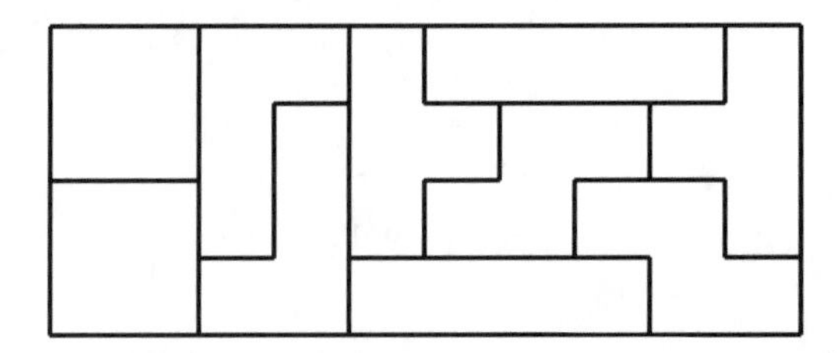

R120. We've seen that in each case there must be at least one pair of parallel lines. We can show that, in fact, in each case there must be exactly one pair of parallel lines.

Consider the first case. If more than one of the lines t_2, t_3, t_4 were parallel to lines s_2, s_3, s_4, s_5, these lines would intersect in ≤ 10 intersection points, and the ninth line can't intersect more than 8 other lines, so there would be fewer than 19 intersection points with only two lines through them, a contradiction. If none of the lines t_2, t_3, t_4 is parallel to any of s_1, s_2, s_3, s_4, and if the ninth line is not parallel to one of them, there would be $\leq 12 + 8 = 20$ intersection points with only two lines through them, a contradiction.

In the second case, if more than one of the lines t_1, t_2, t_3, t_4 is parallel to lines s_2, s_3, s_4, s_5, these lines would intersect in pairs in ≤ 18 points, contrary to assumption.

R122. It could be that $(C, B, K) = (35, 38, 39) \to (0, 73, 4) \to (2, 71, 2) \to (0, 73, 0)$. But John could be more greedy with $(C, B, K) = (35, 38, 39) \to (16, 19, 58) \to (35, 0, 39) \to (16, 19, 20) \to (7, 10, 29) \to (16, 1, 20) \to (7, 10, 11) \to (2, 5, 16) \to (7, 0, 11) \to (2, 5, 6) \to (0, 3, 8) \to (3, 0, 5) \to (0, 3, 2) \to (2, 1, 0) \to (1, 0, 1) \to (0, 1, 0)$. In fact, he could leave any odd number of baseball cards between 1 and 73.

R124. One construction is to follow the pattern established in the following examples for $n = 2, 4, 6, 8$.

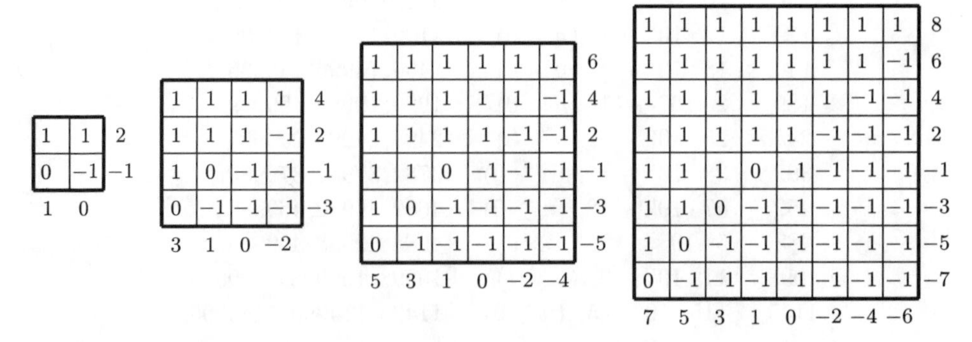

R125. One construction is to follow the pattern established in the following example for the 5×6 board.

1	1	1	1	1	−1	4
1	1	1	1	−1	−1	2
1	1	1	−1	−1	−1	0
1	1	−1	−1	−1	−1	−2
1	−1	−1	−1	−1	−1	−4
−5	−3	−1	1	3	5	

We've seen that the $n \times n$ grid with entries -1, 0, 1 can have distinct row and column sums when n is even. But what about the odd case? It's not hard to show there is no solution for $n = 3$, 5, 7, but larger values aren't so immediate. For example, it's not clear in the grid below that the numbers can't be shunted around so that the 2 and a 0 couldn't be replaced by a couple of 1's.

1	1	1	1	1	1	1	1	1	9
1	1	1	1	1	1	1	1	0	8
1	1	1	1	1	1	1	0	0	7
1	1	1	1	1	0	0	0	0	5
1	1	2	1	0	0	−1	−1	−1	2
0	1	0	−1	−1	−1	−1	−1	−1	−5
1	0	−1	−1	−1	−1	−1	−1	−1	−6
1	−1	−1	−1	−1	−1	−1	−1	−1	−7
−1	−1	−1	−1	−1	−1	−1	−1	−1	−9
6	4	3	1	0	−1	−2	−3	−4	

To help resolve this question, we posted it on the Web where it drew considerable response. Exhaustive computer searches showed there were no solutions for the cases $n = 9, 11, 13, 15, 17$. Finally, several days later, Fred Galvin posted a beautiful proof that there could be no solution when n is odd. His proof was extended by Michael Kleber and others. Their theorem is quite striking:

An $n \times m$ matrix (that is, a grid of n rows and m columns), $n \leq m$, with entries -1, 0, 1 and all row and column sums distinct, exists only when

(i) *$n = m$ and is even, in which case the missing sum is n or $-n$, or*

(ii) *$m = n + 1$, in which case the missing sums are m and $-m$.*

R130. We can fence any rectangular shape $a \times b$, where $a \geq 3$ (and $a + b = 473$). For example, for $3 \leq a \leq 22$, (and $b = 473 - a$), we can use lengths 1 and $a - 1$ at one end and a at the other, and $23-a, 32, 33, \ldots, 43$ and the rest at the other two sides, except for $a=12$ and $a=22$. In these cases we can use 2 and $a-2$ in place of 1 and $a-1$. For the near square (236×237), we can use 32, 39, 40, 41, 42, 43 and 24, 33, 34, 35, 36, 37, 38 for the sides, and 17, 23, 25, 26, 27, 28, 29, 30, 31 and $1, 2, \ldots, 15, 16, 18, 19, 20, 21, 22$ for the ends. The cases $23 \leq a \leq 235$ are not difficult because there are plenty of small pieces.

R138. There are 56 ways of arranging the squares into a sheaf of six. We will denote a fold in the left vertical line by LU and LD, for up and down folds, respectively. Similarly, RU and RD will denote up and down folds about the right vertical axis, and BU, BD, will denote up and down folds of the bottom half about the horizontal line through the middle. See the next page for a description of the following table.

LU, RU, BU	546312	LD, RD, BD	213645
LU, RU, BD	312546	LD, RD, BU	645213
LU, RD, BU	654123	LD, RU, BD	321456
LU, RD, BD	123654	LD, RU, BU	456321
RU, LU, BU	564132	RD, LD, BD	231465
RU, LU, BD	132564	RD, LD, BU	465231
RU, LD, BU	456321*	RD, LU, BD	123654*
RU, LD, BD	321456*	RD, LU, BU	654123*
LU, BU, RU	365412	LD, BD, RD	214563
LU, BU, RD	541236	LD, BD, RU	632145
LU, BD, RU	631254	LD, BU, RD	452136
LU, BD, RU_T	163254	LD, BU, RD_T	452361
LU, BD, RD_T	125634	LD, BU, RU_T	436521
LU, BD, RD	125463	LD, BU, RU	364521
RU, BU, LU	145632	RD, BD, LD	236541
RU, BU, LD	563214	RD, BD, LU	412365
RU, BD, LU	413256	RD, BU, LD	652314
RU, BD, LU_T	341256	RD, BU, RD_T	652143
RU, BD, LD_T	325416	RD, BU, LU_T	614523
RU, BD, LD	325641	RD, BU, LU	146523
BU, LU, RU	361452	BD, LD, RD	254163
BU, LU, RU_T	136452	BD, LD, RD_T	254631
BU, LU, RD	145236	BD, LD, RU	632541
BU, LD, RU	365214	BD, LU, RD	412563
BU, LD, RD_T	521364	BD, LU, RU_T	463125
BU, LD, RD	521436	BD, LU, RU	634125
BU, RU, LU	143652	BD, RD, LD	256341
BU, RU, LU_T	314652	BD, RD, LD_T	256413
BU, RU, LD	365214*	BD, RD, LU	412563*
BU, RD, LU	145236*	BD, RU, LD	632541*
BU, RD, LD_T	523146	BD, RU, LD_T	641325
BU, RD, LD	523614	BD, RU, LU	416325

There are six ways to arrange the folds: LRB, LBR, RLB, RBL, BRL, and BLR. For each of these we have the options of making an up fold or a down fold. In addition, as we have seen in the previous problem, there are some cases when a fold can be tucked into a pocket, and when this is done, we will add a subscript T on that fold. The preceding table lists all possible folds and the arrangements they produce. Asterisks (*) indicate repetitions.

> **Rikki-Tikki-Tavi** took an $m \times n$ piece of paper, divided it into mn unit squares, labelled the squares in the top row from left to right, 1 to n, from $n+1$ to $2n$ in the second row, $2n+1$ to $3n$ in the third, and so on. By making $(m-1)+(n-1)$ folds about the $m-1$ horizontal lines and $n-1$ vertical lines, a stack of mn squares in a sheaf is formed. By varying the sequence of folds and tucks, in how many different orders can the squares be arranged? We saw that the answer was 56 in the 2×3 case, and for 2×2 the answer is 8. The $1 \times n$ case is the classical stamp-folding problem.

R140. The sum of all the numbers with eight digits is their average value multiplied by the number of terms,

$$\left(\frac{10000000 + 99999999}{2}\right) \times (9 \times 10^7) = 109999999 \times 45 \times 10^6.$$

We now want to sum the 9^8 eight-digit numbers which do not contain a zero, and we'll do this by considering each place in turn. First, among these numbers, 9^7 start with a 1 (in the ten-millions place), 9^7 start with a 2, 9^7 start with a 3, and so on, up to 9. Therefore, the sum of the first digits is $9^7 \times 1 + 9^7 \times 2 + \cdots + 9^7 \times 9 = 45 \times 9^7$.

Using the same argument, the sum of the second digits (the millions place) is 45×9^7, and the same for the sum of the digits in the other six places. Therefore the sum of the eight-digit numbers which do not contain a zero is

$$\begin{aligned} T &= (45 \times 9^7) \times 10^7 + (45 \times 9^7) \times 10^6 + \cdots + (45 \times 9^7) \times 10^0 \\ &= 45 \times 9^7 \times 11111111 \end{aligned}$$

Now $9^7 \times 45 \times 11111111$ divided by $45 \times 10^6 \times 109999999$ is 0.483128184, which is less than 1/2. Consequently, the eight-digit numbers without a zero add up to less than those that have zeros!

R141a. Obviously, the units digit must be even and divisible by 5, so it must be 0; the tens digit must be even. The last four digits must be divisible by 3 (add to a multiple of 3) and 8. The final eight digits must be divisible by 7, and once they are set in place the rest is guaranteed. There are many solutions; one such is 9875643120.

R141b. It is clear that $b_5 = 0$ or 5. Other conditions are that b_2, b_4, b_6, and b_8 are even, $b_1 + b_2 + b_3 \equiv 0 \pmod 3$, $b_3 + b_4 + b_5 \equiv 0 \pmod 4$, $b_7 + b_8 + b_9 \equiv 0 \pmod 9$, $b_6 + b_7 + b_8 \equiv 0 \pmod 8$, and $b_1 b_2 b_3 b_4 \equiv b_5 b_6 b_7 \pmod 7$. By systematically considering the cases, it can be shown that 0 must be used as one of the digits. Then it is not difficult to find examples; one such is 783204165.

R141c. From $C_5 = 0 \pmod 5$, we conclude that $c_9 = 0$ or 5, but not the latter because C_2 is even. Then, because $1+2+\cdots+9 = 45$ is divisible by 9 and $c_1+c_2+\cdots+c_8 \equiv C_9 \equiv 0 \pmod 9$, the missing digit must be 9. Because C_3 and C_6 are divisible by 3, c_3 and c_6 must be 3 and 6 in some order. With this beginning it is easy to find all the solutions:

173856420	156473280	753216480	756813240
873156420	876513240	716543280	516783240
713526480	176583240	746153280	586713240
273156480	756183240		

R143. On the third day, the boys played $(15+22+25)/2 = 31$ games. Because each boy played at least once in every pair of successive games, Victor must have played (and lost) in each even numbered game (in particular, Victor lost the sixth game). The other two boys must have played against each other in the odd-numbered games, including the antepenultimate game.

R146. You may not be able to get more than nine of the ten dimes. You can check this yourself by considering the distribution of coins shown below.

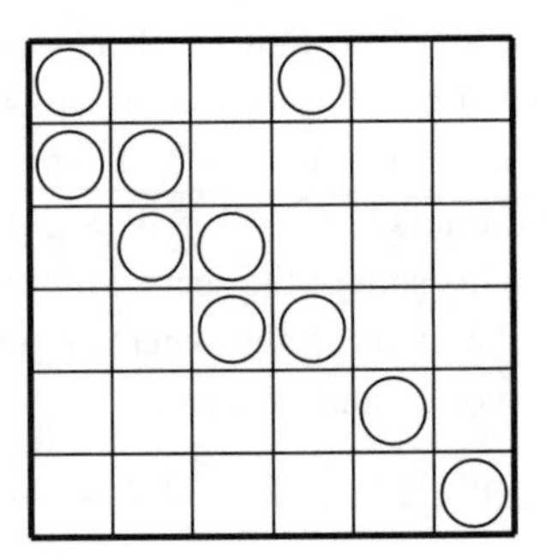

For Rikki-Tikki-Tavi. I have placed a dime in k of the mn squares of an $m \times n$ board. You may choose r rows and c columns and take all the coins you find in them. What are the minimal values of r and c for which you can be assured of getting all k coins, independent of how I distribute them?

R157. If there are more than 4 board members the numbers of men and women need not be equal.

W

M W

W W

W W

M

R159. Yes, you can specify all three weighings before you start; namely, weigh AB against CD, AC against BD, AD against BC.

Four coins contain at least one dud. If they contain both, then (just) one of the three weighings will balance, and the duds are the pair in one scale or the other of the balanced weighing. If AC $=$ BD, then the duds are either A and C, or B and D. Which they are depends on the other two weighings. There are four cases:

If AB $>$ CD and AD $>$ BC then A is heavy and C is light.
If AB $>$ CD and AD $<$ BC then B is heavy and D is light.
If AB $<$ CD and AD $>$ BC then D is heavy and B is light.
If AB $<$ CD and AD $<$ BC then C is heavy and A is light.

This accounts for 12 of the twenty (5×4) cases. In the other eight, there will be no balance, and E is one of the duds. In the three weighings, just one of the coins will always be on the heavy side, or on the light side. For example, if AB $>$ CD, AC $<$ BD, AD $<$ BC, then B is heavy and E is light.

If we don't know that the combined weight of the two duds is equal to two good coins, then a weighing may separate the 20 cases into two tens, which cannot be sorted by two more weighings, which can have at most $3^2 = 9$ outcomes.

R161. If the number of the teams participating in the tournament is only 7, the same conclusion cannot be made, as shown in the following table in which each team has won 3 games (a 1, or a 0, in row i and column j, indicates that team i beat, or lost to, respectively, team j).

	1	**2**	**3**	**4**	**5**	**6**	**7**	**Score**
1	–	1	1	0	1	0	0	3
2	0	–	1	1	0	1	0	3
3	0	0	–	1	1	0	1	3
4	1	0	0	–	1	1	0	3
5	0	1	0	0	–	1	1	3
6	1	0	1	0	0	–	1	3
7	1	1	0	1	0	0	–	3

R162. Each pentomino covers squares that add to 15, and so covers an odd number of odd numbers. Have fun checking that this is the only solution:

5	2	3	2	0	4	0	5	4	2
2	3	4	1	9	1	4	3	3	1
4	6	6	1	2	5	2	1	5	6
7	0	4	2	1	3	2	4	0	3
2	0	2	4	7	4	0	8	1	4
2	4	6	2	1	2	1	6	0	2

Without rotations and reflexions, the 6×10 rectangle can be tiled with the twelve different pentominoes in 2339 ways. Most of the solutions come in sets which have certain sub-blocks that can be exchanged or reflected into each other. For example, in this solution (disregarding the "sum" requirement), the N and Y can be interchanged to give a different solution; or the W and F. Also, the W, F, and L can be rearranged to give another solution. Combinations of these changes will generate seven other solutions (that is, each solution "belongs" to a set of eight others). It turns out that 2339 solutions can be reduced to 804 essentially different sets.

The twelve pentominoes will also tile a 3×20 rectangle (in 2 ways, 1 set), a 4×15 rectangle (in 368 ways, 19 sets), a 5×12 rectangle (in 1010 ways, 33 sets). All the solutions are described in detail in a colorful Web site produced by Adrian Smith:

`http://www.snaffles.demon.co.uk/pentanomes/pentanomes.html`

For an introduction to the enormous literature about pentominoes, see:

`http://sue.scs.uvic.ca/ cos/inf/misc/PentInfo.html`

The classic reference to polyominoes is Solomon Golomb's book, *Polyominoes*, Scribners, New York, 1965.

R163. Not counting rotations and reflexions as different, there are 10 different solutions. These can be grouped into 4 different classes, with those in the same class distinguished from each other by internal rotations, reflexions, or swaps (the shaded areas show the region that has been changed from a preceding solution in its class).

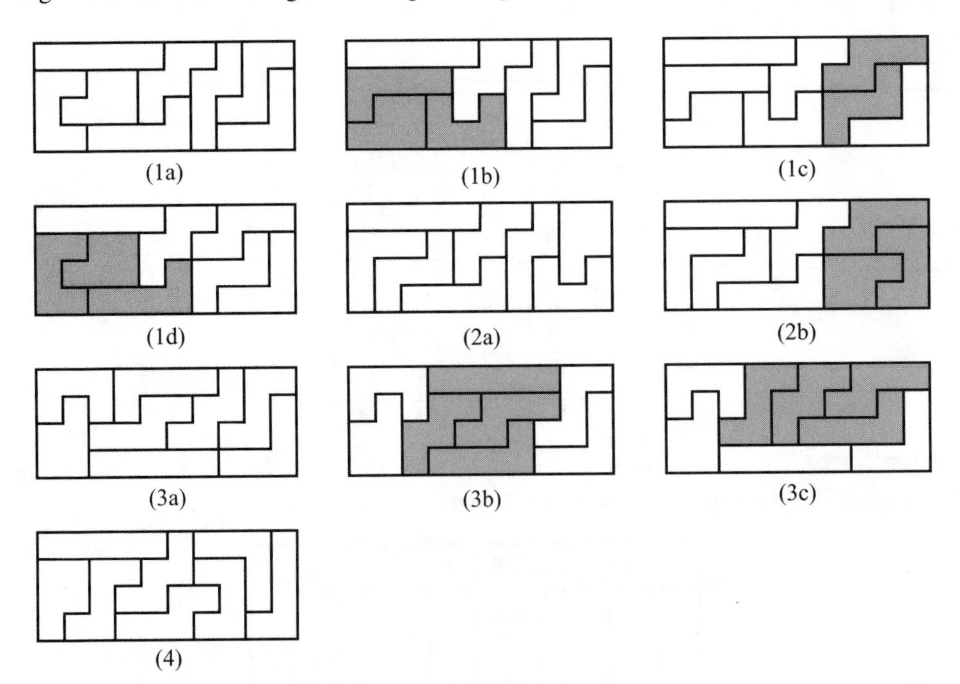

R165. We can assume that $A \leq B \leq C$. It's clear that $A+B+C$ must be a multiple of 6. It is also necessary that $C \leq 3A$, $C \leq A+B$, and $B+C \leq 5A$, because if C is all threes, A must have at least as many ones, $A+B$ will also be all threes, whereas if

A is all ones, $B+C$ will be all fives. Thus A, B, C satisfy the following properties P:

$$\begin{array}{ll} \text{(i)} & 0 < A \leq B \leq C \\ \text{(ii)} & A+B+C \text{ is a multiple of } 6 \\ \text{(iii)} & C \leq \min\{3A, A+B, 5A-B\} \end{array}$$

We claim that with the single exception of 2, 2, 2, these conditions are also sufficient. We'll prove these by induction on the sum $A+B+C$.

Suppose that P holds for A, B, C where $A+B+C=6$. Then A, B, C is 2, 2, 2 or 1, 2, 3. The second of these can be attained as a stack, but not the first. Suppose that P holds for A, B, C, where $A+B+C=12$. If $A=1$, the condition $C \leq 3A$ implies $C \leq 3$, but then $A+B+C \leq 1+3+3=7<12$, a contradiction. Therefore $A \geq 2$. Also $A \leq 4$, and the only solutions to $A+B+C=12$ with $2 \leq A \leq 4$ are $2,2,8$; $2,3,7$; $2,4,6$; $2,5,5$; $3,3,6$; $3,4,5$; $4,4,4$. The condition $C \leq 3A$ eliminates the first two of these, and the last five are attainable as a stack of two triangles.

Now suppose that P holds for A, B, C where $A+B+C=6k$, $k>2$.

Case 1. $A=B=C$. Here $3A=A+B+C=6k$, so A is an even integer larger than 2, and we've already shown that such a triple is attainable (see the solution to the original problem).

Case 2. $A=B<C$. Suppose $C=A+d$, $d>0$. Then $A+B+C=3A+d=6k$, so d is a positive multiple of 3. Therefore $C \geq A+3$, so $A-3 \leq C-6$. It is easy to check that P holds for $A-3, B-3, C-6$, so by induction, it is attainable as a stack of triangles. Adding $1,2,3$ and $2,1,3$ to the stack gives A,B,C.

Case 3. $A<B=C$. Suppose $C=A+d$, $d>0$. Then $A+B+C=3A+2d=6k$, so d is a positive multiple of 3. Therefore $C \geq A+3$, so $A-1 \leq C-3$. Again it is easy to show that $A-1, B-3, C-2$ satisfies P, so by induction, it is attainable, and therefore, so is A,B,C.

Case 4. $A<B<C$. Here it is easy to show that $A-1, B-2, C-3$ satisfies P, so it is attainable by induction, and so is A,B,C.

This completes the induction, and the proof.

If the conditions for a solution A,B,C are satisfied, there will be nonnegative integers a,b,c,d,e,f such that

$$a(1,2,3)+b(1,3,2)+c(2,1,3)+d(2,3,1)+e(3,1,2)+f(3,2,1)=(A,B,C),$$

or equivalently,

$$a+b+2c+2d+3e+3f=A$$

$$2a+3b+c+3d+e+2f=B$$

$$3a+2b+3c+d+2e+f=C$$

With some algebraic manipulation, these equations can be put into the form

$$a=2d+2e+3f+(C-A)-(A+B+C)/6$$

$$b=-2d-e-2f-C+(A+B+C)/2$$

$$c=-d-2e-2f+A-(A+B+C)/6$$

To find particular solutions to the problem, we need only find nonnegative d, e, f, that will make a, b, c nonnegative as well.

R176. There are 30 different "elevator sequences" that start at the basement level.

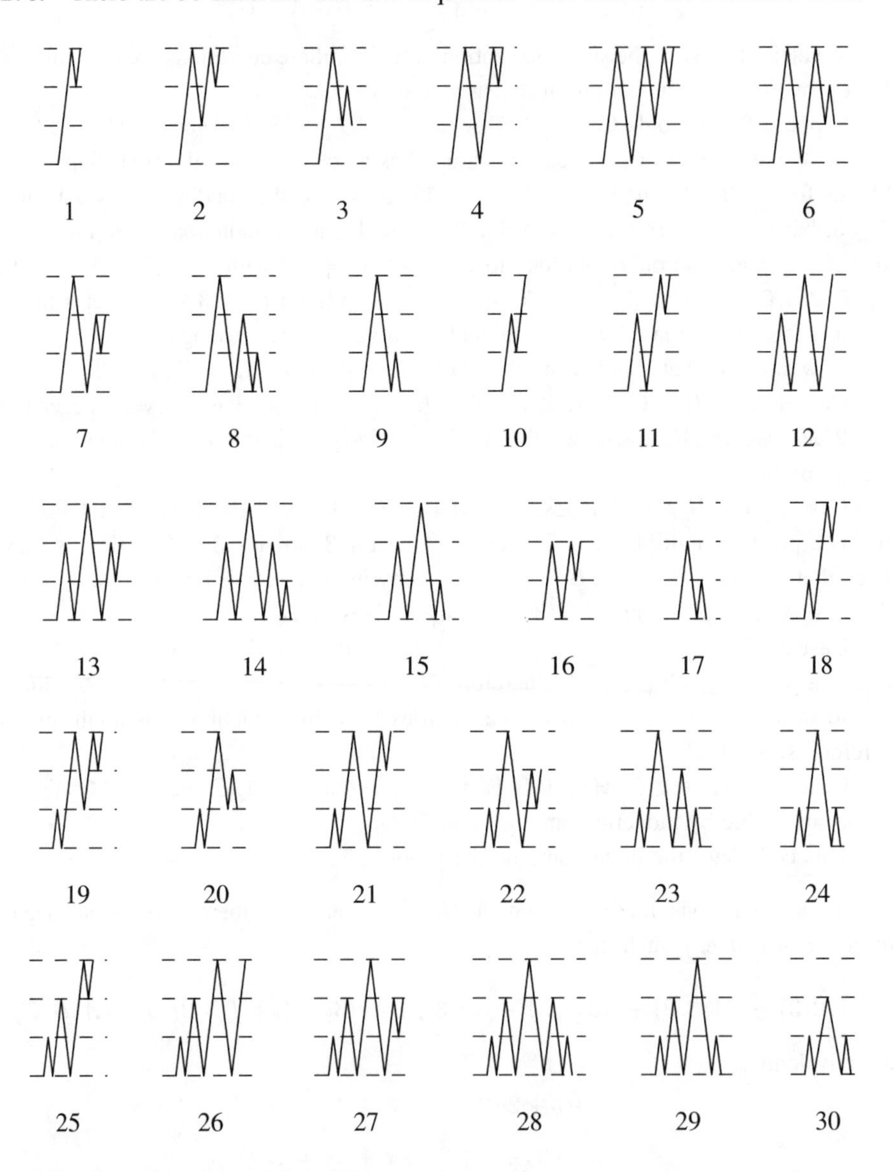

On the facing page is a tree that shows the result of pushing A and B buttons to produce the elevator sequences. A move to the left corresponds to an A button, to the right is a B button. The numbers at the ends of the branches correspond to the order in which the sequences are generated using the backtracking algorithm (the order of the sequences in the preceding display).

For Rikki-Tikki-Tavi. Find a general formula for the number of elevator sequences for a building with k floors?

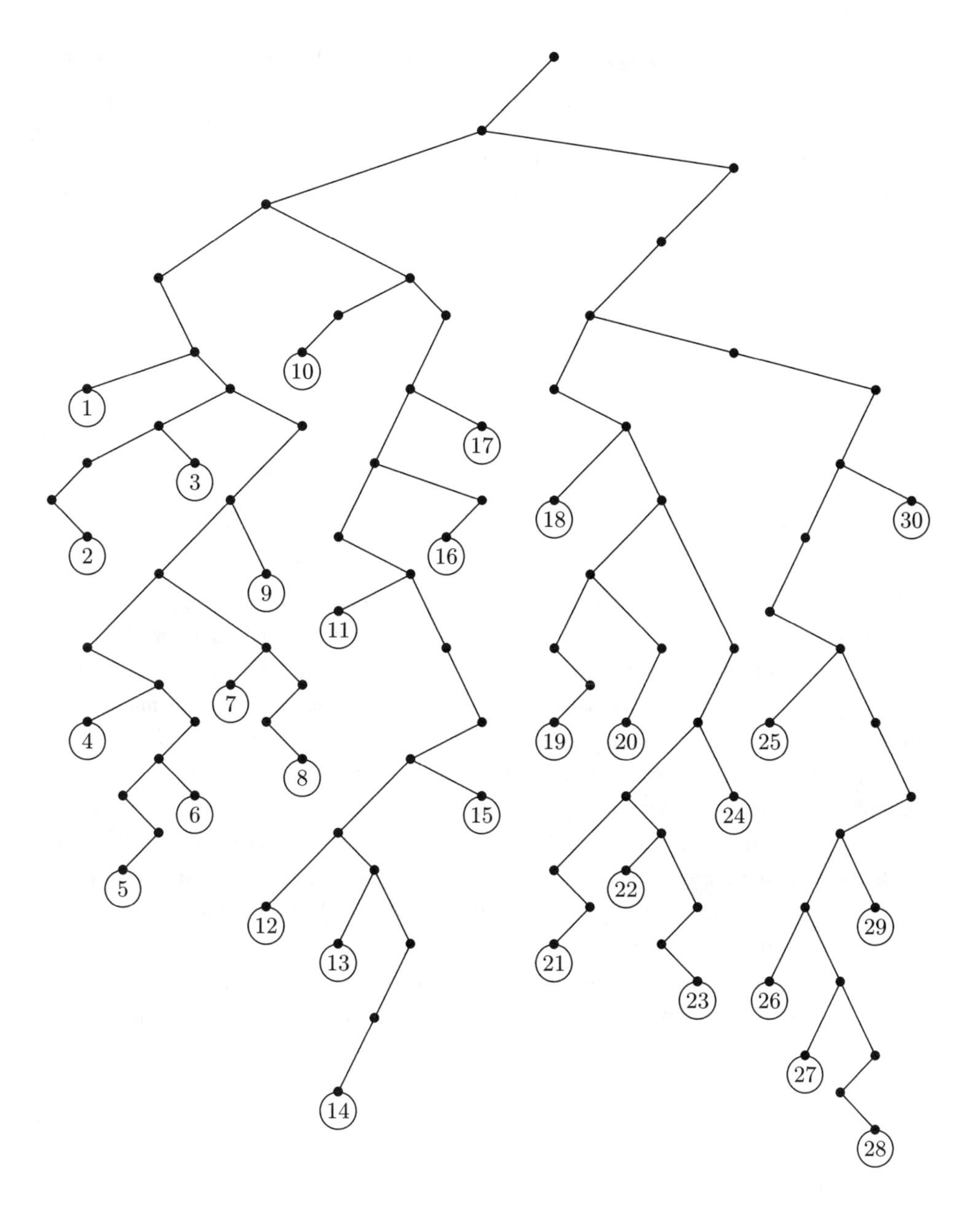

R177. Let $\mathcal{S}$ be the set of all grids whose squares in the bottom row and rightmost column are white. There are $2^9 = 512$ grids in $\mathcal{S}$, since each of the other nine squares can be black or white.

It is clear that a grid not in $\mathcal{S}$ is equivalent to some grid in $\mathcal{S}$, for by choosing row operations we can make the last column completely white, and then using column operations we can make the bottom row completely white. The resulting grid is an element of $\mathcal{S}$.

We'll now prove that no two grids in $\mathcal{S}$ are equivalent. First, it should be clear that row and column operations on a grid can be carried out in any order with the same result. This is because the final color of a square only depends on how many times it is changed, and it doesn't matter, for that square, whether it's changed by a row operation or by a column operation.

So suppose we have a grid, say A, and we perform a row operations on row 1; b, c, and d row operations on rows 2, 3, and 4, respectively; e, f, g, h column operations on columns $1, 2, 3, 4$, respectively; and that with these operations, we reach another grid, say B, which is also in $\mathcal{S}$. That is,

A =

	a	b	c	d	
4	a1	a2	a3		h
3	a4	a5	a6		g
2	a7	a8	a9		f
1					e
	1	2	3	4	

$\Longleftrightarrow$

	a	b	c	d	
4	b1	b2	b3		h
3	b4	b5	b6		g
2	b7	b8	b9		f
1					e
	1	2	3	4	

= B

where the letters in the grids denote the colors of the squares (either black or white).

Because the lower-right corner remains white under the transformation, $d + e$ is an even number. Similarly for the squares $(4,2), (4,3), (4,4)$ and $(1,1), (2,1), (3,1)$ which show that $d+f, d+g, d+h$ and $a+e, b+e, c+e$ are also even; that is, $x+e$ and $d+y$ are even, where $x = a$ or b or c and y is f or g or h. Thus, $(x+e)+(d+y)+(d+e)$ is even, so $x + y$ is even for any of the other nine squares, which means that the corresponding squares in the two grids have to have the same color.

The preceding also shows that each element, A, of $\mathcal{S}$ is equivalent to $2^7 = 128$ grids, because there are seven squares in the bottom and rightmost columns that can be independently colored either black or white depending on which of the four row and three column operations one makes on A.

Here's another way to think about the problem. Number the squares of the board from 1 to 4 across the top row, 5 to 8, 9 to 12, in the next two rows, and 13 to 16 in the bottom row. We can represent the coloring of the board as a 16-tuple, $(x_1, x_2, \ldots, x_{16})$, where x_i is 0 or 1 depending upon whether square i is white or black, respectively. For example,

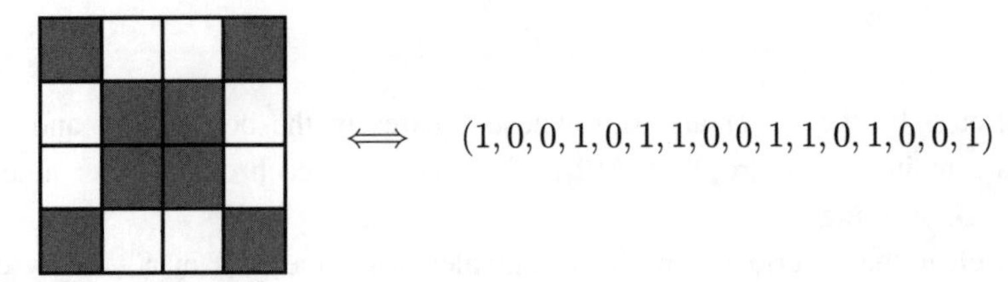

The different colorings of the board correspond to the 2^{16} 16-tuples of 0's and 1's.

In a similar way, we can represent the row and column operations by 16-tuples of 0's and 1's. Specifically, represent the row operation that changes the colors in the top row by $(1,1,1,1,0,0,0,0,0,0,0,0,0,0,0,0)$, the operation that changes colors in the first column by $(1,0,0,0,1,0,0,0,1,0,0,0,1,0,0,0)$, and so forth. The point of this representation is that the board coloring that results from applying operation x to board y is just the nim-sum of x and y, $x \oplus y$ (see Nim-sum in the Treasury).

For readers familiar with group theory, these 2^{16} board colorings (16-tuples of 0's and 1's) form a group $\mathcal{G}$ under nim-addition, and the 8 basic operations generate a subgroup $\mathcal{H}$ with $2^7 = 128$ elements (as a vector space, $\mathcal{H}$ has dimension 7). Two boards are equivalent if and only if they are in the same coset of $\mathcal{H}$, and the number of essentially different grids is equal to the number of elements in the factor group $\mathcal{H}/\mathcal{G}$, which is $2^{16}/2^7 = 2^9$. (For an $n \times n$ board the number of essentially different grids is $2^{n^2}/2^{2n-1} = 2^{(n-1)^2}$.)

R178. Let $S(n)$ be the smallest nonnegative integer that can be extracted by this process when starting with $1, 2, \ldots, n$. Here is a table of minimal values for $n = 1, 2, \ldots, 20$.

n	$S(n)$	Example
1	1	
2	3	
3	0	$((1,2),3)$
4	16	$(((1,2),3),4)$
5	15	$((((2,3),5),1),4)$
6	63	$(((1,4),(3,5)),(2,6))$
7	8	$(((((4,5),6),(2,7)),1),3)$
8	0	$((((4,5),7),(2,6)),((1,3),8))$
9	3	$(2,(1,(((6,7),((3,4),8)),(5,9))))$
10	1	$(((((((4,5),9),6),(8,10)),2),3),7),1)$
11	0	$((((((3,7),(9,11)),6),(8,10)),(1,2)),(4,5))$
12	0	$((((((1,3),7),(8,10)),(((5,6),9),(11,12))),2),4)$
13	1	$((((((((3,7),(9,11)),6),(8,10)),5),(12,13)),2),4),1)$
14	3	$(((((((5,10),(11,14)),(7,8)),((4,6),(12,13))),3),9),1),2)$
15	0	$((((((1,4),(7,8)),((2,5),(10,11))),((3,6),(13,14))),9),(12,15))$
16	4	$(((((0,9),(12,15)),4),16),2)$, where the 0 comes from combining $0 = ((1,3),8) = ((6,7),13) = ((5,10),(11,14))$
17	3	$((((((0,6),(8,10)),3),9),1),2)$, where $0 = ((4,7),(16,17)) = (((13,14),15),((11,12),5))$
18	3	$(((((6,7),11),(8,10),((0,3),(16,17))),1),2)$, where $0 = ((4,12),(14,18)) = (5,9),(13,15))$
19	4	$(((0,4),(3,5)),2)$, where $0 = ((1,6),(17,18)) = ((8,13),(16,19)) = (((11,12),(14,15)),((9,10),7))$
20	0	$(((((0,2),3),6),12),(0,5))$ where $0 = ((1,4),(7,8)) = ((9,15),(16,20)) = (((11,14),(17,18)),((10,13),19))$

These results were obtained by Dean Hickerson, with some help from David Wilson and Michael Kleber. Hickerson then extended the table and proved that the answers from 8 onwards are periodic with period twelve! (Dean Hickerson & Michael Kleber, "Reducing a set by subtracting squares," *Journal of Integer Sequences*, **2** (1999), Article 99.1.4 (html document) (electronic))

R181. The tiling can still be carried out in exactly the same way. Given the quadrilateral,

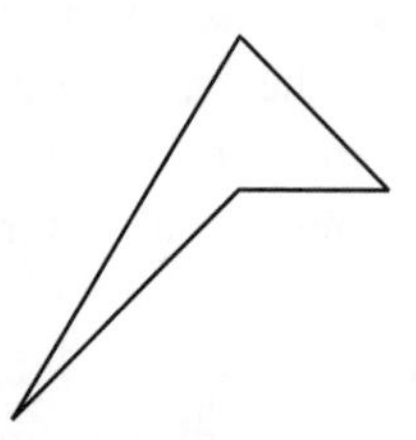

make a hexagon by gluing together two of the given quadrilaterals along a common edge, where one of them has been rotated through 180 degrees about the midpoint of an edge, and then fill the plane by translating the hexagon accordingly.

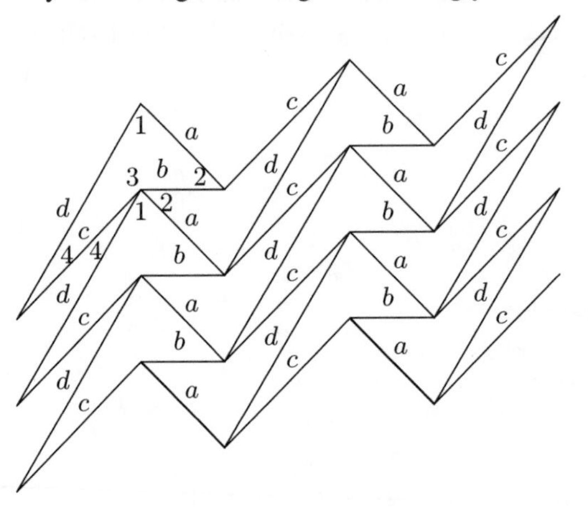

The quadrilateral can even be crossed,

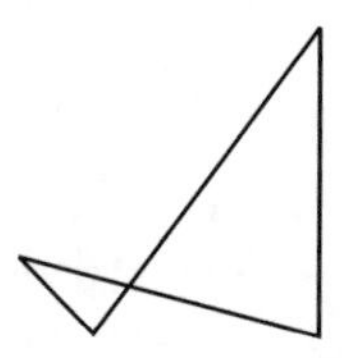

provided that the areas of the two parts are considered to be of opposite sign. If the larger part is positive, then the plane is covered just once positively (as illustrated in the following figure).

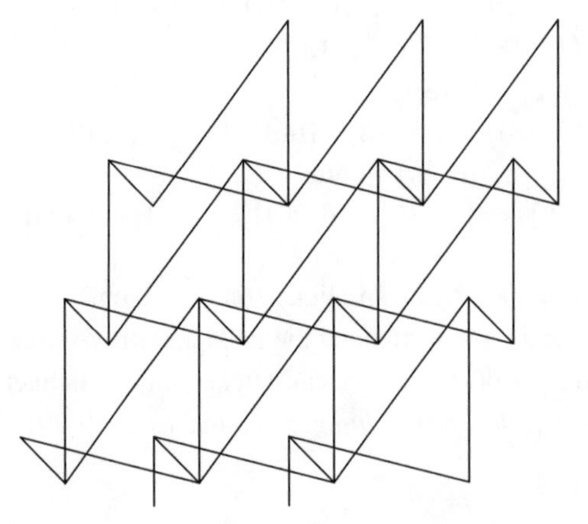

If the areas of the two crossed parts are equal, so that the total area of the quadrilateral is zero, then, when you attempt to tile, the quadrilaterals just form a strip with zero area; that is, any area that appears to be covered is in fact covered the same number of times negatively as positively.

R182. In base 2 the sum of the digits of a square can be any number:

$$(2^n-1)^2 = 2^{n+1}(2^{n-1}-1)+1 = \underbrace{111\ldots1}_{n-1}\underbrace{000\ldots0}_{n}1_2$$

so 17592186044415^2, when written in base 2, has digital sum 44.

In base 3, you can verify that $(3^n-1)^2$ and $(3^n-2)^2$ are respectively

$$\underbrace{222\ldots2}_{n-1}1\underbrace{000\ldots0}_{n-1}1_3 \quad \text{and} \quad \underbrace{222\ldots2}_{n-2}12\underbrace{000\ldots0}_{n-2}11_3$$

with digital sums $2n$ and $2n+1$. For example, 31381059608^2, when written in base 3, has digital sum 44.

But in base 4, we can 'cast out threes', just as we cast out nines in base 10 (see **Divisibility tests** in the Treasury) and discover that the digital sum can only be of shape $3n$ or $3n+1$, and not 44, for example.

In base 5 the sums are of shape $4n$ or $4n+1$. For example, 9765622^2 is

$$4444444434000000014_5$$

with digital sum 44.

In base 6 digital sums can be of shape $5n$, $5n+1$ or $5n+4$. For example, 1679608^2 is 555553200000144_6 with digital sum 44 again.

Base 7 (and 13, 16, ...) is like base 4 and base 10, and does not permit sums $3n+2$. For bases 8 and 9 there are the examples

$$524283^2 = 3777754000031_8 \quad \text{and} \quad 177146^2 = 88888300001_9$$

both with digital sum 44.

R187. It's clear by checkering either board that each piece has to make an even number of moves. A pair of pieces (for example, a and A) can't swap in 4 $(=2+2)$ moves because they would need to use the same intermediate (greek) square. So the number of moves is at least $4\times(2+4)=24$. In fact, in order to get at least two of the four corner pieces out of their corners, we must use the middle greek bridges, β and γ, three times each, twice in one direction, and once in the other direction. This means that at least four pieces will have to make an extra side step, back and forth, (to allow for meeting or passing, the latter to exchange the order of the pieces traversing the bridge in the same direction), entailing an extra eight moves, a total of at least 32.

There are several ways of doing it in 32 moves. In this solution the pairs A-a and D-d swap in the minimum of 6 moves each, but B-b and C-c each take two extra moves.

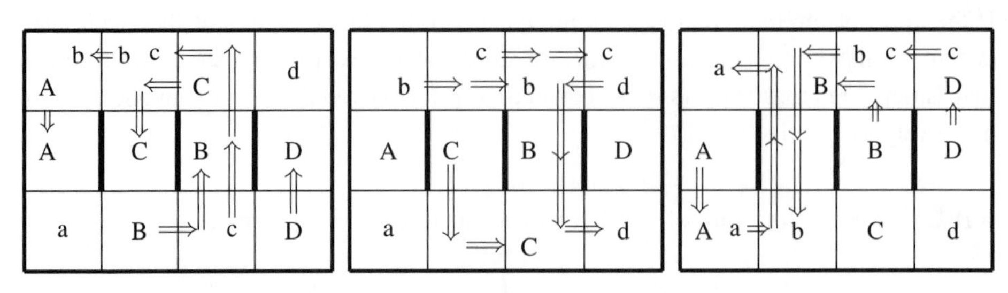

First 10 moves Second 10 moves Final 12 moves

Variations are easy to find; here's another, slightly different.

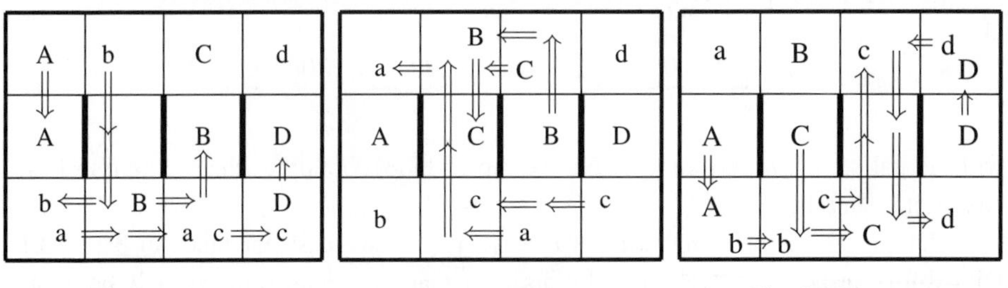

First 10 moves Second 10 moves Final 12 moves

For Rikki-Tikki-Tavi. It's not possible to do this puzzle with 3 pairs of knights on a 3×3 board, but it can be done with $5, 6, 7, \ldots$ pairs of knights on a $3 \times 5, 3 \times 6, 3 \times 7, \ldots$ board. What is the least number of moves in each case?

R191. Suppose Miss Toshiyori's phone number is x. Then

$$400000 + x = 4(10x + 4), \ 39x = 399984, \ x = 10256.$$

If you don't know the number of digits, say k, then we have

$$\begin{aligned} 4 \times 10^k + x &= 4(10x + 4) \\ 39x &= 3\underbrace{99\ldots99}_{k-2}84 \\ 13x &= 1\underbrace{33\ldots33}_{k-2}28. \end{aligned}$$

Now 13 and 3328 are each divisible by 13, and so is 333333 (see **Divisibility tests** in the Treasury), so $k-2 = 3, 9, 15, \ldots, k \equiv 5 \bmod 6$, and

$$x = 10256, \ 10256410256, \ 10256410256410256, \ 10256410256410256410256, \ldots.$$

If you replace 4 by 5 throughout, you get

$$5 \times 10^k + x = 5(10k + 5), \ 49x = 5(10^k - 5)$$

and want the k-digit number 199...99 to be divisible by 49. The smallest x has 41 digits: 10204081632653061224489795918367346938775. The next one has 83 digits—append a 5 and repeat the earlier 41 digits! Can you explain the pattern of digits? Is it because 49 is a bit less than a half of a hundred?

Rikki-Tikki-Tavi will want to try other digits and multiples. He may be able to get some help by reading pages 157–171 of John H. Conway & Richard K. Guy, *The Book of Numbers*, Copernicus, Springer, 1995, 1998.

R192a. Any arrangement of coins in the first row of a 6×6 board leads to a solution, and we will see later that this is true for any even-sided square board. Note that there are 2^6 ways of placing the coins on the first row, 2^3 of which are symmetrical. These 8 lead to four solutions with an axis of symmetry—if the first row is symmetrical, so are all others, in particular the last, which turns out (see later, under **R192b**) to be the complement of the first. We may think of the rows as binary numbers, with the coins as ones and empty squares as zeros. In the four solutions, the first & last (top) rows are 0 & 63, 12 & 51, 20 & 43, 33 & 30, each pair adding to 63.

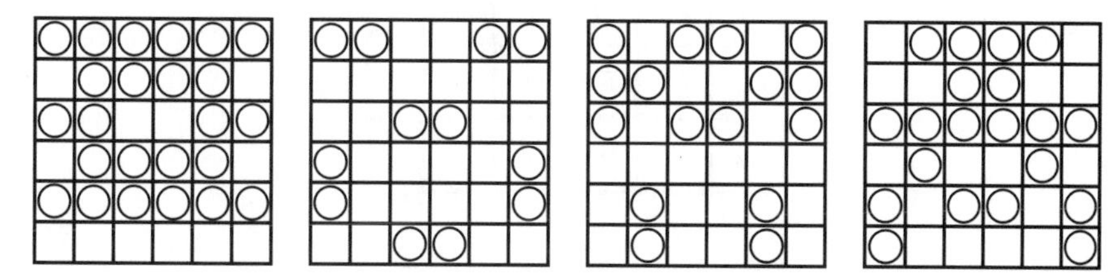

The remaining 56 arrangements comprise 28 reflected pairs, 4 of which (complements as well as reflexions), can be seen as the side edges of the four solutions we've already found, 21 & 42, 25 & 38, 7 & 56, 11 & 52, still adding to 63.

The other 24 pairs occur as the 24 edges of six solutions, together with their reflexions. It's easy to see that any two different starting patterns will yield different edges, and that we've already found all the symmetries—you can't have diagonal symmetry because the corner squares each have just one neighbor.

We've found all the solutions: $(4 \times 4) + (6 \times 8) = 2^6$. Check that all the numbers from 0 to 63 occur. This is easy to do because opposite edges are always complements, adding to $2^6 - 1$. E.g., the first of the remaining six solutions has opposite edges 1 & 62 and 10 & 53. Its reflexion will have opposite edges 32 & 31 and 20 & 43.

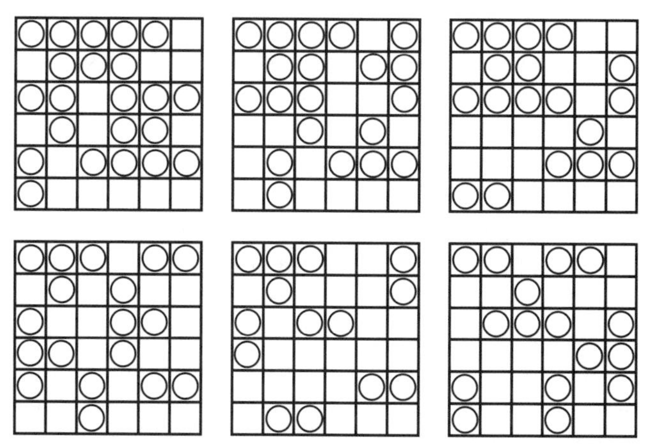

The number of coins may be any even number from 12 to 24. The largest and smallest numbers occur in our first two solutions above.

In general the $2n \times 2n$ board has 2^{n-1} solutions with an axis of symmetry, and $2^{2n-3} - 2^{n-2}$ without symmetry, a total of $2^{2n-3} + 2^{n-2}$ essentially different solutions,

Here is another approach. The letters in the table stand for 0 or 1 according as the square is empty or contains a coin. If we write a string of letters such as bdf we mean their sum modulo 2, and if there's a bar over them, $\overline{bdf}$, then add in an extra 1. For example, if $b = 0$ and $d = f = 1$, then $bdf = 0$ and $\overline{bdf} = 1$. The entries in each row, working up from the bottom, are uniquely determined from that bottom row.

$\overline{f}$	$\overline{e}$	$\overline{d}$	$\overline{c}$	$\overline{b}$	$\overline{a}$
e	$\overline{df}$	$\overline{ce}$	$\overline{bd}$	$\overline{ac}$	b
$\overline{d}$	$\overline{ce}$	$\overline{bdf}$	ace	$\overline{bd}$	$\overline{c}$
c	$\overline{bd}$	$\overline{ace}$	$\overline{bdf}$	$\overline{ce}$	d
$\overline{b}$	$\overline{ac}$	$\overline{bd}$	$\overline{ce}$	$\overline{df}$	$\overline{e}$
a	b	c	d	e	f

Opposite edges of the outer border are reflected complements, as are those in the central square. In the intermediate border, opposite edges are just reflected. As we shall see later, on larger boards, the frames alternately complement or merely reflect. Note that adjacent edges have alternate members complemented.

In the 6×6 board, this means that exactly 10 coins occupy squares in the outer border, and exactly 2 in the central square. There will be a minimum number of 12 coins, achieved with

$$a = b = e = f = 0, \quad c = d = 1,$$

and a maximum of 24 (put $a = b = c = d = e = f = 0$—or take the complementary values in either case).

R192b. Any board which has an even dimension can be 'coined', and here is how to find all solutions. Imagine the board to be checkered and place it with the even dimension horizontal and the white square on the right. We'll tell you all the ways to put coins on the black squares. You can choose two of these and reflect one of them, left to right, and put the reflexion on the white squares.

As before, the letters in the tables stand for 0 or 1 according as the square is empty or contains a coin. Strings of letters are to be added modulo 2, and a bar implies an extra 1.

If you want to coin an $m \times n$ board, where n is even, make a table, n columns wide, building it from the bottom row but paying attention only to the black squares. The $(m+1)$th row must contain no coins. Instructions for making this row all zeros are given on the right. It will be seen that the $(n+1)$th row consists of zeros and above that the pattern continues reflected in a 'frieze pattern' manner. We illustrate with $n = 8$ and 10.

For larger n the pattern continues with consecutive or alternating letters in the equations on the right.

	d		c		c		a	
0		0		0		0		
	$\overline{d}$		$\overline{c}$		$\overline{b}$		$\overline{a}$	$a = b = c = d = 1$
d		$\overline{cd}$		$\overline{bc}$		$\overline{ab}$		$a = c = 1,\ b = d = 0$
	$\overline{cd}$		bcd		abc		$\overline{b}$	$b = 1,\ a \neq c \neq d$
c		$\overline{bcd}$		$abcd$		$\overline{bc}$		$a = b = 1,\ c = d = 0$
	$\overline{bc}$		$abcd$		bcd		$\overline{c}$	$a = c = 0,\ b = d = 1$
b		$\overline{abc}$		$\overline{bcd}$		$\overline{cd}$		$b = 0,\ a \neq c \neq d$
	$\overline{ab}$		$\overline{bc}$		$\overline{cd}$		$\overline{d}$	$a = c = 0,\ b = d = 1$
a		b		c		d		

	e		d		c		b		a	
0		0		0		0		0		
	$\overline{e}$		$\overline{d}$		$\overline{c}$		$\overline{b}$		$\overline{a}$	$a = b = c = d = e = 1$
e		$\overline{de}$		$\overline{cd}$		$\overline{bc}$		$\overline{ab}$		$a = c = e = 0,\ b = d = 1$
	$\overline{de}$		cde		bcd		abc		$\overline{b}$	$a = d = 0,\ b = c = e = 1$
d		$\overline{cde}$		$bcde$		$abcd$		$\overline{bc}$		$a = b = e = 1,\ c = d = 0$
	$\overline{cd}$		$bcde$		$\overline{abcde}$		bcd		$\overline{c}$	$a = b = c = 1,\ d = e = 0$
c		$\overline{bcd}$		$abcde$		$bcde$		$\overline{cd}$		$a = b = c = 0,\ d = e = 1$
	$\overline{bc}$		$abcd$		$bcde$		cde		$\overline{d}$	$a = b = e = 0,\ c = d = 1$
b		$\overline{abc}$		$\overline{bcd}$		$\overline{cde}$		$\overline{de}$		$b = c = e = 0,\ a = d = 1$
	$\overline{ab}$		$\overline{bc}$		$\overline{cd}$		$\overline{de}$		$\overline{e}$	$a = c = e = 1,\ b = d = 0$
a		b		c		d		e		

If both dimensions of the board are odd, then there may not be a solution, although there usually is. We proceed as before, putting capital letters in the white squares and lower case in the black ones, of which there are now one less. The patterns are now slightly different, according as

$$n = 4k + 3 \quad \text{or} \quad 4k + 1.$$

We illustrate with $n = 7$ and $n = 9$. Wider boards will be similar, merely containing longer runs of consecutive letters. Again the patterns repeat after every $2n + 2$ rows, reversing and complementing after every $n + 1$ rows. The $(2n + 2)$th row consists entirely of zeros, but the $(n + 1)$th row contains alternate ones, so that an odd square board can never be coined. The instructions on the right are given only on even-numbered rows; we know that we can coin an even by odd board, and you can check that the odd-numbered rows can always be made zero.

A	a	B	b	C	c	D	
0	0	0	0	0	0	0	OK!
$\overline{A}$	$\overline{a}$	$\overline{B}$	$\overline{b}$	$\overline{C}$	$\overline{c}$	$\overline{D}$	
a	$\overline{AB}$	$\overline{ab}$	$\overline{BC}$	$\overline{bc}$	$\overline{CD}$	c	$a = c = 0,\ b = 1,\ A \neq B \neq C \neq D$
$\overline{B}$	$\overline{ab}$	ABC	abc	BCD	$\overline{bc}$	$\overline{C}$	
b	$\overline{BC}$	$\overline{abc}$	$ABCD$	$\overline{abc}$	$\overline{BC}$	b	$b = 0,\ a \neq c,\ B \neq C,\ A \neq D$
$\overline{C}$	$\overline{bc}$	BCD	$\overline{abc}$	ABC	$\overline{ab}$	$\overline{B}$	
c	$\overline{CD}$	bc	$\overline{BC}$	ab	$\overline{AB}$	a	$a = b = c = 0, A \neq B \neq C \neq D$
$\overline{D}$	c	$\overline{C}$	b	$\overline{B}$	a	$\overline{A}$	
1	0	1	0	1	0	1	A 7×7 square is impossible
D	$\overline{c}$	C	$\overline{b}$	B	$\overline{a}$	A	
$\overline{c}$	$\overline{CD}$	bc	$\overline{BC}$	ab	$\overline{AB}$	$\overline{a}$	$a = b = c = 1, A \neq B \neq C \neq D$
C	$\overline{bc}$	$\overline{BCD}$	abc	$\overline{ABC}$	$\overline{ab}$	B	
$\overline{b}$	$\overline{BC}$	abc	$ABCD$	abc	$\overline{BC}$	$\overline{b}$	$b = 1, a \neq c, B \neq C, A \neq D$
B	$\overline{ab}$	$\overline{ABC}$	$\overline{abc}$	$\overline{BCD}$	$\overline{bc}$	C	
$\overline{a}$	$\overline{AB}$	$\overline{ab}$	$\overline{BC}$	$\overline{bc}$	$\overline{CD}$	$\overline{c}$	$a = c = 1,\ b = 0,\ A \neq B \neq C \neq D$
A	a	B	b	C	c	D	

All $m \times 4k + 3$ rectangles are possible, except when $m = 8(k + 1)r + 4k + 3$ for $r = 0, 1, 2, \ldots$.

1	0	1	0	1	0	1	0	1	A 9×9 square is impossible
E	$\overline{d}$	D	$\overline{c}$	C	$\overline{b}$	B	$\overline{a}$	A	
$\overline{d}$	$\overline{DE}$	cd	$\overline{CD}$	bc	$\overline{BC}$	ab	$\overline{AB}$	$\overline{a}$	$a = b = c = d = 1,\ A \neq B \neq C \neq D \neq E$
D	$\overline{cd}$	$\overline{CDE}$	bcd	$\overline{BCD}$	abc	$\overline{ABC}$	$\overline{ab}$	B	
$\overline{c}$	$\overline{CD}$	bcd	$BCDE$	$\overline{abcd}$	$ABCD$	abc	$\overline{BC}$	$\overline{b}$	No solution for a, b, c, d
C	$\overline{bc}$	$\overline{BCD}$	$abcd$	$ABCDE$	$abcd$	$\overline{BCD}$	$\overline{bc}$	C	
$\overline{b}$	$\overline{BC}$	abc	$ABCD$	$abcd$	$BCDE$	bcd	$\overline{CD}$	$\overline{c}$	$b = c = 1,\ a = d = 0,\ A \neq B \neq C \neq D \neq E$
B	$\overline{ab}$	$\overline{ABC}$	$\overline{abc}$	$\overline{BCD}$	$\overline{bcd}$	$\overline{CDE}$	$\overline{cd}$	D	
$\overline{a}$	$\overline{AB}$	$\overline{ab}$	$\overline{BC}$	$\overline{bc}$	$\overline{CD}$	$\overline{cd}$	$\overline{DE}$	$\overline{d}$	No solution for a, b, c, d
A	a	B	b	C	c	D	d	E	

Continuation of the pattern shows that $m \times (4k + 1)$ rectangles are possible, except when $m = 4r + 1$ for $r = 0, 1, 2, \ldots$.

Barry Cipra found a nice argument for showing the impossibility of coining the above-mentioned exceptions. For example, consider the 5×5 board. Start as above by coloring the checkerboard black and white, with white in all the corners, so that there are 12 black squares. Place "guards" on the white squares so that each black square is adjacent to exactly one guard. Here are two different ways of doing this, one with 5 guards, the other with 4. But the number of coins on black squares adjacent to a guard must be odd, so the number of coins on the black squares of the first diagram is odd, while the number on the second is even, a contradiction!

The argument works for any $4n + 1 \times 4m + 1$ board, since the pattern can be extended in multiples of four in each direction. For example, for the 5×9 board we have the following placements of 8 guards or 7.

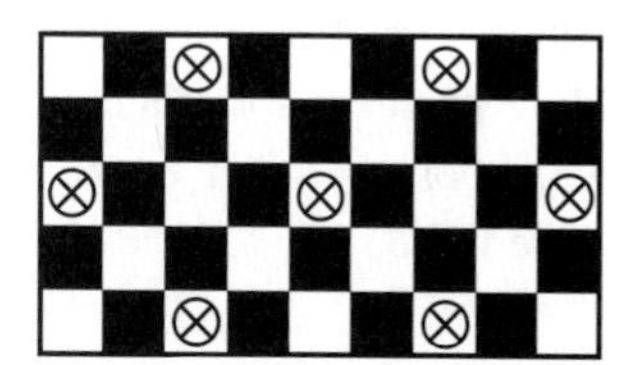

Rather than continuing along these lines for the $4n + 3$ boards, there's a better approach, also discovered by Barry Cipra.

Define a **Cipra guarding** of a checkered rectangular board to be an assignment of guards to (some of the) white squares so that every black square is guarded by an *even* number (possibly zero) of guards. By an **odd Cipra guarding** we mean a Cipra guarding with an odd number of guards; any other is an **even Cipra guarding**.

Cipra's theorem. A rectangular board has an odd Cipra guarding if and only if the board cannot be coined.

To prove this, suppose we have an odd Cipra guarding of a rectangular board that can be coined. Let B denote the set of guarded black squares, and represent the coins by ones and empty squares by zeros. Each guard is adjacent to an odd number of coins so the sum of the numbers on the (black) squares adjacent to a single guard is an odd number. Therefore, the sum of these numbers over all guards is an odd number (the sum of an odd number of odd numbers is odd). On the other hand, each square in B is adjacent to an even number of guards, so that in the previous grand total, each guarded square is counted an even number of times, so the sum is even, giving a contradiction. We conclude that a board with an odd Cipra guarding cannot be coined.

To prove the converse, that the rectangles we know can't be coined do indeed have an odd Cipra guarding, we exhibit such guardings. Suppose that the dimensions are m and n. Then they are either

$m \equiv 1 \bmod 4$, $n \equiv 1 \bmod 4$, and you can place guards on all squares (i, j) where i and j are both odd, $1 \leq i \leq n$, $1 \leq m \leq m$; or

$m \equiv 3 \bmod 4$, $n \equiv 3 \bmod 2m{+}2$, with a zigzag of guards on squares $((2m{+}2)i \pm j, j)$, $1 \leq j \leq m$, $0 \leq i \leq (n{-}m)/(2m{+}2)$.

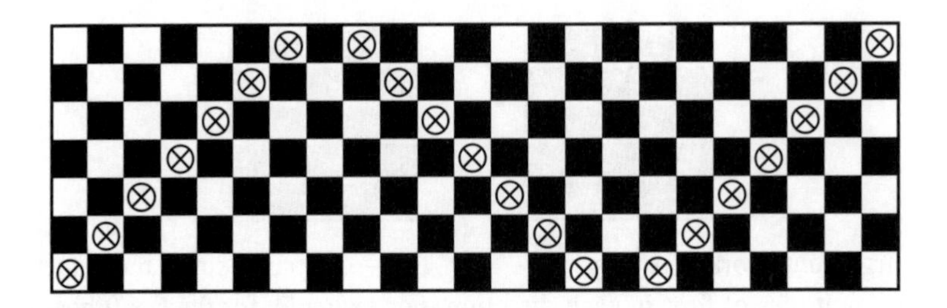

For Rikki-Tikki-Tavi. Which boards have even Cipra guardings? How many Cipra guardings are there for a given rectangular board? And what about the corresponding problem for triangular boards where the corner triangles have only one neighbor, the other edge triangles have two neighbors and the inside triangles have three. What about honeycomb shaped boards?

R198. (a) It can't be done. Even if you were to choose all 20 points of the icosagon, no three points are equally spaced because 20 isn't divisible by 3.

(b) It's possible to choose eight of the twenty points without forming an isosceles triangle. Number the vertices consecutively from 0 to 19, and choose points 0, 1, 3, 4, 10, 11, 13, and 14. The even ones are spaced 4 6 4 6 around the circle (measuring distance by arclength, where adjacent points are separated by one unit), while the odd ones are spaced 2 8 2 8. The distances between the odds and the evens are

	0	4	10	14
1	1	3	9	7
3	3	1	7	9
11	9	7	1	3
13	7	9	3	1

with no two equal distance in the same row or column.

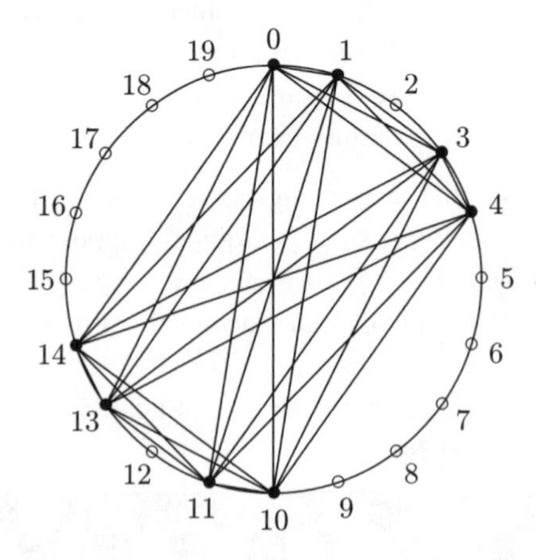

R200. No, it cannot be done. There are 49 equally spaced triples of positions and, amazingly, they fit into a "magic cube of positions" ($27 = 9 \times 3$ rows parallel to an edge; $18 = 6 \times 3$ diagonals parallel to a face and 4 body diagonals)!

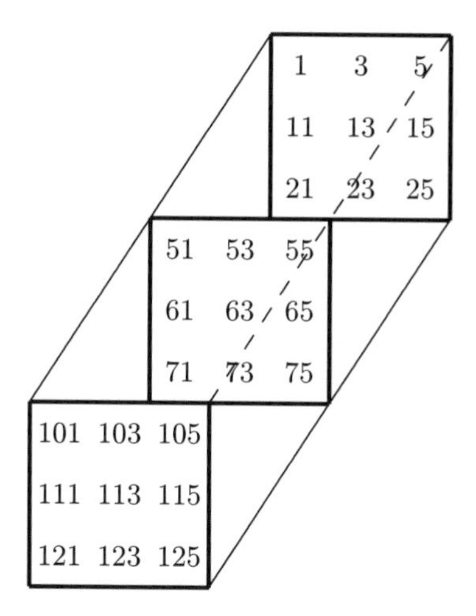

If the numbers 1 through 27 could be placed in these positions the result would be a magic cube in which the numbers in all rows, columns, face diagonals, and body diagonals, would equal $(1+2+\cdots+27)/9 = (27 \times 28)/18 = 42$. However, it is not possible to construct such a magic cube, for think of the cube as three parallel layers (three parallel to the front face, or three parallel to the right face, or three parallel to the top face). Add up the three rows in two of the three layers, and add to that the row sum, the column sum, and two diagonal sums from the third layer. If m is the middle number in that third layer, we will have added $1+2+\cdots+27+3m = 378+3m$. On the other hand, we have also added 10 magic lines, so the sum is also equal to $10 \times 42 = 420$. Equating these, $378 + 3m = 420$, yields $m = 14$. This means that 14 has to be in the middle of every face, which is impossible.

R202. The board below is the smallest one that guarantees a win for the first player.

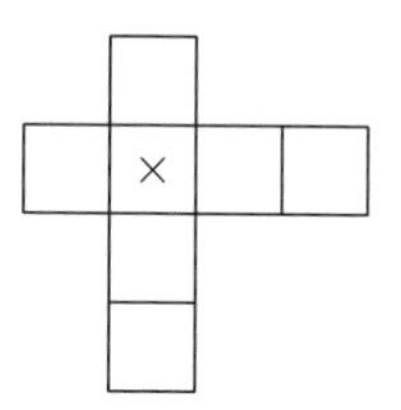

The first player can force a win by playing in the cell marked with an $\times$.

Here are two unusual boards with seven cells. The first player can force a win by playing in cell A, B, C, or D.

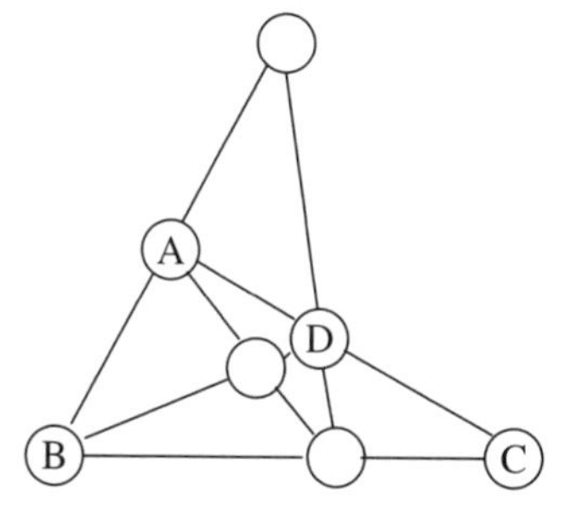

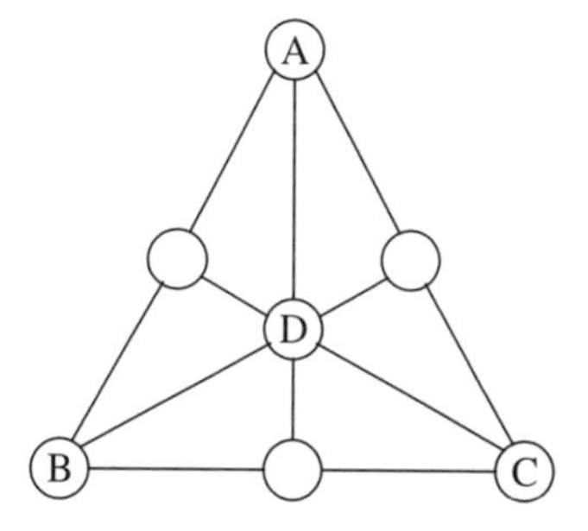

R203. Label the rows of the board $1, 2, \ldots, 8$ from bottom to top, and the columns $\mathrm{a}, \mathrm{b}, \ldots, \mathrm{h}$ from left to right.

On the first board, a winning move is to mark squares c7 and d6, and follow this with the Tweedledum and Tweedledee strategy (or, more subtly, by b8, c8, or d8, e8). On the second board, a winning move is b8, c8, d8. On the third board, a winning move is c8, d8, followed by opponent-like moves: respond to a play in one of the double spaces with the same play in the second double space, and a play in one of the single spaces with a play in the other single space.

On the fourth board, a winning move is g6 and h6, followed by the Tweedledum and Tweedledee strategy, which works unless there is a play (eventually) at b3 or g3, in which case the response is g3 or b3, respectively. Other winning moves are g8, h8, or h6, h7, h8.

For Rikki-Tikki-Tavi. Which player has a winning strategy for this game (on rows, columns, and diagonals) on an rectangular board with r rows and c columns, when r and c are both even?

R205. The numbers on the chessboards denote the successive moves of a rook circuit that starts on square 1 and visits each square of the board. The squares in each pair $(1,2), (3,4), (5,6), \ldots, (63,64)$ are on the same row or column, so the first player wins by moving to square $n+1$ if the rook is on square n and n is odd, or to square $n-1$ if the rook is on square n and n is even. (In a similar way, the first player can win by pairing the squares $(64,1), (2,3), (4,5), \ldots, (62,63)$.)

On the left board, the rook moves at least four squares each time. On the right board, the winning strategy would be difficult to detect because half the first player's moves are vertical and half are horizontal; moreover, their lengths vary from 1 to 7.

50	52	38	40	49	51	53	39
28	30	32	34	59	29	31	33
62	16	18	20	61	63	17	19
4	6	8	10	3	5	7	9
47	45	43	41	48	46	44	42
57	55	37	35	58	56	54	36
27	25	23	21	60	26	24	22
1	15	13	11	2	64	14	12

14	27	19	29	13	28	34	12
43	42	6	52	41	5	53	4
15	37	9	60	36	59	35	10
44	45	20	31	22	32	33	21
56	26	7	55	23	57	54	3
16	38	18	30	40	17	39	11
64	46	62	61	47	63	49	48
1	25	8	51	24	58	50	2

R207. The more general game in which any number of beans may be taken from one heap or equally from both heaps is known as Wythoff's Game. This was later rediscovered by Rufus Isaacs in the form called Wyt Gueens (see Berlekamp, Conway, Guy, *Winning Ways for your Mathematical Plays*, 2nd edition, A. K. Peters, 2000, pp. 59–60, 74–75). It has connexions with the golden ratio and Fibonacci numbers. The P-positions, those positions that can be won by the previous player, are pairs of corresponding numbers (the heap sizes) from the two Beatty sequences:

$$\begin{array}{cccccccccccc} 1 & 3 & 4 & 6 & 8 & 9 & 11 & 12 & 14 & 16 & 17 & \cdots \\ 2 & 5 & 7 & 10 & 13 & 15 & 18 & 20 & 23 & 26 & 28 & \cdots \end{array}$$

which differ by

$$\begin{array}{cccccccccccc} 1 & 2 & 3 & 4 & 5 & 6 & 7 & 8 & 9 & 10 & 11 & \cdots \end{array}$$

So in this game, Nicole, who plays first, can win the $(10, 12)$ game by reducing the 12 heap to 6, the number corresponding to 10 in the above pair of sequences. You can check that whatever Pryor does from the $(6, 10)$ position, Nicole can always find a winning reply.

R210a. Here are some solutions with six-fold rotational symmetry.

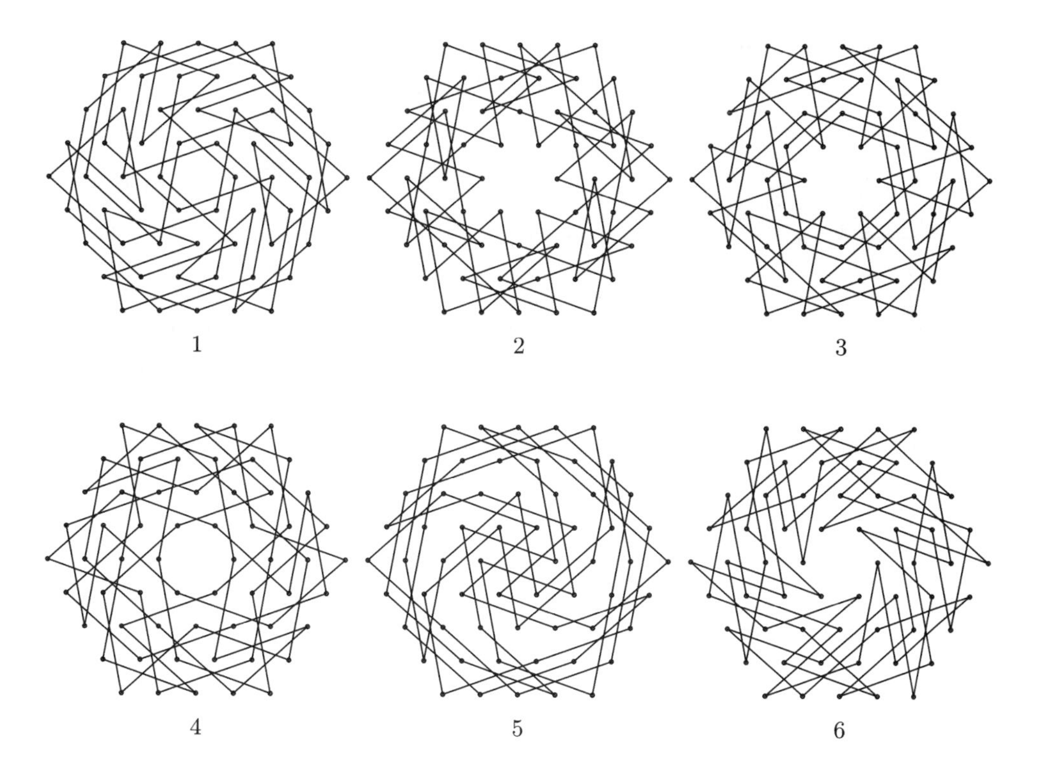

Such tours can be constructed by partitioning the hexagon into six congruent triangular regions, as indicated below left, and identifying them as a single graph with ten vertices (below right) where two vertices are connected if and only if they are connected with a single move in the original hexagonal grid. For example, in the graph at the right, a is connected to vertices g, i, and f because in the original hexagonal grid, one can go to these places in a single move.

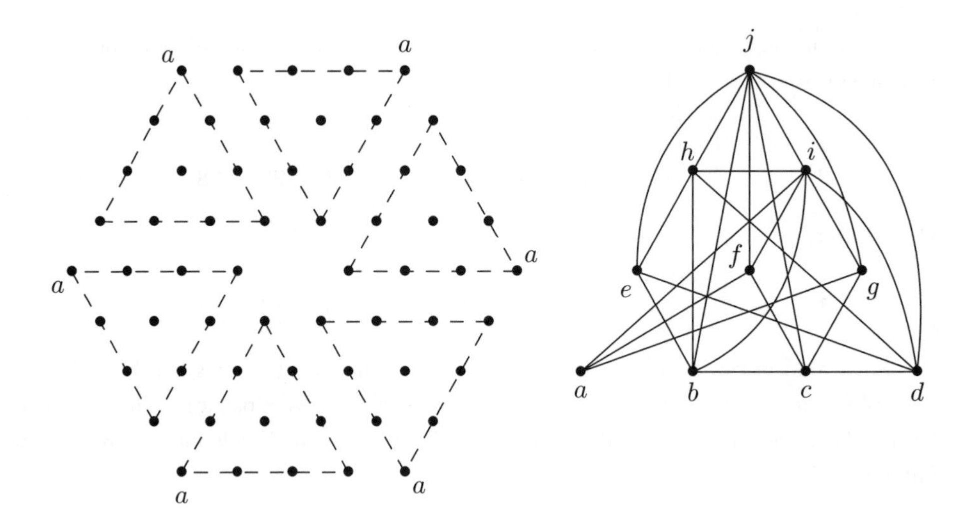

The first ten moves of a six-fold symmetric tour must visit the vertices a, b, c, d, e, f, g, h, i, j in some order. Thus, a necessary step in getting a six-fold symmetrical solution to the 5-hex problem is to get a Hamiltonian circuit on the previous graph. The preceding solutions were generated from the following circuits.

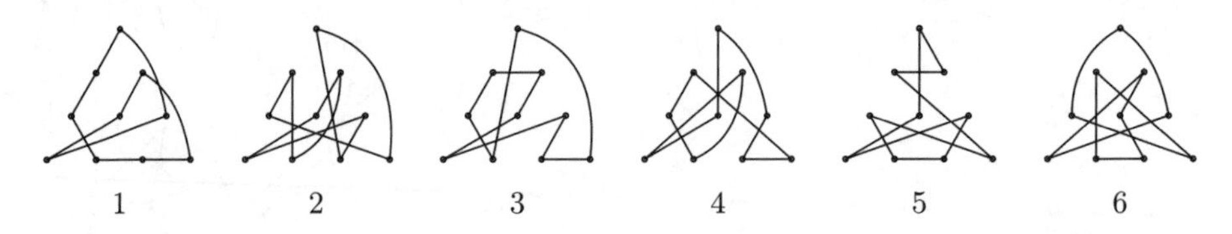

Each of these ten-move Hamiltonian circuits generate a number of different paths of length ten on the original hexagonal grid, because certain moves, such as the move from vertex j to vertex e, can be carried out in more than one way on the original hexagonal grid. But the ones we want are those that start at a corner and terminate on an adjacent corner. A hand count found 111 such Hamiltonian circuits on the triangular graph, and these generated 149 different six-fold symmetric circuits on the hexagonal grid.

R210b. Here are six-fold symmetric solutions for hexagonal boards with 5, 6, 7, 8, and 9 vertices on a side.

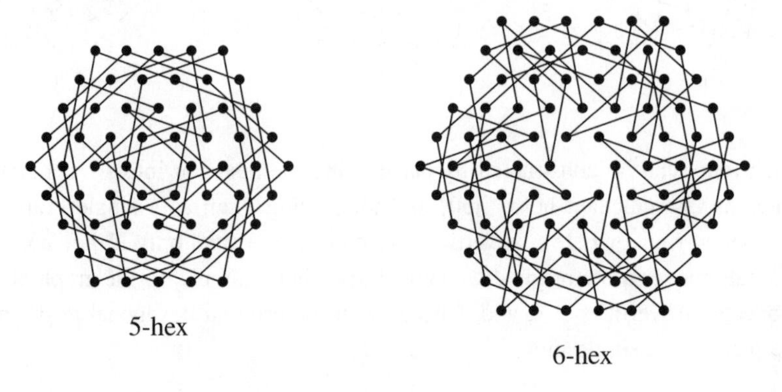

5-hex

6-hex

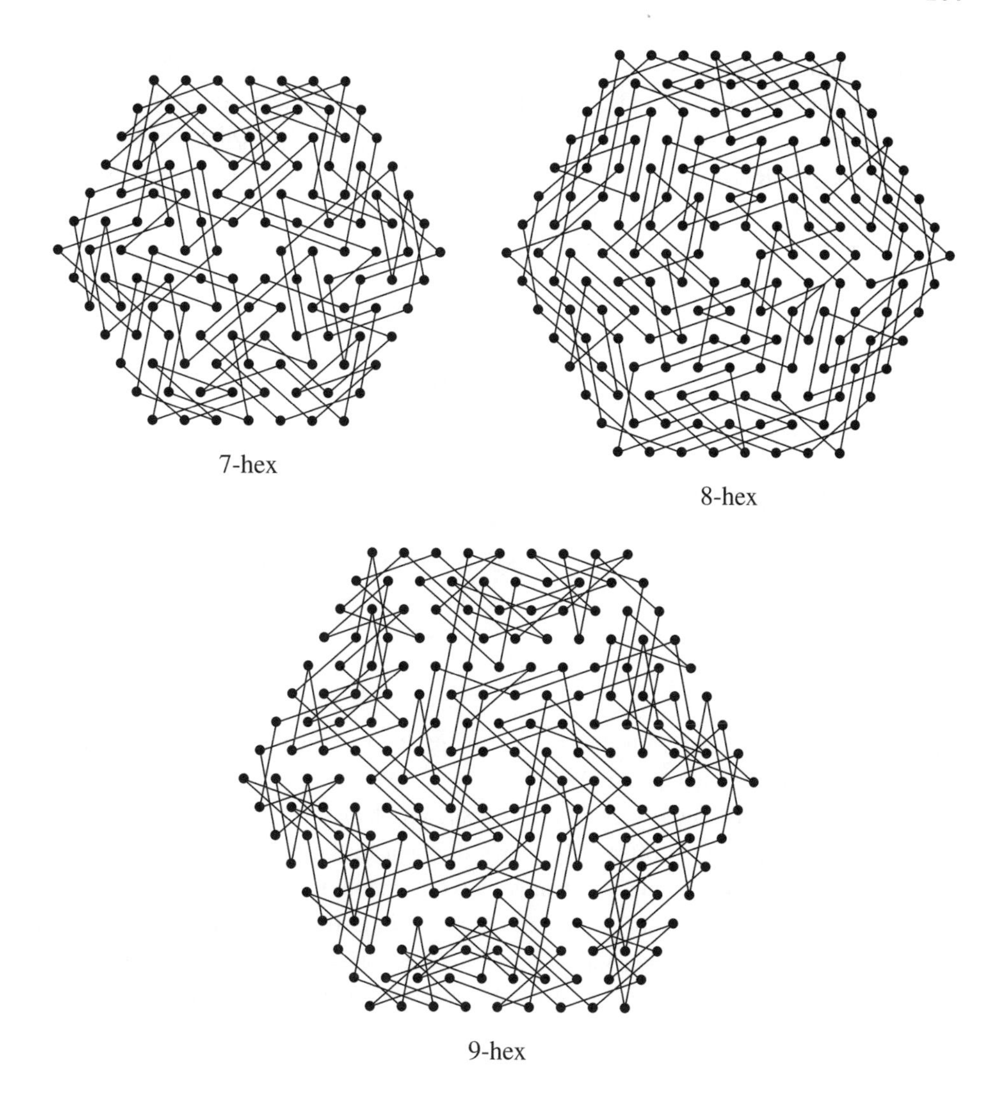

7-hex

8-hex

9-hex

To get symmetric solutions for larger boards, adjoin six loops, such as the one shown below, by deleting edges AB and CD and adding edges AC and BD. Notice that the dotted portion of the loop, which begins at S and ends at T can be adjusted to make the border fit any hexagonal grid with at least four vertices on each edge.

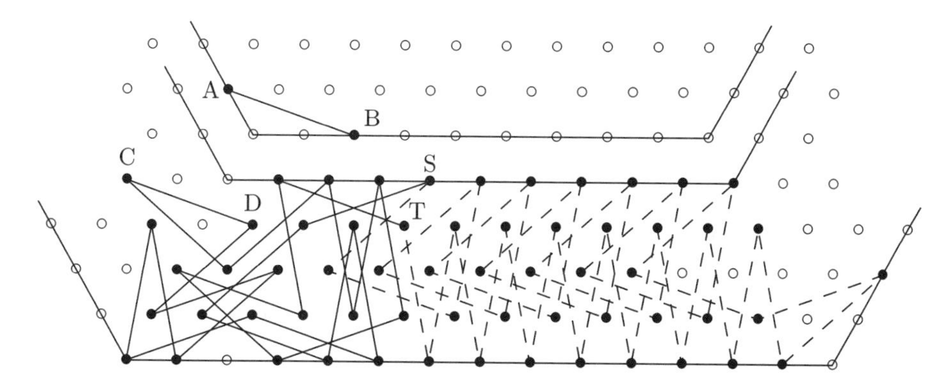

For Rikki-Tikki-Tavi. Circuits for hexagonal grids which contain the center point cannot have six-fold symmetry, but can have bilateral symmetry, and these exist for boards with more than three vertices on a edge. Other interesting touring problems of this sort can be formed by replacing the hexagon with other shapes, such as triangles, diamonds, parallelogram, or combinations of these shapes. One can also consider (m, n) toad hops instead of $(1, 2)$ hops.

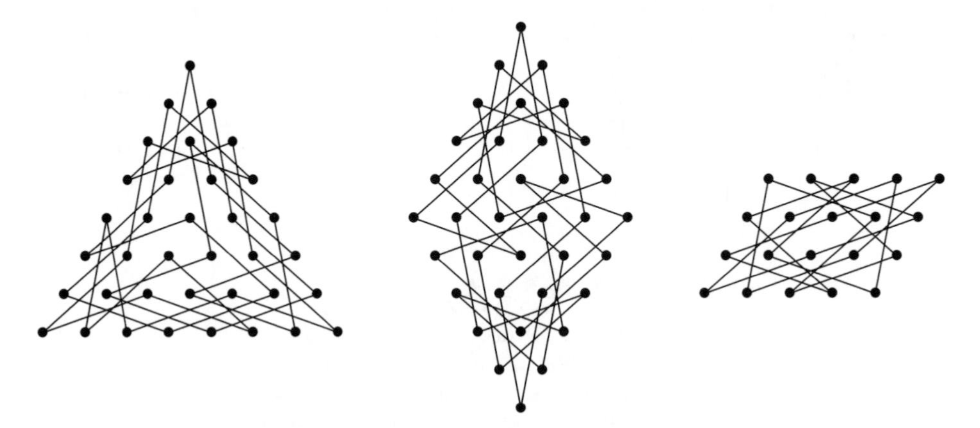

R211. We shall see that with this method of packing the tennis balls, we can get an extra ball in if the box has length 238.

The problem is essentially a two-dimensional problem of packing circles, and in such problems, it's best to chop off a strip of width equal to the radius round the boundary and then 'spread points' (centers of tennis balls) in what's left.

So we're packing (pairs of) equilateral triangles in a strip of unit width, i.e., packing unit diamonds, whose long diagonal, AB, is $\sqrt{3}$. The horizontal component, AD, of such a diagonal is $\sqrt{2}$, by Pythagoras. How close can we put the next diamond?

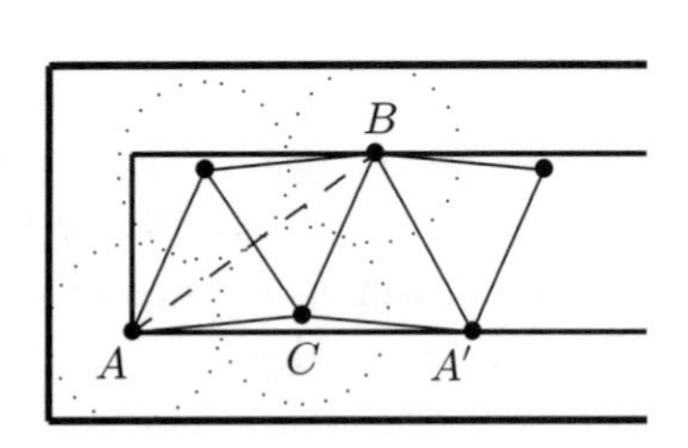

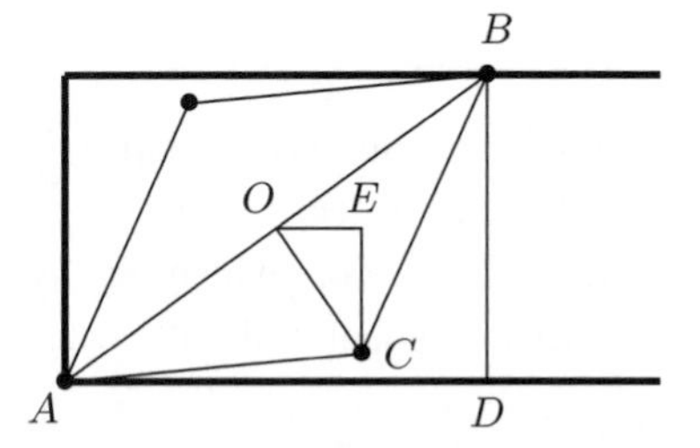

The center, O, of the first diamond has coordinates $(\sqrt{2}/2, 1/2)$ (taking A to be the origin). Triangles COE and ABD are similar and of shape $(1, \sqrt{2}, \sqrt{3})$ so that $OE/OC = BD/BA = 1/\sqrt{3}$. Since $OC = 1/2$ we have $OE = \sqrt{3}/6$ and the distance between diamonds is $AA' = 2(\frac{1}{2}AD + OE) = \sqrt{2} + (\sqrt{3}/6) = 1.991563832 < 2$. It follows that we can pack $4n + 1$ tennis balls in a $1 \times 2 \times 2n$ box provided $1 + (\sqrt{2} + \sqrt{3}/3)\, n < 2n$, or equivalently, when

$$n > 1/(2 - \sqrt{2} - \sqrt{3}/3) = 30 + 21\sqrt{2} + 17\sqrt{3} + 12\sqrt{6} = 118.5372255.$$

So, with $n = 119$ we can pack 477 balls in a $1 \times 2 \times 238$ box!

R212. We have seen that a necessary condition for a tiling with these pieces is that the 1×1 tile be placed on a square labeled 1. But the label must also be 1 had we numbered the squares in the rows of the board from right to left instead of left to right. There are only four squares which are numbered 1 in both of these cases (see the figure at the left). Each of these four squares for the 1×1 tile allow a tiling with the remaining 1×3 tiles. The figure at the right shows one such tiling.

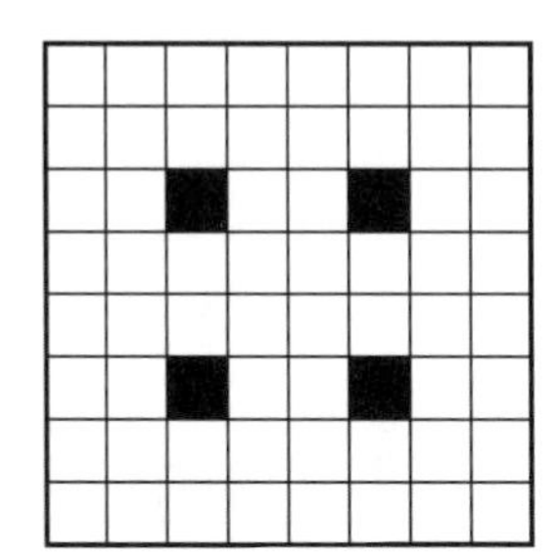

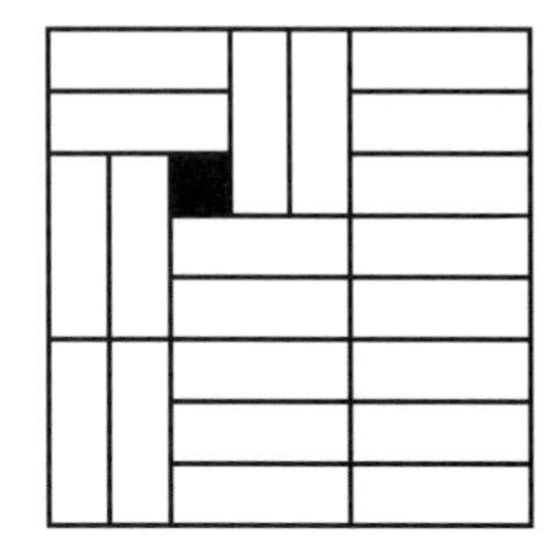

R213. Yes, provided you include $343 = 7^3$ and $686 = 2 \cdot 7^3$, because 686! contains $98 + 14 + 2 = 114$ factors of 7, while 86! contains $12 + 1$, and $114 - 13 = 101$.

R217. The white squares of the chessboard lie along the seven (falling) diagonals labeled 3, 5, 7, 9, 11, 13, 15 as shown in the following diagram.

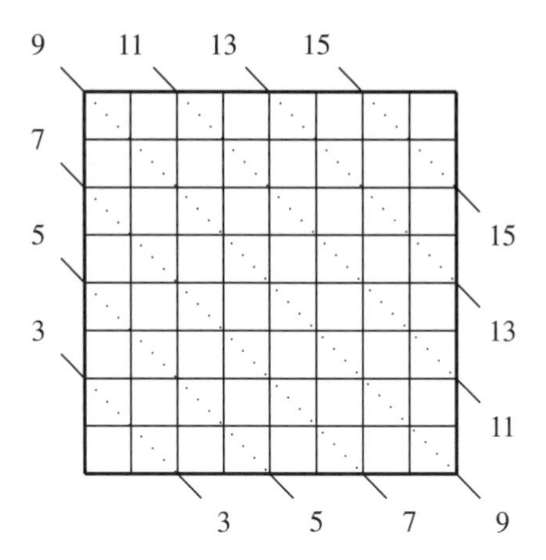

The coordinate sums of the squares on the same (falling) diagonal are each equal to the label assigned to the corresponding diagonal. For example, square $(1, 2)$ lies on diagonal 3, square $(5, 8)$ is on diagonal 13, and so forth.

If it were possible to place eight queens on the white squares so that no two of them attack one another, these eight queens would have to be on eight different diagonals. But there are only seven diagonals, so such a placement is impossible.

Suppose it were possible to place six queens on the white squares so that no two of them attack one another. Then all but one of these seven diagonals will have a queen on it. In particular, at least one of the two diagonals 3 and 15 will have a queen on it. Thus, there must be a queen on one of the squares $(1, 2)$, $(2, 1)$, $(7, 8)$, or $(8, 7)$. By symmetry we may assume that there is a queen on the square $(1, 2)$.

Suppose, first, that diagonal 15 does not have a queen on it. Then there is only one place to put the queens on diagonals 3, 7, and 9, and only two places for a queen on diagonal 11, but in either case, there is no square left for a queen on diagonal 13. See the diagram below on the left.

Thus, if there is a solution with six queens on white squares, diagonals 3 and 15 must each have one of the queens.

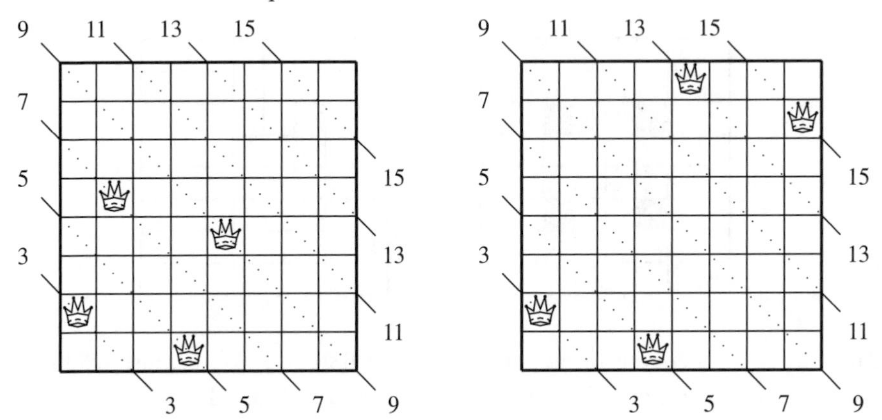

Suppose that diagonals 5 and 13 also each have queens on them. They must be placed as shown in the above diagram on the right. But now there is no place for a queen on diagonals 7, 9, or 11, contrary to the fact that two of these diagonals must be occupied by a queen.

Thus, only one of the diagonals 5 and 13 can have a queen on it, and we may suppose that it is diagonal 5. Now we'll show that either diagonal 7 or diagonal 11 doesn't have a queen on it. For suppose they both do. Then the board looks like the following diagram.

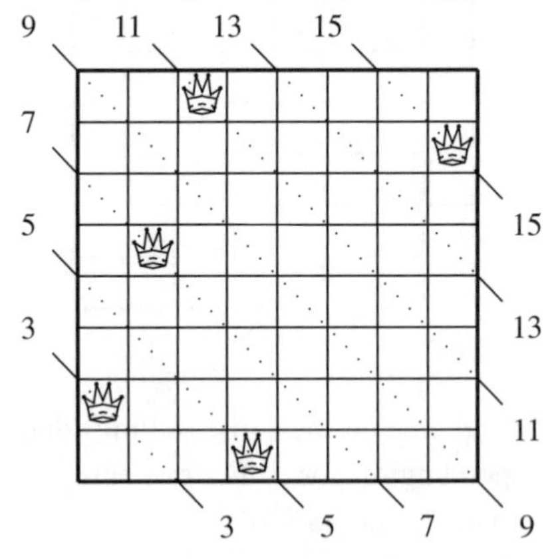

But now it is impossible to find a place for a queen to cover diagonal 9.

This completes the proof that no nonattacking queens placement can have six queens on the white squares.

Suppose there were either 0 or 2 queens on the white squares. Then 8 or 6 queens would be on the black squares. Now rotate the board 90° and exchange the colors on all the squares. This will result in 8 or 6 nonattacking queens on the white squares, which we've seen can't happen. Therefore, we can't have 0 or 2 queens on the white squares either.

There are 92 different ways of placing 8 queens on the 8×8 board so that no two of them will attack any other (if we eliminate rotations and reflexions of the board there are 12 essentially different solutions). In each of these, exactly four of the queens are on the white squares and four are on the black. Here's one solution.

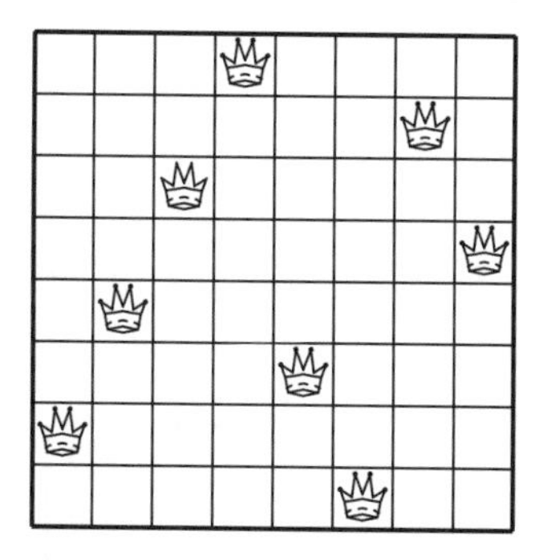

R218. The configuration is made up of nine small equilateral triangles, and if we denote their common sum by m, then $9m = 1+2+\cdots 27 = (27)(28)/2$, so $m = 42$ (the meaning of the universe, according the "The Hitchhiker's Guide to the Galaxy").

Suppose there is such a magic triangle and label some of the cells as shown.

```
            A
          *   *
         b     c
       d   x  y   e
     *             *
    f   *       *   g
   h     *     *     i
  j  p  r  *  *  s  q  k
```

First, by considering triangles bdx and cye, we'll show that $x + y = 2A$. We have $b+d = 42 - x$ and $c+e = 42 - y$. Also, adding $A+b+c = 42$ and $A+d+e = 42$ we have $2A+(b+d)+(c+e) = 84$, so from the preceding, $2A+(42-x)+(42-y) = 84$, or $x + y = 2A$, as claimed.

Similarly, $r+s = 2A$ (using triangles fjr and gsk) and $p+q = 2A$ (using triangles hjp and iqk).

It follows that

$$\begin{aligned} 84 &= (x+p+s)+(y+r+q) \\ &= (x+y)+(p+q)+(r+s) \\ &= 2A+2A+2A \\ &= 6A \end{aligned}$$

and $A = 14$. Because the same argument can be used at each of the corners, the magic property forces all the corner squares to have the same value, namely 14. Because this is not allowed, the triangle cannot be magic.

There's another way of thinking about the problem, similar to that used in the second solution to the original problem. First, replace the stars by numbers so that each of the

37 triangles has its corners in arithmetic progression.

1
2 3
4 7
5 6 8 9
10 19
11 12 20 21
13 16 22 25
14 15 17 18 23 24 26 27

Next, place the 27 numbers in a cube so that every row and diagonal is an arithmetic progression.

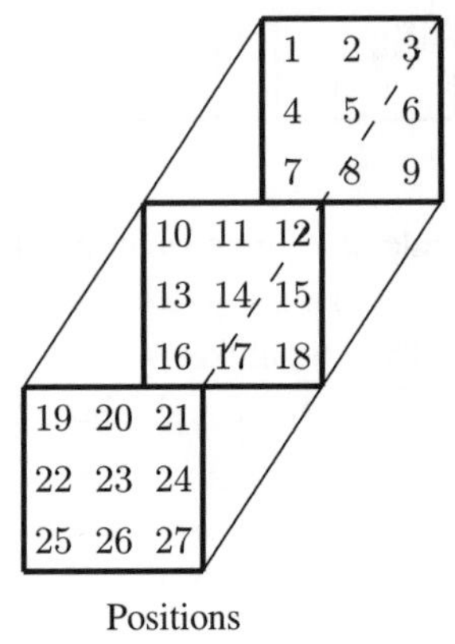

Positions

It is not possible to arrange the numbers $1, 2, \ldots 27$ in a cube so that every row, column, and diagonal adds to the same number (see **R200** for a proof). However, if we disregard the face diagonals, such cubes do exist (and they are called magic cubes). Here is an example.

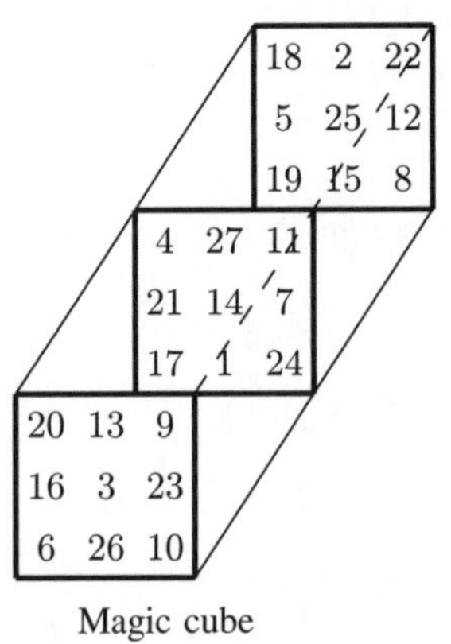

Magic cube

In a magic cube the 27 rows and columns, as well as 10 other lines through the middle (the four body diagonals and the six lines joining midpoints of opposite edges), a total of 37 lines, each sum to the magic number 42. If we place these numbers in the original triangle according to the directions in the positions cube (for example, 18 goes to position 1, 2 to position 2, 22 to position 3, and so on), we get the following triangle.

18
2 22
5 19
25 12 15 8
4 20
27 11 13 9
21 17 16 6
14 7 1 24 3 23 26 10

We haven't succeeded in getting all triangles to add to 42, and it can't be done. The reason is that 14 necessarily goes into position 14 (regardless of the magic cube, the same argument as that in **R200** shows that 14 must go into the middle square, and regardless of how the arithmetic progressions are put into a cube, 14 must go into the middle square) but the other numbers can be rotated and these correspond to different rotations of the cube. The upshot is that we would have to have 14 at each of the corners of the triangle (as we saw in the first proof), and that isn't allowed. In this triangle, there are six failures: $10, 1, 19$; $10, 3, 20$; $10, 7, 22$; and $18, 21, 6$; $18, 25, 8$; $18, 27, 9$ (these are the face diagonals of the magic cube).

There are 14 equilateral triangles that are not parallel to the first 37. Four of them, $5, 15, 22$; $6, 13, 23$; $11, 15, 16$; $12, 13, 17$ do have the right magic sum. That's the best that can be done: 31 out of 37, or 35 out of 51.

For Rikki-Tikki-Tavi. Extend the figure to $81, 243, 729, \ldots$ places, getting closer and closer to the "Sierpiński Gasket" (which is what inspired the problem). How many equilateral triangles are there are each stage? How many of them can you label so that their sums are all the same?

R219. Yes. Number the 21 people from 0 to 20, and for the first committee, choose persons numbered 0, 2, 7, 8, and 11. For the second committee, choose 1, 3, 8, 9, and 12, and for the kth committee, choose k, $k+2$, $k+7$, $k+8$, and $k+11$, where the numbers here are taken modulo 21. This gives 21 committees and it is clear that each pair of people serves together on just one committee.

R230. Let G be a graph whose vertices are the invited speakers, and draw a directed edge from A to B if A went to B's talk. By assumption, there are 200 such edges. But the number of different pairs of vertices is $(20 \times 19)/2 = 190$. Therefore at least ten pairs of vertices are connected by two edges (A to B and B to A), meaning that at least ten pairs of speakers heard each other's talks.

R232a. If b and c have a common factor $d > 1$, then we can only change the relative numbers of coins in two bowls by a multiple of d. But if b and c have no common factor (larger than 1), then we can find positive integers m and n such that $cm - bn = 1$ (see **Diophantine equations** in the Treasury), and then, by making m consecutive moves, we can add $cm = bn + 1$ coins, $n + 1$ to the bowl from which we started, and n to each of the other bowls. By repeated applying this procedure as in the solution of the original problem, we can eventually get the same number of coins in each bowl.

R232b. For seven bowls, and indeed for any odd number of bowls, any distribution can be achieved in which the total number of coins is even. To see this we first show how to add coins to two adjacent bowls. Put coins in alternate bowls until one complete tour has been made, then remove one coin from each adjacent pair. For example, with 11 bowls, put coins in bowls 1 and 3, 5 and 7, 9 and 11, and then take coins from bowls 3 and 5, 7 and 9, leaving coins in adjacent bowls 11 and 1. With 13 bowls put coins in 1 and 3, 5 and 7, 9 and 11, 13 and 2, and take from 3 and 5, 7 and 9, 11 and 13, leaving coins in adjacent bowls 1 and 2.

This algorithm enables us to move one coin to the next bowl; for example, suppose we wish to move a coin from bowl 1 to bowl 2. Use the above method to put coins in 2 and 3 and then remove 1 and 3. We can now achieve any distribution with an even number of coins. Put the right total number of coins in any way you like, then move coins one by one until you get the desired distribution.

R234. Let s, a, b, c, d be as in the solution, so that the perimeters are $p = s + a + b$, $q = s + b + c$, $r = s + c + d$ and $p - q + r = s + d + a$. It is more convenient algebraically to orient the square as shown, where there is no loss of generality in assuming that P is in the triangular octant as indicated.

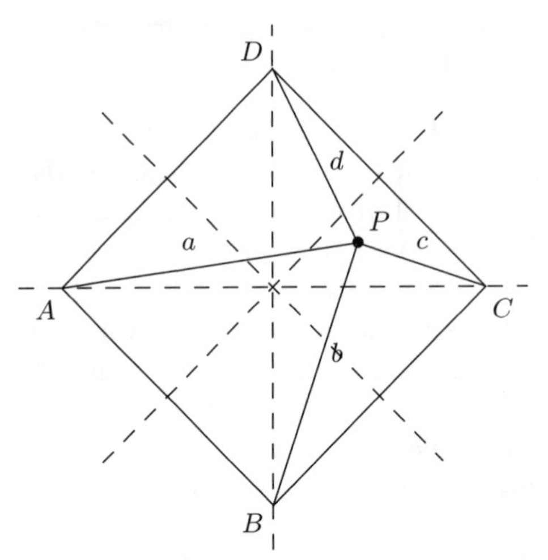

Given the location of P, $a \geq b \geq d \geq c$, so that $p \geq q \geq r$. Write $a - c = p - q = u$, $b - d = q - r = v$ and $p + q = w$.

Then, with some tedious algebra, we can write a, b, c, d in terms of s, u, v, w:

$$a = \frac{(2s - u - w)^2 - 2u^2 + v^2}{4w - 8s} \qquad b = \frac{(2s - v - w)^2 + u^2 - 2v^2}{4w - 8s}$$

$$c = \frac{(2s + u - w)^2 - 2u^2 + v^2}{4w - 8s} \qquad d = \frac{(2s + v - w)^2 + u^2 - 2v^2}{4w - 8s}$$

and $A = (-t, 0)$, $B = (0, -t)$, $C = (t, 0)$, $D = (0, t)$, and $P = (x, y)$, where $t = s/\sqrt{2}$.

Then Pythagoras's theorem tells us that

$$a^2 = (x + t)^2 + y^2, \quad b^2 = x^2 + (y + t)^2, \quad c^2 = (x - t)^2 + y^2, \quad d^2 = x^2 + (y - t)^2$$

and

$$a^2 + c^2 = 2(x^2 + y^2 + t^2) = b^2 + d^2.$$

We can use the cosine formula to find the cosines of angles PBA and PBC. These are complementary so the sum of the squares of their cosines is 1:

$$(s^2 + b^2 - a^2)^2 + (s^2 + b^2 - c^2)^2 = 4s^2 b^2.$$

Substitute the expressions for a, b, c, d into this last equation to get an equation for the length of the side of the square, s, in terms of the perimeters p, q, r (actually in terms of u, v, w).

$$\begin{aligned}32s^6 \; - \; 32s^4w^2 \; + \; 16s^3w(w^2 - u^2 - v^2) \; - \; 2s^2(w^4 - 10w^2(u^2 + v^2) \; + \\ + \; 5(u^2 - v^2)^2) \; - \; 8sw(w^2(u^2 + v^2) \; - \; (u^2 - v^2)^2) \; + \\ + \; w^4(u^2 + v^2) \; - \; 2w^2(u^2 - v^2)^2 \; + \; (u^2 + v^2)(u^2 - v^2)^2 \; = \; 0.\end{aligned}$$

The point P is a point of intersection of the two hyperbolas $|a-c| = u$ and $|b-d| = v$ whose foci are A, C and B, D respectively. They have equations

$$\frac{x^2}{u^2/4} - \frac{y^2}{(2s^2 - u^2)/4} = 1 \qquad \frac{y^2}{v^2/4} - \frac{x^2}{(2s^2 - v^2)/4} = 1$$

and intersect in the points

$$x^2 = \frac{u^2(2s^2 - v^2)(2s^2 - u^2 + v^2)}{8s^2(2s^2 - u^2 - v^2)} \qquad y^2 = \frac{v^2(2s^2 - u^2)(2s^2 + u^2 - v^2)}{8s^2(2s^2 - u^2 - v^2)}.$$

If we solve the sixth degree equation for s, using the values $u = 1.4$, $v = 1$, $w = 9$ from the problem, we get six real roots, and the corresponding values for a, b, c, d, x, y are

s	a	b	c	d	x	y
−10.7452	8.3147	8.1305	6.9147	7.1305	0.7015	0.5021
−1.2446	3.5514	3.3932	2.1514	2.3932	2.2680	1.6438
1.3901	2.2164	2.0935	0.8164	1.0935	1.0799	0.8106
1.6003	2.1085	1.9912	0.7085	0.9912	0.8713	0.6589
4.4627	−2.4991	3.7364	−3.8991	2.7364	0.7097	0.5128
4.5367	3.9506	−2.7873	2.5506	−3.7873	0.7093	0.5124

Although the perimeters add up correctly, the first two rows have a negative square side, which we can't draw, and the last two rows have negative distances from P to two corners. But the third and fourth rows work out and correspond to the figures (drawn to scale) as shown.

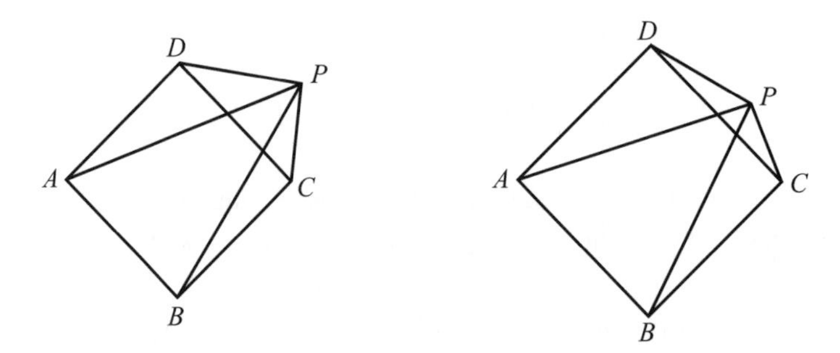

For Rikki-Tikki-Tavi. Find the relationship between the lengths of the perimeters, p, q, r and the numbers of solutions, the size of the square, and whether the points P are inside or outside the square. For example, if one perimeter is very different from the other two, then it's clearly impossible: how different is that?

There's a fairly old unsolved problem here as well. It isn't known if you can find a point which is an integer distance from all four corners of an integer-sided square. See Richard K. Guy, *Unsolved Problems in Number Theory,* 2nd edition, Springer, New York, 1994, Section D19.

R236. Subdivide the square into four congruent squares, and place a sentry at the center of each of these subsquares.

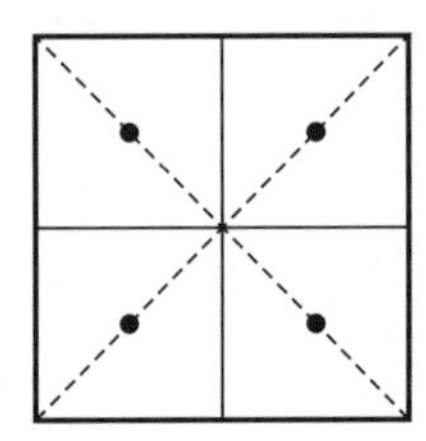

By looking at the corners and the center, it is clear that this is the optimal placement of the four sentries.

In the case of five sentries, divide the square into rectangles as shown, in such a way that $SH = HR = 4$ (all distances are in rods), $PK = KL = LQ = 8/3$, and $XH = XK (= KJ = JQ)$.

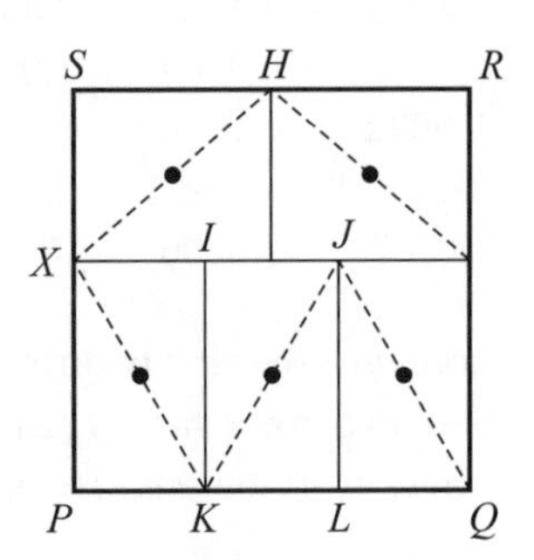

Let $a = SX$. Then, because $XH = XK$,

$$\begin{aligned} 4^2 + a^2 &= \left(\frac{8}{3}\right)^2 + (8-a)^2 \\ 16 + a^2 &= \frac{64}{9} + 64 - 16a + a^2 \\ 16a &= \frac{64}{9} + 48 \\ a = \frac{496}{144} &= \frac{4}{9} + 3 = 3.44444\overline{4} \end{aligned}$$

A point in the square can be no further than

$$\frac{\sqrt{a^2+4^2}}{2} = 2.639327\ldots$$

from one of the sentries and this is less than $2\sqrt{2} = 2.828427125\ldots$, so the mayor is wrong.

The optimal placement of sentries is an arrangement much like the previous example (see figure below) except that the common diagonal is chosen to equal the diameter of the circumcircle of triangle ABC. In this optimal case, no point in the square is more than 2.609284... rods from a sentry (the diameter, 5.21857..., is the smaller real root of the equation $1024\,x^6 + 2048\,x^5 + 1024\,x^4 - 655360\,x^3 + 360448\,x^2 - 11010048\,x + 111411200 = 0$). Our proof of optimality involves the kind of mathematics that Swinnerton-Dyer says that no gentleman would do in public.

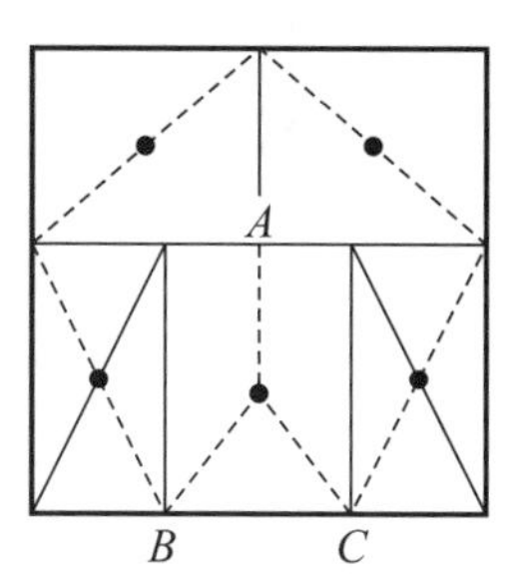

R243. We can interpret the problem in a graph-theoretic setting. Consider the animals as vertices of a graph, and draw an edge of the graph between two animals if they are compatible. Since each animal is incompatible with at most three others, it is compatible with at least four, so that the valence of each vertex of our graph is at least four. Now there's a theorem due to Gabriel Dirac, stepson of the famous physicist, which says that if the minimum valence of a graph on n vertices is at least $n/2$, then the graph contains a Hamilton circuit (see **Dirac's Theorem** and **Hamilton circuits** in the Treasury). So four pairs of consecutive animals on this circuit are compatible and can share a cage.

This is a very short solution, but won't satisfy most readers because it is **nonconstructive**: it doesn't tell you which animals to put in which cages. We'll elaborate under **Graphs** and **Hamilton circuits** in the Treasury.

R244. Suppose, to make things concrete, we have nine guests, labeled 0, 1, ... , 8, and that they are seated as shown (the placecards labels are inside the circle, the guest numbers on the outside). (Note that there is no loss of generality in supposing that the placecards are placed consecutively around the circle, for if not, simply renumber the labels on the placecards and guests in a consistent manner.)

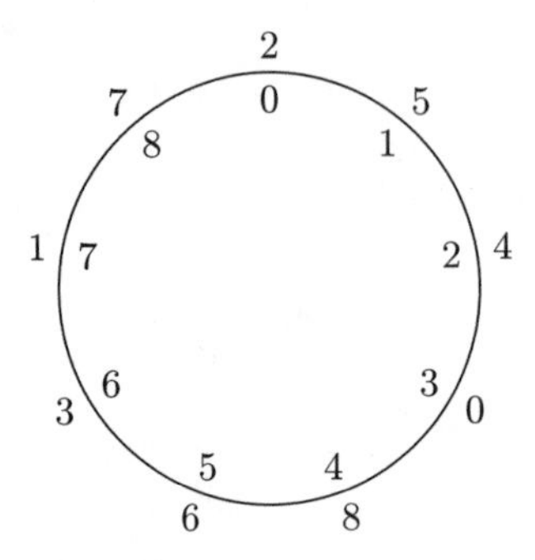

We can represent this seating arrangement by marking appropriate squares on a 9×9 grid. A square (a, b) is marked with an $\times$ if guest a is at the seat with placecard b.

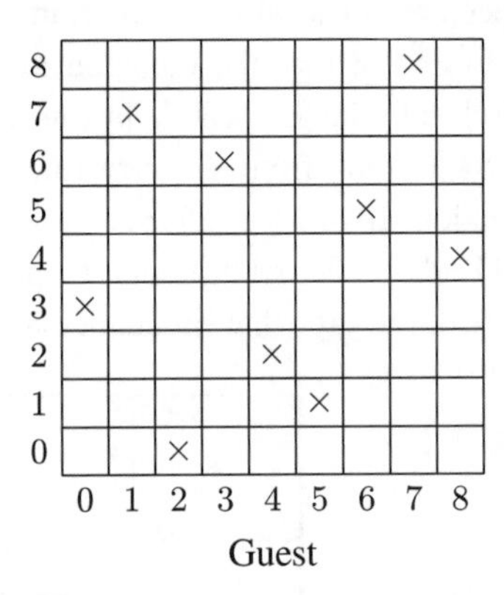

In this example, none of the guests is properly seated. Geometrically, this means that there is no marked square on the main diagonal (the rising diagonal from the lower left corner to the upper right corner).

Now consider the rising diagonals that start on the left and upon reaching an edge, wrap around to the opposite side and continue. The following diagram shows two such diagonals, labeled 2 and 6.

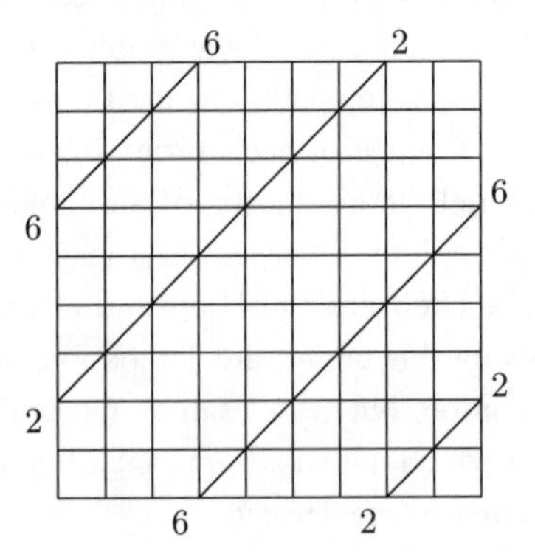

There are exactly 9 such rising diagonals, which we can label with the numbers of the rows in which they start at the left edge.

When each guest moves one seat to the right, the number on the placecard in front of them goes down by 1, except for the guest in position 0, whose placecard number goes up by 8. On the 9×9 grid, the marked squares all move down one row, except for the mark in the bottom row, which moves up to the top row. Squares on the same rising diagonal will remain on the same rising diagonal after the move and the number of their diagonal will go down by 1, unless it is diagonal 0, in which case it goes up by 8.

In the original seating, no guests were properly placed. That means that there were no marks on the 0 diagonal. Thus, one of the other rising diagonals must have at least two marks. Indeed, diagonals 3, 5, and 7 each have two marks. Thus, after 3 shifts (or 5 shifts, or 7 shifts) to the right, diagonal 3 (or 5, or 7, respectively) will be in the 0 diagonal position, and two of the guests will be properly placed.

This argument is general, and 9 can be replaced by any integer larger than 1.

R246a. The boards below show that for each k between 0 and 13 (inclusive) there is a array of twelve squares that can be covered with six dominoes in exactly k different ways.

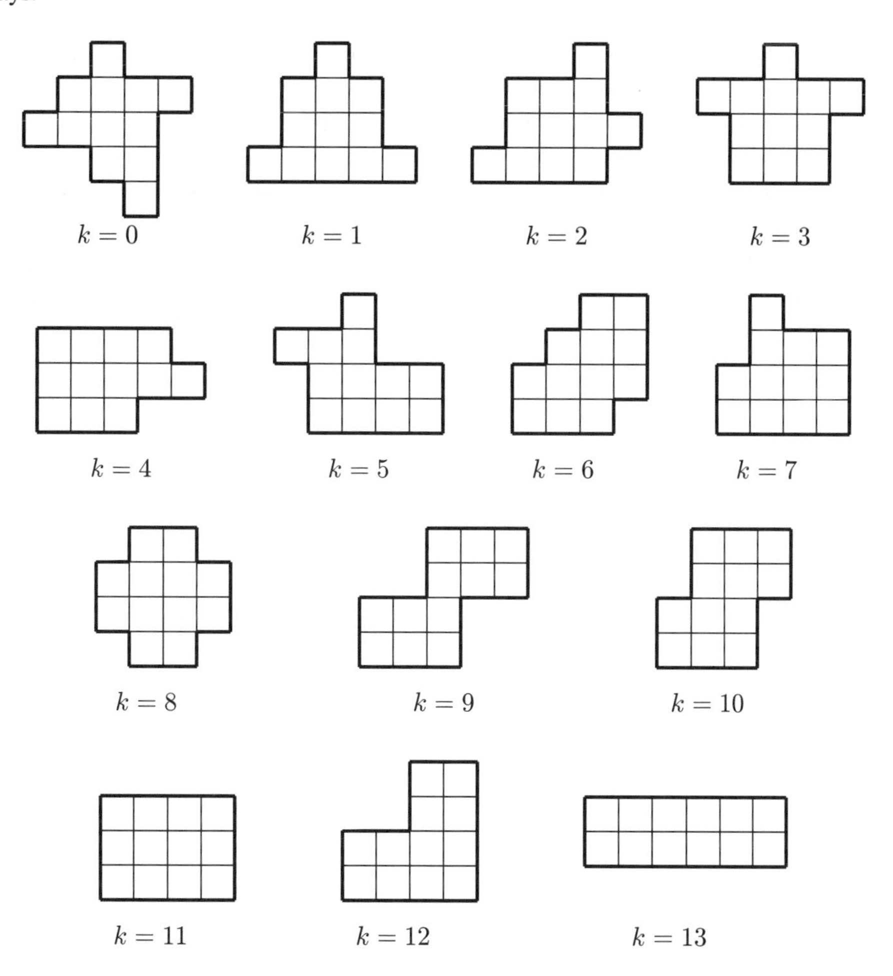

R246b. We know that the number of ways to tile a 2×8 rectangle with dominoes is $T_8 = u_9 = 34$. To see how many ways there are to tile a 4×4 square, slice it equatorially and note that there must be an even number of dominoes crossing the equator:

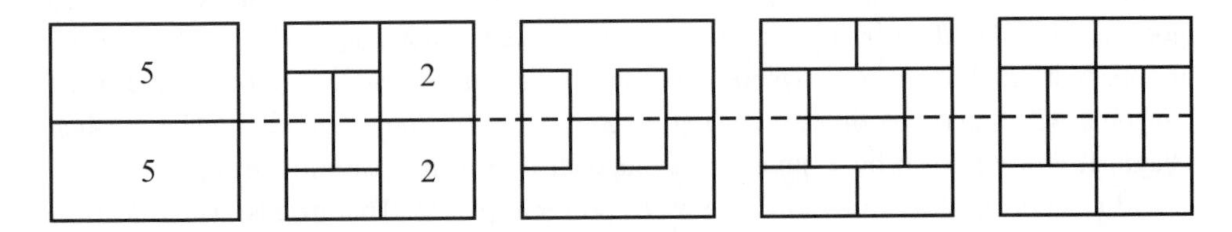

The numbers in the regions are the numbers of ways of tiling them with dominoes. Note that the second square can be reflected, left to right, so that, by the **Fundamental counting principle** (see the Treasury), the total number of tilings is

$$(5 \times 5) + (2 \times 2 \times 2) + 0 + 1 + 1 = 35, \quad \text{winning by one!}$$

R248. There is no solution for $t = 15$. The brute force way to verify this is to use a computer and the backtracking algorithm to test all possibilities. But we can use our results for $t = 10$ to prove our claim!

A solution mod 15 would have to coincide, modulo 5, with a mod 10 one for the first 4 rows (a mod 15 solution gives a mod 5 solution, and a mod 10 solution gives a mod 5 solution), so the fourth row, mod 5, is 1 1 4 4 or 2 2 3 3 or one of their reversals. If the fifth row is $abcde$, then, modulo 5, $a + b \equiv b + c$ and $c + d \equiv d + e$, so that $a \equiv c \equiv e$. On the other hand the fifth row must comprise one member from each residue class modulo 5, a contradiction.

If the modulus is not a triangular number (or even if it is), then it may still be possible to find trapezoids, instead of triangles, which have the same property. For example, 12 can be arranged as

```
    *  *  *
  *  *  *  *
*  *  *  *  *
```

There are solutions, and they come in sets of four (multiply each of the following by 1, 5, 7, or 11).

```
    0  6  9           1  0  2            1  0  7            3  0  9
  1  11  7  2       5  8  4  10        3  10  2  5        1  2  10  11
5  8  3  4  10    6  11  9  7  3    11  4  6  8  9     5  8  6  4  7
```

For Rikki-Tikki-Tavi. Investigate modular trapezoids and modular triangles for other moduli. Find as many solutions as you can for each modulus. Are there any moduli whose residues cannot cannot be arranged as a trapezoid? Find examples where each residue occurs twice, or thrice.

For more on this, see Marc Roth & Harry Nelson, Counting certain triangles of modular sums, *J. Recreational Math.*, **30**(1999–2000) pp. 23–28.

R250. Consider a round-robin tournament between four players, A, B, C, D, and suppose that it is possible that every player could win a gold medal. For this to happen, no player could win 3 games, and every player must win at least one. Altogether they play $\binom{4}{2} = 6$ games, and one of them, say A, must have won two (the scores have to be 2, 2, 1, and 1). So, suppose that A beats B and C, and that B beats C. For C to received a gold medal, it is necessary that C $\to$ D $\to$ B. Therefore, the only player B had defeated is C and C was beaten by A. Therefore B can't have received a gold medal (there is no X such that B $\to$ X $\to$ A). So, not all players can win a gold medal in a tournament with 4 players.

For a tournament with 5 or more players, however, it is possible that each of them could win a gold medal. The patterns of results, shown here for $n = 9$ and $n = 10$ are general and can be used to prove the claim for any odd, or even, integer ≥ 5.

	0	1	2	3	4	5	6	7	8
0	–	1	1	1	1	0	0	0	0
1	0	–	1	1	1	1	0	0	0
2	0	0	–	1	1	1	1	0	0
3	0	0	0	–	1	1	1	1	0
4	0	0	0	0	–	1	1	1	1
5	1	0	0	0	0	–	1	1	1
6	1	1	0	0	0	0	–	1	1
7	1	1	1	0	0	0	0	–	1
8	1	1	1	1	0	0	0	0	–

	0	1	2	3	4	5	6	7	8	9
0	–	1	1	1	1	1	0	0	0	0
1	0	–	1	0	1	0	1	0	1	0
2	0	0	–	1	0	1	0	1	0	1
3	0	1	0	–	1	0	1	0	1	0
4	0	0	1	0	–	1	0	1	0	1
5	0	1	0	1	0	–	1	0	1	0
6	1	0	1	0	1	0	–	1	0	1
7	1	1	0	1	0	1	0	–	1	0
8	1	0	1	0	1	0	1	0	–	1
9	1	1	0	1	0	1	0	1	0	–

A 1 in the grid means that the player in that row defeated the player in that column. The pattern for an odd number of players, say $2n + 1$ players, is that player k defeats players $k+1, k+2, \ldots k+n$, where we work modulo $2n+1$. Since player $k+i$ defeats player $k+i+n$ for $i = 1, 2, \ldots n$, player n will "indirectly" beat players $k+n+1, k+n+2, \ldots, k+2n$, and will therefore win a gold medal.

For an even number of players, say $2n$ players, the zeroth player defeats players $1, 2, \ldots, n$. This completely determines the top row and leftmost column. Now remove this row and column and consider the remaining $2n-1 \times 2n-1$ table (retaining the original row and column labels). In the k-th row, place a 1 in columns $k+1, k+3, \ldots, k+(2n-3)$ where we work modulo $2n-1$, and write $2n-1$ if the remainder is 0. Player $k+(2i+1)$ defeats player $k+(2i+1)+1 = k+2(i+1)$, and therefore, k defeats or indirectly defeats everybody except possibly the zeroth player which we have yet to check. The zeroth player has defeated players 1, 2, $\ldots, n$, and players $2n-1$ and $2n-2$ were defeated by one of these players. In fact, for each player k, $k = 2, 3, \ldots, 2n-2$, players $2n-1$ and $2n-2$ were defeated by one of the players who beat k. But both $2n-1$ and $2n-2$ defeated the zeroth player, and therefore every player has either beaten, or indirectly beaten, every other player (including the zeroth player). This means that each player should receive a gold medal.

R252. The analysis of the solution shows that $3(k-1) \leq 2n$, where n is the number of lines in each of the three directions. It follows that $k \leq 2n/3+1$. If $n = 3r-1$ we can't put more than $2r$ coins on the grid, and if $n = 3r$ or $3r+1$ we can't put more than $2r+1$ coins. But we can put that many. For example, in the first case, put coins on the points with coordinates $(0, r, 2r-1)$, $(1, r+1, 2r-3)$, $(2, r+2, 2r-5), \ldots,$ $(r-2, 2r-2, 3)$, $(r-1, 2r-1, 1)$ and $(r+1, 0, 2r-2)$, $(r+2, 1, 2r-4), \ldots,$ $(2r-1, r-2, 2)$, $(2r, r-1, 0)$. In the other two cases we may put the coins on $(0, r, 2r)$, $(1, r+1, 2r-2)$, $(2, r+2, 2r-4), \ldots,$ $(r-1, 2r-1, 2)$, $(r, 2r, 0)$, $(r+1, 0, 2r-1)$, $(r+2, 1, 2r-3), \ldots,$ $(2r-1, r-2, 3)$, $(2r, r-1, 1)$. Here are pictures for the first few values of n.

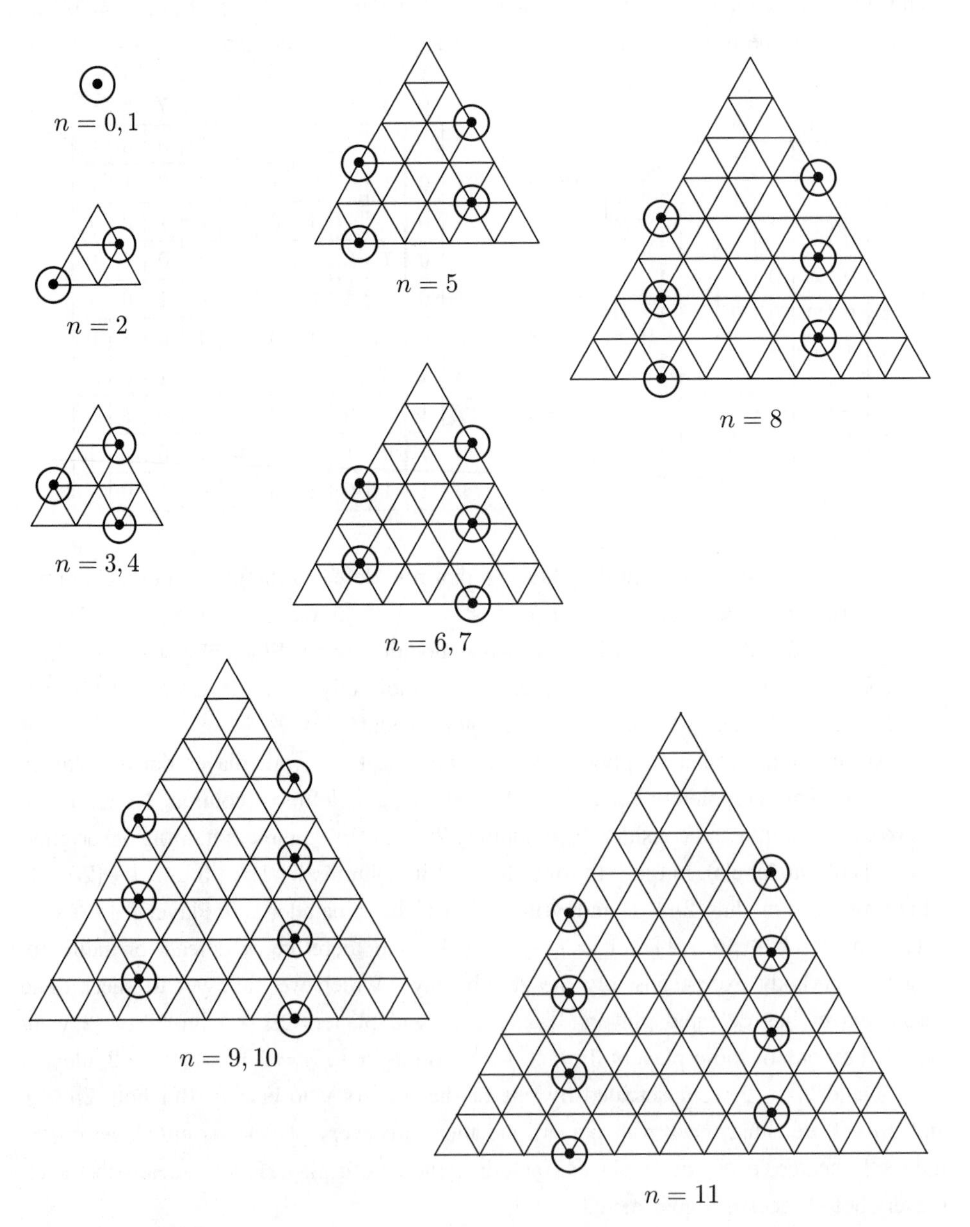

The Treasury

Technical Terms, Techniques, Tricks of the Trade

Algebra. A great many problems are easy to formulate and solve using the notation of elementary algebra. Two useful formulas are

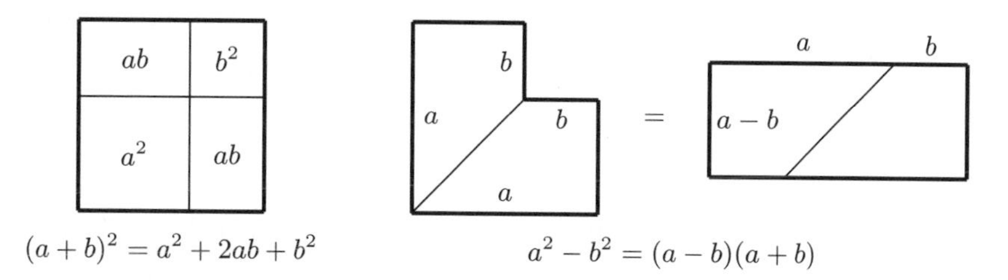

$(a+b)^2 = a^2 + 2ab + b^2$ $\qquad$ $a^2 - b^2 = (a-b)(a+b)$

These both have generalizations to higher dimensions. For the first, see **Binomial coefficients** and **Binomial theorem**. The second, useful in **P79**, is related to **Geometric series**:

$$a^3 - b^3 = (a-b)(a^2 + ab + b^2)$$
$$a^4 - b^4 = (a-b)(a^3 + a^2b + ab^2 + b^3)$$
$$a^5 - b^5 = (a-b)(a^4 + a^3b + a^2b^2 + ab^3 + b^4)$$
$$a^6 - b^6 = (a-b)(a^5 + a^4b + a^3b^2 + a^2b^3 + ab^4 + b^5)$$

and so forth. Difference of squares is used to find the formula for **Pythagorean triples**.

Simple equations are helpful in **P7**, **P28**, **P52**, **P68**, **P69**, **P92**, **P93**, **P94**, **P96**, **P101**, **P134**, **P135**, **P139**, **P165**, and **P167**. Other examples of the use of algebra are listed under **Cryptarithms**, **Farey series**, **Fermat's little theorem**, and **Inequalities**.

Algorithms. Sets of instructions for carrying out a procedure or solving a problem. The oldest and most famous is the **Euclidean algorithm**. This is the idea behind **P41**, **P74**, and **P75**.

Other algorithms are used for sorting numbers (**P13**), finding a counterfeit coin (**P14** and **P159**), arranging people (**P153**, **P160**, **P166**), arranging binary objects (pennies and dimes in **P39**, zeros and ones in **P59**, heads and tails in **P90**, black and white squares in **P177**), choosing rows and columns (**P146**), placing coins (**P253**), and several for emptying cookie jars (**P34**), and four arithmetic operations (**P255**). See also **Backtracking**, **Euler circuits**, **Games** and **Hamilton circuit**.

Archimedean solids. These are polyhedra whose faces are all regular polygons, though possibly of differing numbers of edges. In an Archimedean solid, the faces that surround a given vertex are congruent to those that surround any other vertex, and these may be identified by a sequence of numbers corresponding to the number of sides of the respective faces. Thus, a cube has the form $4,4,4$, written as 4^3, and a tetrahedron is 3^3; an octahedron, dodecahedron, and icosahedron are $5^3, 3^4$, and 3^5, respectively. There are infinitely many **prisms** described by 4^2n, and **antiprisms**, 3^3n. A diagrammatic version of the square antiprism is shown in **R111**. There are just thirteen others, the truncated tetrahedron, 6^23, also shown in **R111**, and 6^24, 6^25, 8^23, 10^23, 3^44, 3^45, $(34)^2$, $(35)^2$, 3435, 3436, 3536, and 4^33. **Rikki-Tikki-Tavi** will want to make models of them all.

Arithmetic series. One of the most important formulas in mathematics is that for the sum of the first n positive integers:

$$1+2+3+\cdots+n=\frac{n(n+1)}{2}$$

The formula for $n=7$ is used in **P21**; for $n=9$ in several, including **P47**, **Q141**; for $n=13$ and 14 in **P38**; for $n=15$ and 16 in **P123** and **P244**; and for $n=43$ in **P130**. See also **P21**, **P38**, **P47**, **P123**, **P130**, **Q141**, **P150**, **P244**.

Should you forget the formula, it's easy to derive, as Gauss did when he was a small boy. Write the sum in increasing and in decreasing order, as shown.

$$\begin{array}{lcccccccccccc} S_n & = & 1 & + & 2 & + & 3 & + & \cdots & + & n-1 & + & n \\ S_n & = & n & + & (n-1) & + & (n-2) & + & \cdots & + & 2 & + & 1 \end{array}$$

Adding yields

$$2S_n \;=\; (n+1) \;+\; (n+1) \;+\; (n+1) \;+\; \cdots \;+\; (n+1) \;+\; (n+1),$$

so that $2S_n=n(n+1)$ and the formula follows. Indeed, twice a triangular number is a so-called **pronic** number and this is illustrated in Figure (i) under **Triangular numbers**.

More generally, an arithmetic series is a sum in which each term after the first is obtained from the previous one by adding or subtracting a fixed constant (the **common difference**)

$$a+(a+d)+(a+2d)+(a+3d)+\cdots+\big(a+(n-1)d\big).$$

Note that in order to have n terms the nth will have to be $a+(n-1)d$. This is what we call Theorem 0 of Mathematics:

Theorem 0. *There is one less piece of wire than there are telegraph poles.*

To sum these n terms, it's easiest to do the same as Gauss

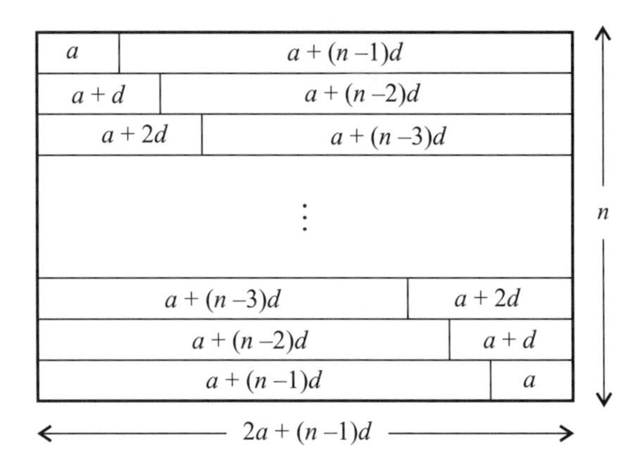

and notice that it's n times the average of the first and last terms,

$$n \times \left(\frac{a + (a + (n-1)d)}{2}\right) = \frac{1}{2}n\bigl(2a + (n-1)d\bigr).$$

Arrangements. Problems that require objects to be grouped in particular ways to satisfy stated conditions. We are arranging numbers in a row or round a circle in **P30**; people round a table in **P174**; heads and tails in **P177**; families in apartments in **P190**; coins on a grid in **P192**; queens on a chessboard in **P217**; and animals in cages in **P243**. Also, see **Fundamental counting principle** and **Permutations**.

Associativity. When we add three numbers, $a + b + c$, it doesn't matter if we do it as $(a + b) + c$ or as $a + (b + c)$. This is known as the associative law. On the other hand, $a - b - c$ is ambiguous, because $(a - b) - c$ is not the same as $a - (b - c)$.

In **P237a**, **P237b** and **P242**, we are replacing the pair (a, b) with $a+b$, $a+b+ab$ and $a + b \pmod{13}$ respectively, and in each case the operation is associative, and we finish with the same answer, but in **P178**, where we replace (a, b) with $a^2 - b^2$, it matters very much which way we associate the operations. For example, $\{5, \{3, 2\}\} \to \{5, 5\} \to 0$, but $\{\{5, 3\}, 2\} \to \{16, 2\} \to 252$.

In **S142**, we must associate the two operations in each row and column properly: although it doesn't matter in the top rows of the solutions, the bottom row of solution (a) must read $(4 \times 3) + 4 = 16$ and not $4 \times (3 + 4)$.

Backtracking. A method of systematically listing all alternatives by identifying the branches on a tree using a depth-first approach. Systematic listmaking is often useful in problems of arrangements and configurations, but it is expensive of time, and when there are thousands of cases to consider, it is only feasible on a computer. One usually tries to avoid the tediousness of backtracking by using ingenuity (for example, see **P105**).

As an example of backtracking, consider how one might make a list of all the subsets of $\{a, b, c\}$. Begin by considering a (will it go into the subset, or not?), then b (will it go in the subset, or not), and then c (will it go into the set, or not?). Here's the tree of possibilities (a bar over the element means that element is not chosen):

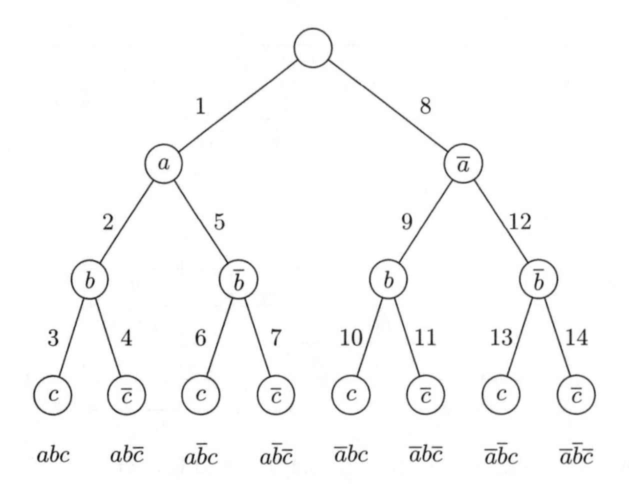

Each branch of the tree corresponds to a subset, and we can list the subsets by traversing the branches in the order specified along the edges (start at the top and at each juncture go to the left until you reach a terminal point, then backtrack and take the leftmost choice that has not yet been taken and proceed as before). This algorithm produces the list of subsets in the following order:

1. $\{a, b, c\}$
2. $\{a, b\}$
3. $\{a, c\}$
4. $\{a\}$
5. $\{b, c\}$
6. $\{b\}$
7. $\{c\}$
8. $\{\,\}$

where $\{\,\}$ denotes the empty set, the set with no elements.

Problem **Q176** gives a detailed illustration of the backtracking algorithm. Systematic listmaking is often useful in problems of arrangements and configurations.

Binary notation. In binary notation, numbers are written as sums of powers of 2, and expressed in base 2 using only 2 digits, 0 and 1. For example, $26 = 16 + 8 + 2$, which is abbreviated by 11010, where it is understood that

$$11010 = 1 \cdot 2^4 + 1 \cdot 2^3 + 0 \cdot 2^2 + 1 \cdot 2^1 + 0 \cdot 2^0 = 16 + 8 + 2 = 26$$

in base 2. Note that $2^1 = 2$ and $2^0 = 1$.

The first ten integers written in binary notation are

$$1,\ 10,\ 11,\ 100,\ 101,\ 110,\ 111,\ 1000,\ 1001,\ 1010.$$

Binary notation is useful in problems involving halving or doubling, and in situations where there are just two states or two kinds of objects. In **Q8** the parrot is uttering two kinds of noise and doubling the length of his word; in **P13** we wish to arrange numbers with their doubles in different camps; in **P43** the number of cells is doubling; and in

Q90 the pennies are in one of two states. **P63** is a cryptarithm in binary notation, and in **R192a** we use binary numbers to identify coin arrangements. Compare **Decimal notation** and **Positional notation**.

Binomial coefficients. The coefficients in the expansion of the binomial expression $(x+y)^n$. The first few powers of $(a+b)^n$ are

$$(a+b)^0 = 1$$

$$(a+b)^1 = a+b$$

$$(a+b)^2 = a^2 + 2ab + b^2$$

$$(a+b)^3 = a^3 + 3a^2b + 3ab^2 + b^3$$

$$(a+b)^4 = a^4 + 4a^3b + 6a^2b^2 + 4ab^3 + b^4$$

$$(a+b)^5 = a^5 + 5a^4b + 10a^3b^2 + 10a^2b^3 + 5ab^4 + b^5$$

(see **Binomial theorem**), so the binomial coefficients are

$$\begin{array}{ccccccccccc} & & & & & 1 & & & & & \\ & & & & 1 & & 1 & & & & \\ & & & 1 & & 2 & & 1 & & & \\ & & 1 & & 3 & & 3 & & 1 & & \\ & 1 & & 4 & & 6 & & 4 & & 1 & \\ 1 & & 5 & & 10 & & 10 & & 5 & & 1 \\ & & & & & \vdots & & & & & \end{array}$$

This triangular array of numbers is called **Pascal's triangle**.

For a nonnegative integer n and each integer k, $0 \le k \le n$, the coefficient of $x^{n-k}y^k$ is equal to $\dfrac{n!}{k!(n-k)!}$, which we denote by $\binom{n}{k}$. This symbol is read "n choose k" because it counts the number of ways of choosing k objects from a collection of n different objects. To see this, we know that a particular subset of k different objects can be arranged $k!$ different orders (see the argument in **Fundamental counting principle**). If there are S ways of choosing k objects from n different objects, then the total number of arrangements of length k with elements chosen from among n different objects is $S \times k!$. On the other hand, this number is equal to $n!/(n-k)!$ (see **Arrangements**). Therefore, $S \times k! = n!/(n-k)!$, and it follows that

$$S = \frac{n!}{k!(n-k)!} = \binom{n}{k}.$$

This interpretation of the binomial coefficients makes it easy to verify relationships between them. For example, Pascal's triangle suggests bilateral symmetry: $\binom{n}{k} = \binom{n}{n-k}$. To see that this is indeed the case, the left side of the identity counts the number of ways of choosing k objects from among n different objects, but this is the same as the number of ways of choosing $n-k$ objects (to leave out) from the n objects.

It is also easy to see that

$$\binom{n}{k} = \binom{n-1}{k} + \binom{n-1}{k-1}.$$

To see this, keep your eye on one particular object, say X. The number of ways of choosing k from n is $\binom{n}{k}$, and these sets of k can be classified into those that don't include X and those that do. The number which don't include X is $\binom{n-1}{k}$ (choose all k from the other $n-1$), whereas the number that do include X is $\binom{n-1}{k-1}$ (choose $k-1$ objects different from X and then adjoin X to the set).

The addition formula makes it easy to extend Pascal's triangle row by row. For example, the next row in Pascal's triangle (shown above) is

$$1 \quad 6 \quad 15 \quad 20 \quad 15 \quad 6 \quad 1$$

which means that $(x+y)^6 = x^6 + 6x^5y + 15x^4y^2 + 20x^3y^3 + 15x^2y^4 + 6xy^5 + y^6$.

In **R219** there are $\binom{21}{2} = 21 \times 20/2 = 210$ pairs of people on 21 comittees, each committee containing $\binom{5}{2} = 10$ pairs, so each pair serves on exactly one committee.

Binomial theorem. A general formula for the expansion of $(x+y)^n$.

$$(x+y)^n = \binom{n}{0}x^n + \binom{n}{1}x^{n-1}y + \binom{n}{2}x^{n-2}y^2 + \cdots + \binom{n}{k}x^{n-k}y^k + \cdots + \binom{n}{n-1}xy^{n-1} + \binom{n}{n}y^n$$

To prove this, multiply out the product

$$(x+y_1)(x+y_2)\cdots(x+y_n),$$

by taking all possible products of n terms, one from each binomial factor, to get x^n (choose x from each factor) + $x^{n-1}(y_1 + y_2 + \cdots y_n)$ (choose y from one of the factors and x from each of the others) + $x^{n-2}(y_1y_2 + y_1y_3 + \cdots y_2y_3 + \cdots + y_{n-1}y_n)$ (choose y from two of the factors and x from each of the others) + and so on. Now drop the subscripts from each y, and collect the terms. (See **Binomial coefficients**).

Coloring. A method for showing that an arrangement or configuration is impossible. In **S104** we have black and white numbers; in **S118** black and white cubes; in **S121** black and white halves of dominoes. There are two colors for the positions of chips in **S123** and for positions of disks in **S128b** and **S129**, and two colors of triangle in **S152**. We use three colors in **S155c** and **S212**, while for the T- and L-tetrominoes in **S238** we use two colors, and for the Z- and I-tetrominoes and in **S156** we are effectively using four colors. The fact that the T-tetromino has three squares of one color and only one of the other is also crucial in **P119b**.

Combinatorial geometry. This refers to problems related to arrangements and configurations of geometric objects. For example, **P46**, **P120**, and **P251** are about intersections of straight lines: how many triangles are formed in the first, the number of intersections in the second, and their labeling in the third. **P111** is about touching circles, and **P112** about the intersection of the edges of a polygon.

Commutativity. In a binary operation such as addition or multiplication the order of the summands or factors doesn't matter: $a+b=b+a$, $a\times b=b\times a$. This is known as the commutative law. Subtraction and division are not commutative: in general, $a-b\neq b-a$, $a/b\neq b/a$.

Configurations. This is a general term and covers geometrical configurations of points, lines, circles, etc., such as in **P46**, **P111**, and **P252**, as well as arrays of numbers: **magic squares**, **latin squares** (**P116**, **P131**), other rectangular arrays (**P77**, **P124**, **P125**, **P144**) and more general configurations (**P93**, **P94**, **P100**, **P147**, **P151**, **P170**). Sol LeWitt's puzzle is an interesting example of a configuration and here's a related problem.

Those squares of LeWitt, with the lines,
Will make a great many designs.
But can you bestow,
In each column and row,
Two each of four kinds of inclines?

That is, can you arrange the sixteen pieces of the LeWitt puzzle described in **P127** in a square so that each row and column will have exactly two horizontal lines, two vertical lines, two up diagonals and two down diagonals? (For Barry Cipra's ingenious insight, see **Magic squares**.)

A famous configuration is the **Fano configuration**, although it was discovered by Kirkman, much earlier than Fano. It is the simplest nontrivial *projective geometry*, comprising seven points and seven lines. As this is not a Euclidean geometry, we can't represent all the lines by straight lines; one of them is a circle. In this geometry, each line contains three points, each pair of points is on a single line, and each pair of lines intersect in a single point. There are two quite distinct labelings:

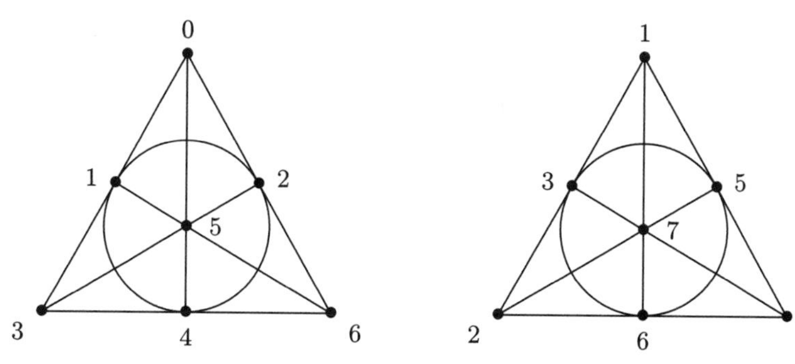

The first labeling is done by cycling the **difference set** $\{0,1,3\}$ modulo 7; that is, the lines are 013, 124, 235, 346, 450, 561, and 602.

The second labeling is done by **nim-addition**: the lines are triplets which nim-add to zero and so are heap sizes of P-positions in the game of **Nim**: 123, 145, 167, 246, 257, 347, 356. We've arranged the labeling so that the only nim-sum (see **Nim-addition**) which isn't an ordinary sum (the first two add to the third) corresponds to the only line which isn't straight.

Each labeling leads to many things, for example, the first gives Heawood's 7-color map on the torus (see Gerhard Ringel, *Map Color Theorem*, Springer-Verlag, 1974, p. 3, Fig. 1.4), and the second is the basis for the Hamming code (see **Error-correcting codes**). The table of scores in **R161** is an incidence matix for the Fano configuration!

Congruences. See **Modular arithmetic**.

Contradiction. See **Proof by contradiction**.

Continued fraction. See **Euclidean algorithm** and (linear) **Diophantine equation**.

Cosine formula. An extension of the Pythagorean theorem to arbitrary triangles with angles A, B, C and opposite sides a, b, c. It enables you to calculate the length of the third side of a triangle if you know the other two and the angle between them:

$$a^2 = b^2 + c^2 - 2bc \cos A$$

or to calculate the cosines of the angles if you know the three sides:

$$\cos A = \frac{b^2 + c^2 - a^2}{2bc}.$$

Counterfeit coins. There are many weighing problems of the kind we see in **P14 P159**, and **P256**. Here are some places where you can read more. First Freeman Dyson's beautiful ternary solution to the famous 12-coin problem, which swept the world in WWII, and Cedric Smith's versification of it:

Freeman J. Dyson, Note 1931, The problem of the pennies, *Math. Gaz.*, **30**(1946) pp. 231–233.

Blanche Descartes, The twelve coin problem, *Eureka*, **13**(1950) pp. 7, 20.

Cedric Austen Bardell Smith, The counterfeit coin problem, *Math. Gaz.*, **31**(1947) pp. 31–39.

and many references can be found in

Richard K. Guy & Richard J. Nowakowski, Coin-weighing problems, *Amer. Math. Monthly*, **102**(1995) pp. 164–167.

Coverings. These may involve overlapping, as in **P235** where pairs of circles overlap each other and the boundary of the big circle, and in **P236**, where the sentries' beats overlap; or they may be *exact* in the sense that each bit of area is covered by just one object (except that the usual convention is that the boundaries may be shared). In this latter case they may also be regarded as exact packings or dissections or **tilings**.

Cryptarithms. Puzzles in which digits are replaced by letters or symbols, and where the original digits must be found. These can be deliberately set, as **P45**, **P63**, **P142**, **P191**, **P215**, **P229**, and **P234**, or indirectly set as in **P53**, **P62**, and **P71**.

Decimal notation. The base 10 notational system for representing integers (and real numbers). For example, 2468 is shorthand for

$$2 \times 10^3 + 4 \times 10^2 + 6 \times 10^1 + 8 \times 10^0.$$

See **P107**, **P108**, **P140**; compare **Binary notation** and **Positional notation**, and see also **Digits**, **Divisibility**, and **Divisibility tests**.

Degree. An overworked word in mathematics. A unit of measurement of angle, the largest exponent in a polynomial, the size of the set that is rearranged in a permutation group, and many other things, apart from its frequent literary use. Where it refers to the number of edges incident with the vertex of a graph, we prefer to use the word **valence**, especially as this coincides with its important use in chemistry.

Difference sets. Sets of numbers whose differences comprise a complete set of nonzero residues (see **Residues**). For example, the differences of $\{0, 1, 3\}$ are ± 1, ± 2, ± 3, all the nonzero residues modulo 7, which we used to label the Fano configuration (see **Configurations**). Other examples are $\{0, 1, 4, 6\}$ mod 13 and $\{0, 2, 7, 8, 11\}$ mod 21, used in **S219**.

Digits. Problems in which the digits of the number must satisfy stated conditions. See **P24**, **P53**, **P62**, **P71**, **P84**, **P117**, **P214**, **P249**, **P254**; see also **Cryptarithms**, **Divisibility** and **Divisibility tests**.

Directed graphs. See **Graphs**

Diophantine equations. Equations in which we are only interested in integer solutions. The most famous occurs in Fermat's Last Theorem which states that if $n > 2$ then the Diophantine equation $x^n + y^n = z^n$ has no nonzero integer solutions. Fermat, around 1637, thought that he had a proof of this, but the first accepted proof, by Andrew Wiles, came only in 1995. For $n = 2$ there are infinitely many integer solutions: see Pythagorean triples.

The simplest Diophantine equation is the linear one, $ax + by = c$. If a and b have a common factor which doesn't divide c, then we can't have any integer solutions. On the other hand, if the equation does have solutions, then so does the equation $ax + by = d$ where d is a multiple of c (just take the corresponding multiples of x and y). So we need only consider the equation $ax + by = 1$, with a and b having no common factor greater than 1, since $a(cx) + b(cy) = c$ gives solutions to our original equation.

A neat way to solve $ax + by = 1$ (which always has solutions if a is prime to b, but not otherwise) is to write a/b as a *continued fraction*. For example, let us solve $37x + 13y = 5$. Write

$$\frac{37}{13} = 2 + \frac{11}{13}, \qquad \frac{13}{11} = 1 + \frac{2}{11}, \qquad \frac{11}{2} = 5 + \frac{1}{2}$$

and we get

$$\frac{37}{13} = 2 + \frac{1}{\frac{13}{11}} = 2 + \frac{1}{1 + \frac{2}{11}} = 2 + \frac{1}{1 + \frac{1}{\frac{11}{2}}} = 2 + \frac{1}{1 + \frac{1}{5 + \frac{1}{2}}}.$$

To save space this is written as

$$\frac{37}{13} = 2 + \frac{1}{1+}\,\frac{1}{5+}\,\frac{1}{2} \quad \text{or} \quad \frac{37}{13} = 2 + \frac{1}{1+}\,\frac{1}{5+}\,\frac{1}{1+}\,\frac{1}{1}.$$

The *convergents* to this continued fraction are obtained by chopping off its tail:

$$\frac{2}{1},\quad 2+\frac{1}{1}=\frac{3}{1},\quad 2+\frac{1}{1+}\,\frac{1}{5}=\frac{17}{6},\quad 2+\frac{1}{1+}\,\frac{1}{5+}\,\frac{1}{1}=\frac{20}{7},\quad 2+\frac{1}{1+}\,\frac{1}{5+}\,\frac{1}{1+}\,\frac{1}{1}=\frac{37}{11}$$

You can discover several interesting properties of these. The important one for our purposes is that neighboring fractions are close to one another in the following sense:

$$2\times1-3\times1=-1,\quad 3\times6-17\times1=1,\quad 17\times7-20\times6=-1,\quad 20\times13-37\times7=1.$$

The last of these shows that $37(-7)+13(20) = 1$, so that $(x, y) = (-7, 20)$ is a solution of $37x + 13y = 1$. All other solutions are found by adding multiples of 13 to x and subtracting the same multiples of 37 from y: $x = -7 + 13t$, $y = 20 - 37t$, where t is any integer, positive, negative or zero. The solutions of our original equation $37x + 13y = 5$ are obtained by multiplying by thest solutions by 5: $x = -35 + 65t$, $y = 100 - 185t$.

You'll find these ideas useful in **Q41** (water buckets), **S73** (curious coins), **S74** and **S75** (curious elevators), **Q111** (Euler's formula), **Q165** (triangle stacking), **Q191** (Miss Toshiyori's phone number), **R218** (Sierpinski gasket), and **Q232** (coins in bowls).

Discriminant. See **Algebra**.

Dissections. See **Tilings**.

Divisibility. We say that an integer m is **divisible** by n, or that n is a **divisor** of m, or that m is a **multiple** of n, if there is an integer a such that $m = an$. Note that every integer divides 0, but we cannot divide m by zero since $a \times 0 \neq m$ (unless m is zero, and 0/0 is meaningless).

If $N = p_1^{n_1} p_2^{n_2} \cdots p_k^{n_k}$ and $M = p_1^{m_1} p_2^{m_2} \cdots p_k^{m_k}$ are the prime factorizations of N and M ($n_i \geq 0, m_i \geq 0$), then M divides N if and only if $m_i \leq n_i$ for each i. See **Q17**, **P29**, **P33**, **P53**, **P110**, **P133**, **P134**, **P135**, **P150**, **P216**, **P242**, **P247**; also, **Fundamental theorem of arithmetic**.

Divisibility tests. See **P141** and other problems mentioned below.

Divisibility by 2 and 5.

Since 10 is divisible by 2 and 5, a number $10b + a$, ending with the digit a, will be divisible by 2 or 5 just if the last digit is divisible by 2 or 5.

Divisibility by 4 and 8, 25 and 125.

In the same way, since 100 is divisible by 4 and 25, a number will be divisible by 4 or 25 just if its last two digits form a number divisible by 4 or 25. Similarly for

8 and 125 and the last three digits of a number and so on. For example, 522680 is divisible by 8, but not by 16, and it is divisible by 5, but not by 25.

Divisibility by 3 and 9.

The arithmetical check of 'casting out the nines' consists in using the sum of the digits of a number in place of the number itself. See **P47**, **Q117**, **P183**, **P184**, **P185**, **P208**. The reason that this works is because in our decimal notation

$$\ldots edcba \qquad \text{means} \qquad \cdots + 10000e + 1000d + 100c + 10b + a$$

and if we take away the sum of the digits

$$\cdots + e + d + c + b + a$$

we are left with $\cdots + 9999e + 999d + 99c + 9b$, which is clearly divisible by 9 (and 3). So, when we divide the sum of the digits by 9, we will get the same remainder as when we divide the original number by 9. If the remainder is zero, the number is divisible by 9; if it is 3 or 6, the number is divisible by 3, but not by 9. Is 9876543210 divisible by nine?

Divisibility by 11.

For divisibility by 11 (see **P239**), just add alternate digits and look at the difference between the two sums

$$\ldots edcba \qquad \text{means} \qquad \cdots + 10000e + 1000d + 100c + 10b + a.$$

If we take away the alternating sum of the digits

$$\cdots + e - d + c - b + a$$

we are left with $\cdots + 9999e + 1001d + 99c + 11b = 11 \times (\cdots + 909e + 91d + 9c + b)$, so the alternating sum gives the same remainder on division by 11 as the original number. For example, $9 - 8 + 7 - 6 + 5 - 4 + 3 - 2 + 1 = 5$, so that when you divide 987654321 by 11 you get remainder 5.

Divisibility by 7, 11, and 13.

Because $1001 = 7 \times 11 \times 13$ and $999 = 27 \times 37$, we can give a test which gives the remainder when we divide a large number by any of these numbers. Partition the large number into three-digit numbers: $\ldots\ onm\ lkj\ ihg\ fed\ cba$ and add together alternate ones $\cdots + onm + ihg + cba$ and $\cdots + lkj + fed$. The difference between these sums,

$$\cdots + onm - lkj + ihg - fed + cba,$$

differs from the original number,

$$\cdots + onm \times 10^{12} + lkj \times 10^9 + ihg \times 10^6 + fed \times 10^3 + cba$$

by

$$\cdots + onm \times 999999999999 + lkj \times 1001000000 + ihg \times 999999 + fed \times 1001$$

$$= 1001 \times \big(\cdots + onm \times 999000999 + lkj \times 1000000 + ihg \times 999 + fed\big),$$

that is, by a multiple of $1001 = 7 \times 11 \times 13$, so, if we divide this difference by $7, 11$ or 13, we get the same remainder as would be given by the original large number. For example, it was recently claimed that 3 775 283 249 699 was prime, but

$(3+283+699)-(775+249)=985-1024=-39$ is divisible by 13, and so is the original number.

Divisibility by 27 and 37.

If, in the preceding argument for divisibility by 7, 11, 13, we take the *sum* of the three-digit numbers in the partition, then this differs from the original number by

$$\cdots + onm \times 999999999999 + lkj \times 999999999 + ihg \times 999999 + fed \times 999 + 0$$

which is a multiple of $999 = 27 \times 37$. For example, it was also claimed (see last paragraph) that 3 775 283 249 931 was prime, but $3+775+283+249+931=2241$ and $2+241=243$ are divisible by 27, and so, therefore, is the original 'prime'. (Check the numbers 3 775 283 249 933 and 3 775 283 249 937 from the same source.)

Divisors. If a number n is the product of two numbers, say $d \times e = n$, then we say that d and e are **divisors** of n and that n is a **multiple** of d, and of e. For example, the divisors of 120 are 1 & 120, 2 & 60, 3 & 40, 4 & 30, 5 & 24, 6 & 20, 8 & 15, 10 & 12, sixteen in all. We can count the number of divisors without writing them all out by factoring the number, $120 = 2^3 \times 3 \times 5$; then any divisor contains a power of 2: $2^0=1$, $2^1=2$, $2^2=4$, or $2^3=8$, and a power of 3: $3^0=1$ or $3^1=3$, and a power of 5: $5^0=1$ or $5^1=5$; that is, $4 \times 2 \times 2 = 16$ possibilities in all. In general, if

$$n = p^a \times q^b \times r^c \times \cdots$$

is the prime factorization of n, then the number of divisors of n is

$$(a+1) \times (b+1) \times (c+1) \times \cdots .$$

In **R17** we noticed that the number of divisors of $105 = 3 \times 5 \times 7$ is $(1+1) \times (1+1) \times (1+1) = 8$, that is 105 has 8 odd divisors, so that there are 16 representations of 105 as a sum of consecutive integers; 15 of them have more than one term, as we have seen.

Error-correcting codes. These are important devices whereby we can transmit messages (of zeros and ones) over 'noisy' channels, which occasionally change 0- or 1-bits into 1 or 0, yet still retrieve the original message at the other end.

They do this by using only certain 'codewords' which are recognizable as being different not only from one another, but from any non-codewords received with one altered bit. For example the Hamming code of 'words' of seven 'letters' uses only 16 out of the $2^7 = 128$ possible strings of seven zeros and ones. Each codeword has 7 neighbors (change just one of the seven digits), which, in the Hamming code, comprise exactly the 112 non-codewords. That is to say, each seven-letter word is either a codeword or differs from a codeword in only one place.

It's easy to remember the codewords. Write down all sets of different numbers less than 8 whose nim-sum is even (see **Nim-addition**). For example, 1, 2, 3 is such a set. These sets of numbers correspond to 'P-positions' in the game of Nim (positions where you'd like your opponent to be the next one to play). The complete set for heap sizes less than 8 is: 0, 123, 145, 167, 246, 257, 347, 356, 1247, 1256, 1346, 1357, 2345, 2367, 4567, 1234567.

There are quick ways to remember these. Martin Roller and John Conway looked at an ordinary die and said: take all subsets of 'pairs of opposite faces' and 'triples at the corners', appending 7 if necessary to make the sum even. Alternatively, you can look at the second labeling of the Fano configuration (see **Configurations**) and take either no points, the three points on a line, the four points not on a line, or all seven points.

Then these sets of numbers are the positions of the 1-bits in the 16 codewords:

	1	2	3	4	5	6	7
1.	0	0	0	0	0	0	0
2.	1	1	1	0	0	0	0
3.	1	0	0	1	1	0	0
4.	1	0	0	0	0	1	1
5.	0	1	0	1	0	1	0
6.	0	1	0	0	1	0	1
7.	0	0	1	1	0	0	1
8.	0	0	1	0	1	1	0
9.	1	1	0	1	0	0	1
10.	1	1	0	0	1	1	0
11.	1	0	1	1	0	1	0
12.	1	0	1	0	1	0	1
13.	0	1	1	1	1	0	0
14.	0	1	1	0	0	1	1
15.	0	0	0	1	1	1	1
16.	1	1	1	1	1	1	1

You'll see that the 'distance' between any two codewords is at least three—that is any two codewords differ in at least three places (their nim-sum has at least three 1's), so, if you alter a digit in a codeword it is still at distance 2 from any other codeword, so you can discover where the error is (in that position where it differs from the closest codeword).

For example, let's solve **P253** using this. Ask seven questions: 'Is it in the set $\{2,3,4,9,10,11,12,16\}$?', 'Is it in the set $\{2,5,6,9,10,13,14,16\}$?', 'Is it in the set $\{2,7,8,11,12,13,14,16\}$?' and so on, where the sets are the labels of the codewords which have a 1 in the first position, 1 in the second position, 1 in the third position, and so on.

Suppose I've thought of 13 and tell a lie when you ask the third of these questions: No, Yes, No (a lie), Yes Yes, No, No, i.e., 0 1 0 1 1 0 0. The nim-sum of the 'Yes' positions, 2, 4, 5, is 3, which is not zero, so this is not a codeword. The nearest codeword is got by changing position 3 from a 'No' (0) to a 'Yes' (1); the answers should have been 0 1 1 1 1 0 0, the codeword labeled 13.)

Here's a problem that's been very popular lately, which in essence goes back to a 1987 'voting with abstention' puzzle of Richard Beigel and Steven Rudich.

Three players enter a room and a red or blue hat is placed on each person's head. The color of each hat is determined by a coin toss, with the outcome of one coin toss having no effect on the others. The players can see the other players' hats but not their own. No communication of any sort is allowed, except for an initial strategy session before the game begins. Once they have had a chance to look at the other hats, the players must simultaneously guess the color of their own hats or pass. The group shares a hypothetical $3 million prize if at least one player guesses correctly and no players guess incorrectly.

How should the players play to maximize their chances of winning the prize? One approach is to have two players agree to pass and the third respond with "Red." Using this strategy, the chances of winning are 1 in 2.

Can they do better? Yes! Agree to call blue hats 1 and red hats 0. There are eight three-letter "words:"

$$000,\ 001,\ 010,\ 011,\ 100,\ 101,\ 110,\ 111.$$

Take the codewords to be 000 and 111. Each three-letter word is either a codeword or differs from a codeword in just one place. The group strategy is to bet that the three hats will not be of the same color (that is, will *not* correspond to a codeword), so they plan their strategy accordingly. They agree to keep mum if they see two different colors, but to say 'red' if they see two blue hats or 'blue' if they see two red hats. With this strategy their chance of winning is 3 out of 4.

The same game can be played with any number of players. The general problem is to find a strategy for the group that maximizes its chances of winning the prize. Suppose that you are one of seven players. Can you use the Hamming code to maximize your chances? Perhaps you'd like to think about it, and try it out on your friends before reading further.

First, they number themselves from 1 to 7 and, as above, agree to call blue hats 1 and red hats 0. Chances are that the hats are *not* arranged in a codeword. Suppose the seven players see

1	$\times$	0	1	1	0	1	0	$346 = 1$
2	0	$\times$	1	1	0	1	0	$346 = 1$
3	0	0	$\times$	1	0	1	0	$46 = 2$
4	0	0	1	$\times$	0	1	0	$36 = 5$
5	0	0	1	1	$\times$	1	0	$346 = 1$
6	0	0	1	1	0	$\times$	0	$34 = 7$
7	0	0	1	1	0	1	$\times$	$346 = 1$

They do the nim-sums of the blue hat positions shown on the right. If anyone gets zero, they know that their hat's being zero would make a codeword, so they guess 1 (blue). If anyone gets their own number, (for example, the first player), they know that their hat's being blue would make a codeword (here 1 0 1 1 0 1 0), so they guess 'red'. For anything other than 0 or the player's own number, keep quiet. If the word is not a codeword, exactly one person will make a guess, the others will pass. If the word is a codeword, all of them will guess wrong. Since codewords only occur $1/8$ of the time, this gives a 7 to 1 chance of winning.

Euclidean algorithm. The method, mentioned by Euclid, of finding the greatest common divisor of two numbers, by dividing one into the other, then dividing the remainder into the previous divisor and repeating until the remainder is zero. Suppose we wanted to find the greatest common divisor (a, b) of $a = 9555$ and $b = 1001$. Then we successively do the divisions $a = b \times q + r$, where q is the quotient and r, the remainder, becomes the next divisor, $b \to a$, $r \to b$:

$$9555 = 1001 \times 9 + 546$$
$$1001 = 546 \times 1 + 455$$
$$546 = 455 \times 1 + 91$$
$$455 = 91 \times 5 + 0.$$

and $(a, b) = (b, r) = \ldots = (91, 0) = 91$.

This is the same process as expressing the ratio a/b as a **continued fraction**.

$$\frac{9555}{1001} = 9 + \cfrac{1}{1 + \cfrac{1}{1 + \frac{1}{5}}}$$

Do you see the connexion between the "partial quotients" and the continued fraction (see **Diophantine equation**)?

Euler circuits. If you think of a graph as a map, with the vertices as towns and the edges as roadways connecting them, then a *path* is a route you might travel from one place to another (you may or may not visit the same town more than once, and you may or may not return to the starting point).

A graph is *connected* if there is a path between every pair of vertices. A *circuit* in a graph is a path whose first and last vertices are the same. An *Euler circuit* is a circuit that traverses each edge of the graph exactly once. A graph may have an Euler circuit (as in **P9**, **P19a**) or it may not (as in **P19b**, **P19c**, **P20**, **P31**).

Leonhard Euler proved that a graph has an Euler circuit if and only if the graph is connected and every vertex has even degree. To see why, consider how one might construct an Euler circuit for the following graph.

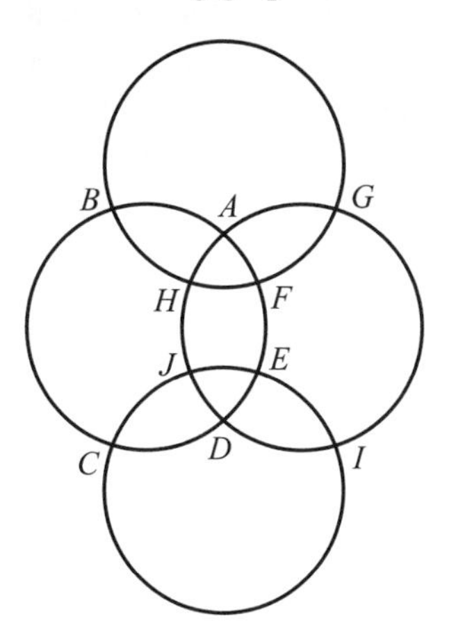

Because we're looking for a circuit, it doesn't matter on which vertex we begin, so suppose we start at point A. Now start out however you wish; you can't get stuck because in arriving at a vertex, you've traversed one edge, but because the degree is even, there has to be another edge from which you can exit.

The only difficulty is that if you haven't been forward looking, you might find yourself back at the starting point before having visited all the edges. For example,

$ABCDEFA\ GIDJHA$. In such a case, you can simply go back and insert side-trip circuits along the way. In this case, for example, when you reach E, insert the circuit, $EICJE$: $ABCDE\ ICJE\ FA\ GIDJHA$. You still haven't covered all edges, so, when you reach F, insert another circuit, $FGBHF$: $ABCDE\ ICJE\ FGBHF\ AGIDJHA$. This is a solution.

If there are just two vertices of odd degree, there is an Euler path (a path that traverses each edge exactly once, but doesn't return to the starting point) that necessarily starts at one of the odd vertices and ends at the other. Again, you can construct such a path in stages. In the following graph, a first attempt might begin at A and continue: $ABF\ GHIJKLMF\ LENCJHDA$. This path, however, hasn't traversed all the edges, so when we first reach vertex B we insert the circuit $BCDEB$ and then continue as before. The result is an Euler path:

$$AB\ CDEB\ FGHIJKLMF\ LENCJHDA.$$

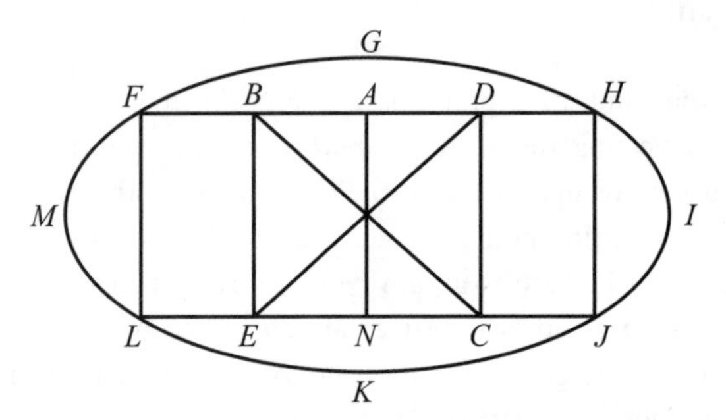

Euler's formula. In any planar graph (a graph that can be drawn in the plane without edge crossings) $V - E + F = 2$, where V is the number of vertices in the graph, E is the number of edges, and F is the number of faces (that is, the regions into which the edges partition the plane, including the region outside the graph). If we think of a planar graph as a map, the faces are the countries together with the surrounding ocean. For example, the first graph shown under **Euler circuits** has 10 vertices, 20 edges, and 12 faces, and it is the case that $10 - 20 + 12 = 2$. In the second graph, $V - E + F = 11 - 22 + 13 = 2$, or, if you count G, I, K, M as vertices, $V - E + F = 15 - 26 + 13 = 2$.

Euler's formula is the main tool in our treatment of **R111**, and it enables us to provide a solution to **P80** which doesn't require you to know how many hexagons there are! Suppose there are h hexagons and p pentagons. Then, as in **S80**, $p = 3h/5$, or $h = 5p/3$. The number of faces is $F = p + h = 8p/3$. If E and V are the numbers of edges and faces, then $2E = 3V = 6h+5p = 15p$ and Euler tells us that $2 = V - E + F = 5p - 15p/2 + 8p/3 = p/6$ so that $p = 12$.

Exponents. Sometimes called powers or indices. The use of a^2 for $a \times a$, a^3 for $a \times a \times a$, etc. is fairly familiar. If you work backwards, dividing by a, this gives a meaning to zero and negative exponents: $a^0 = 1$, $a^{-1} = 1/a$, $a^{-2} = 1/a^2$, and so forth. These definitions are consistent with the rules for dealing with integer exponents:

$$a^m \times a^n = a^{m+n}, \qquad a^m/a^n = a^{m-n}, \qquad (a^m)^n = a^{mn}$$

What meanings can you give to fractional exponents so that these rules still hold?

Factorial. The product of the first n numbers, $1 \times 2 \times 3 \times \cdots \times n$, written $n!$. For example, $1! = 1$, $2! = 2$, $3! = 6$, $4! = 24$, and so forth. If n is zero, the product is empty and it is natural to write $0! = 1$. You'll find that the formulas for binomial coefficients still hold if you use this interpretation.

We sometimes want to know the largest power of a prime in a factorial, for example, the power of 2 in 16! in **P144** and the power of 7 in 600! in **P213**. A formula for computing the largest power of a prime p that divides $n!$ is

$$\left\lfloor \frac{n}{p} \right\rfloor + \left\lfloor \frac{n}{p^2} \right\rfloor + \left\lfloor \frac{n}{p^3} \right\rfloor + \cdots$$

where $\lfloor x \rfloor$ denotes the largest integer less than or equal to the number x (for example, $\lfloor 3.17 \rfloor = 3$).

Fano configuration. See **Configurations**.

Farey sequences. The *Farey sequence of order n* is the increasing sequence of all fractions p/q reduced to their lowest terms and having $0 \le p \le q \le n$. They have several important properties. One is that two consecutive members a/b, c/d, satisfy the relation $bc - ad = 1$. Another is that between two adjacent members, the next entry to be included is their **mediant**

$$\frac{a+c}{b+d}$$

This is the "batting average" way of adding fractions — don't try this at school! It makes it easy to construct a Farey sequence for as far as you have patience:

$$\begin{array}{ccccccccccccccccccc}
\frac{0}{1} & & & & & & & & & & & & & & & & & & \frac{1}{1} \\[6pt]
\frac{0}{1} & & & & & & & & & \frac{1}{2} & & & & & & & & & \frac{1}{1} \\[6pt]
\frac{0}{1} & & & & & & \frac{1}{3} & & & \frac{1}{2} & & & \frac{2}{3} & & & & & & \frac{1}{1} \\[6pt]
\frac{0}{1} & & & & \frac{1}{4} & & \frac{1}{3} & & & \frac{1}{2} & & & \frac{2}{3} & & \frac{3}{4} & & & & \frac{1}{1} \\[6pt]
\frac{0}{1} & & & \frac{1}{5} & \frac{1}{4} & & \frac{1}{3} & \frac{2}{5} & & \frac{1}{2} & & \frac{3}{5} & \frac{2}{3} & & \frac{3}{4} & \frac{4}{5} & & & \frac{1}{1} \\[6pt]
\frac{0}{1} & & \frac{1}{6} & \frac{1}{5} & \frac{1}{4} & & \frac{1}{3} & \frac{2}{5} & & \frac{1}{2} & & \frac{3}{5} & \frac{2}{3} & & \frac{3}{4} & \frac{4}{5} & \frac{5}{6} & & \frac{1}{1} \\[6pt]
\frac{0}{1} & \frac{1}{7} & \frac{1}{6} & \frac{1}{5} & \frac{1}{4} & \frac{2}{7} & \frac{1}{3} & \frac{2}{5} & \frac{3}{7} & \frac{1}{2} & \frac{4}{7} & \frac{3}{5} & \frac{2}{3} & \frac{5}{7} & \frac{3}{4} & \frac{4}{5} & \frac{5}{6} & \frac{6}{7} & \frac{1}{1}
\end{array}$$

To see where to put the next entries, the eighths, look for consecutive denominators which add to 8: $1+7$, $3+5$, $5+3$, $7+1$. This is where to put the mediants

$$\frac{0+1}{1+7} = \frac{1}{8} \qquad \frac{1+2}{3+5} = \frac{3}{8} \qquad \frac{3+2}{5+3} = \frac{5}{8} \qquad \frac{6+1}{7+1} = \frac{7}{8}$$

The two fundamental properties enable us to find bits of Farey sequences without having to write out the whole table. In **R22** we wanted the sequence of order 20 between 3/5 and 13/20. If a/b follows 3/5, then $5a = 3b+1$. Under **Diophantine equations** we learnt how to find the solutions

$$\frac{a}{b} = \frac{2}{3}, \frac{5}{8}, \frac{8}{13}, \frac{11}{18}, \frac{14}{23}.$$

The first is outside our range, and the last has denominator > 20. We can also start from the top: if c/d is the fraction before 13/20, then $13d = 20c + 1$, a solution of which is $c/d = 11/17$. We continue on down with $9/14$, $7/11$, $5/8$, and this is where we came in. Have we found them all?

$$\frac{3}{5}, \frac{11}{18}, \frac{8}{13}, \frac{5}{8}, \frac{7}{11}, \frac{9}{14}, \frac{11}{17}, \frac{13}{20}$$

No! The mediant of 5/8 and 7/11,

$$\frac{5+7}{8+11} = \frac{12}{19}$$

has denominator less than 20, and should be included. It's easy to check that we now have them all, using the two fundamental properties.

> **Rikki-Tikki-Tavi** might wonder if the number of members of a Farey sequence is always a prime number. The first nine are 2, 3, 5, 7, 11, 13, 19, 23, 29. This was pointed out by Leo Moser long ago. But see The Strong Law of Small Numbers, *Amer. Math. Monthly*, **95**(1988) pp. 697–712, and The Second Strong Law of Small Numbers, *Math. Mag.*, **63**(1990) pp. 3–20. By all means make guesses, but don't jump to conclusions!

Fermat's little theorem. This states that if p is a prime, then p divides $a^p - a$ for any number a. This implies that if a is not a multiple of p, then p divides $a^{p-1} - 1$. This gives a neat way of solving **P193b** and **P247**.

For example, in part (a) of **P247**, $2010 = 2 \times 3 \times 5 \times 67$, and by Fermat's little theorem, 67 divides $10^{66} - 1 = 9 \times \underbrace{111\ldots1}_{66}$, so a number that satisfies the conditions is $\underbrace{111\ldots1}_{66}0$ (it is divisible by 3 because the sum of its digits is divisible by 3). For part (b) of the same problem, since 2011 is prime it divides $10^{2010} - 1 = \underbrace{999\ldots9}_{2010}$, which means that it divides $\underbrace{111\ldots1}_{2010}$.

Fermat-Toricelli point. In **R48** we connected 3 houses to a gazebo at this point. To construct it geometrically, draw equilateral triangles externally on the edges of the triangle. Their circumcircles all pass through the F-T point and their centers form another equilateral triangle. This is often called 'Napoleon's theorem'. Napoleon certainly knew some mathematics, but it is doubtful that he discovered this theorem.

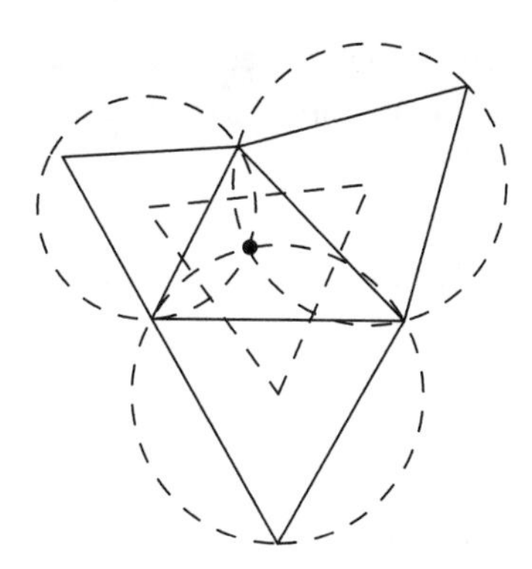

Fibonacci numbers. These are named for Leonardo of Pisa, son of Bonacci (filius Bonacci). They occur frequently in mathematics. Each is the sum of the two previous ones:

$$u_0 = 0, \quad u_1 = 1, \quad u_2 = 1, \quad u_3 = 2, \quad u_4 = 3, \quad u_5 = 5, \quad u_6 = 8, \quad u_7 = 13, \ \ldots$$

They illustrate the best known example, and one of the simplest, of a *recurrence relation*, $u_{n+1} = u_n + u_{n-1}$, which occurs in **P246**. There are dozens of formulas involving the Fibonacci numbers, especially when taken with the companion sequence of *Lucas numbers*

$$v_0 = 2, \quad v_1 = 1, \quad v_2 = 3, \quad v_3 = 4, \quad v_4 = 7, \quad v_5 = 11, \quad v_6 = 18, \quad v_7 = 29, \ \ldots$$

For example, the sum of the first n of either sequence is one less than the term next to the next one. Also $u_{2n} = u_n v_n$. Every third Fibonacci number is even, every fourth one is divisible by 3, every fifth one by 5, every sixth by 8, every seventh by 13, and so on. Whole books have been written about them, and there is a regular journal, the *Fibonacci Quarterly*. Here are some references for further reading.

Verner E. Hoggatt, *Fibonacci and Lucas numbers, Houghton Miflin Mathematics Enrichment Series*, 1969.

N. N. Vorub'ev, *Fibonacci Numbers*, Pergamon Press, 1961.

Fundamental counting principle. If one thing can be done in m different ways and, independently, a second thing can be done in n different ways, then the two things together can be done in mn different ways. We use this to count arrangements, permutations, combinations, choices, and so on. For example, the number of ways we can stand n people in a row is $n!$ (see **Factorial**) because we can choose the first person in n ways, and, for each of these we can choose the second in $n-1$ ways, the third in $n-2$ ways, and so on, giving

$$n \times (n-1) \times (n-2) \times \cdots \times 3 \times 2 \times 1 = n!$$

ways altogether. If we wish to arrange only k of the people in a row, then the number of ways is

$$n \times (n-1) \times (n-2) \times \cdots \times \big(n-(k-1)\big) = \frac{n!}{k!}$$

Notice how Theorem 0 comes in again (see **Arithmetic series**). The subject of combinatorics has been described as the art of counting things in two different ways. For examples, see **Binomial coefficients**.

Fundamental theorem of arithmetic. This states that every positive integer greater than 1 can be factored in exactly one way, apart from the order, as the product of prime numbers.

It is the first essential step in the analysis of the know/don't know problem that we quote at the end of **Logic problems**.

Problems involving the factoring of integers are **P49**, **P79**, **P136**, **P144**, **P171**, **P194**, **P213**. There is a tendency to take the fundamental theorem for granted but there are many number systems in which it does not hold.

For example, if we restrict ourselves to even numbers, then

$$60 = 2 \times 30 = 6 \times 10$$

where 2, 30, 6, 10 are 'primes' (mathematicians say 'irreducible') in the sense that they cannot be expressed as a product of numbers, both of them even. Again, if we adjoin $\sqrt{15}$ to our number system, and consider elements of the form $a + b\sqrt{15}$ where a and b are integers, then

$$63 + 14\sqrt{15} = 7(9 + 2\sqrt{15}) = (6 + \sqrt{15})(8 + \sqrt{15})$$

and each of these four factors is 'irreducible' in that they cannot be expressed as products of numbers of this form. Warning: you *can* factor $6 + \sqrt{15}$ as $(4 + \sqrt{15})(9 - 2\sqrt{15})$, but here $4 + \sqrt{15}$ is a 'unit', corresponding to 1 or -1 in ordinary integers.

Games. The games that we consider in **P201** to **P209** are combinatorial games in which the two players both know what is going on (they have complete information) and there are no chance moves (no rolling of dice or shuffling of cards). If we also assume that the game ends, with the last player winning, then John Conway has developed a remarkable theory to cover them (John H. Conway, *On Numbers and Games*, 2nd edition, A. K. Peters, 2001; and Elwyn R. Berlekamp, John H. Conway & Richard K. Guy, *Winning Ways for your Mathematical Plays*, Academic Press, 1982; 2nd edition, A. K. Peters, 2001).

Such games will have a **winning strategy** for one of the players, who can win regardless of how the opponent plays.

Most of our examples are *impartial* games, in the sense that at any stage of the game, the same options (moves) are available to both players, regardless of whose turn it is to play. See **Nim** for an account of the earlier history of these games.

If the last player is declared the *loser* instead of the winner, then there is still a winning strategy, but the theories, even for impartial games, break down in all but the simplest cases. One of these simple cases is Nim itself. The strategy for Misére Nim (last player *losing*) is the same as for ordinary Nim (last player winning — see **Winning strategy**) until the heap sizes, with one exception, are all 1. Then you reduce the exceptional heap to 1 or 0, leaving an *odd* number of 1-heaps if you're playing Misére Nim, or an even number of 1-heaps is you're playing ordinary Nim.

Geometric series. In connexion with exponential growth, you often come across sequences of numbers each of which is in *constant ratio* with the previous one, such as

$$3,\quad 6,\quad 12,\quad 24,\quad 48,\quad 96,\quad \ldots$$

More familiar ones are 1, 2, 4, 8, 16,... and 1, 10, 100, 1000, 10000,.... Sometimes we need to sum these sequences, and the sum of a sequence is called a *series*.

Experiment shows that $1+2=3$, $1+2+4=7$, $1+2+4+8=15$, which give numbers that are one less than a power of two: $3=2^2-1$, $7=2^3-1$, $15=2^4-1$, and this gives a hint as to how to find the sum, S, of n terms of a geometric series which starts with a and has common ratio r: subtract Sr from S,

$$\begin{array}{rl} S = & a + ar + ar^2 + \cdots + ar^{n-1} \\ Sr = & \phantom{a +{}} ar + ar^2 + \cdots + ar^{n-1} + ar^n \\ \hline S(1-r) = & a \phantom{{}+ ar + ar^2 + \cdots + ar^{n-1}} - ar^n \end{array}$$

then solve for S:

$$S = \frac{a(1-r^n)}{1-r} \quad \text{or} \quad \frac{a(r^n-1)}{r-1}$$

which are the forms one usually uses according as $r < 1$ or $r > 1$. What happens if $r = 1$? (Be careful to notice that the nth term is ar^{n-1}, not ar^n.)

Notice that $a^{n-1} + a^{n-2}b + a^{n-3}b^2 + \cdots + ab^{n-2} + b^{n-1}$ is a geometric series whose sum is $(a^n - b^n)/(a-b)$ and this gives us the generalization of the 'difference of squares', 'difference of cubes' formulas mentioned in **Algebra**:

$$a^n - b^n = (a-b)(a^{n-1} + a^{n-2}b + a^{n-3}b^2 + \cdots + ab^{n-2} + b^{n-1}).$$

Geometry. See **Q48**, **P234**, **P235**; also, **Combinatorial geometry**.

Goldbach conjecture. Every even number bigger than 2 is the sum of two primes. This has been verified for quite large numbers, certainly far enough for us to use it to analyze statement (2) of the know/don't know problem that we mention at the end of **Logic problems**.

Graphs. The vertices in the figure below represent six people. Two vertices are connected by an edge if the people they represent see each other most every day.

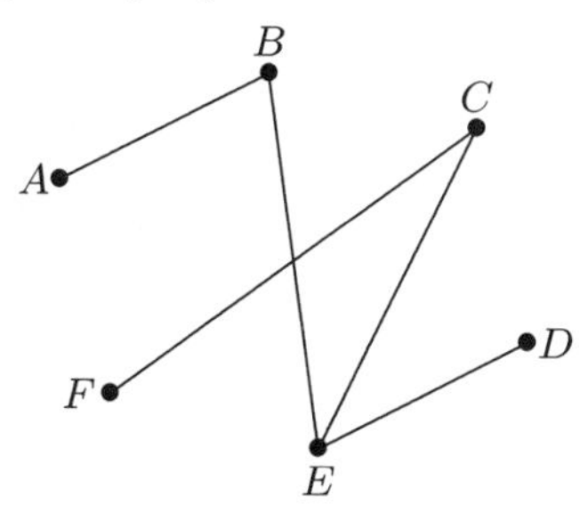

This representation of relationships is an example of a graph. More generally, a **graph** consists of a finite set of points, called **vertices**, together with a finite set of lines (or arcs), called **edges**, each of which joins a pair of vertices.

The **valence** (sometimes called the **degree**) of a vertex is the number of lines that connect to that vertex. In the preceding graph A has valence 1, B has valence 2, and E has valence 3. It's easy to see that the total valence of all the vertices is an even number, since each edge has two ends: this is the **Handshake lemma**. If the valence is the same at each vertex, we call the graph **regular**. For a neat use of graph theory, see the solution to **P107** given under **One-to-one correspondence**.

Graphs are often used as models to represent relationships (edges) between objects (vertices), as exemplied by

	Vertices	Edges
P18, Q18	Venusians, Varmlanders	holding hands
P90	sets of heads and tails	turning moves
P153	knights and ladies	dance partners
P187	squares	knight moves
P197	campers	acquaintances
P220, P221	islands	boat or air trips
P222–225	towns	roads
P226	club members	swapping business cards
P230	speakers, listeners	audience at talk
P243	animals	compatibility
P250	ping-pong players	winning a match

In the last case each edge can carry an arrow, indicating which player beat which, and we have a **directed graph**. We will use the penultimate example, **P243**, to illustrate some other important ideas about graphs.

We're up against some 'hard' problems. One is to decide if two graphs are **isomorphic**, that is, if you can label the vertices of the two graphs so that the labels at the ends of each edge in one graph are matched by the labels of an edge in the other. A second is to decide if a given graph has a **Hamilton circuit**, and a third is to decide if it has a **coloring** (see below) with k colors.

We give two solutions to **P243**, the first by coloring the 'incompatibility' graph, and later, a second, under **Hamilton circuits**, by finding such a circuit in the 'compatibility graph'. The incompatibility graph has animals for vertices and two animals are joined by an edge if they are incompatible with one another. The compatibility graph has the same vertices, but they are joined only when the animals are compatible. These two graphs are examples of **complementary** graphs, each having its edges exactly where the complementary graph does not.

For the first solution we find all regular 3-valent graphs on 8 vertices; each animal is incompatible with just three others. There are just five such graphs—you'll have to take our word for it: for such a small number it's not too hard to make an exhaustive search. Having found them all, it's convenient also to list the complementary graphs, the regular 4-valent graphs on 8 vertices, which we'll use later, in our second solution.

Graphs can be drawn in quite different ways, but we've labelled the vertices in our drawings with the labels 0 1 2 3 4 5 6 7 so that you can check which are isomorphic.

We color the incompatibility graph with four colors, each used twice. When we say **color** a graph it means assigning colors to the vertices so that the ends of each edge get different colors. That is, animals with different colors mustn't be in the same cage. In our notation the four colors (cages) are (0&1), (2&3), (4&5), (6&7).

The five complementary pairs of graphs are:

3-valent (incompatible)	4-valent (compatible)
Kepler's stella octangula (same parity)	Complete bipartite $K_{4,4}$ (opposite parity)
Cube edges (nim-sum 4,5,6)	Cube diagonals (nim-sum 1,2,3,7)
Möbius ladder (add 3 or 4 mod 8)	Square antiprism (add 1 or 2 mod 8)
Double spindle (add 3, nim-add 6)	Crossed cube (add 1 or 4, nim-add 2)
Semitruncated tetrahedron (add 3, sum 8, 04)	Crossed antiprism (add 1, nim-add 2, odd add 4, even nim-add 6)

The parenthesis beneath a name contains a list of the edges in the graph: interpret the entries as follows (see **Hamilton circuits** for the definition of a 1-factor):

add 1	012345670 is a Hamilton circuit	
add 3	036147250 is a Hamilton circuit	
add 2	02460 and 13571 are cycles	
add 4	04, 15, 26, 37 form a 1-factor	
nim-add 6	06, 17, 24, 35 form a 1-factor	
nim-add 2	02, 13, 46, 57 form a 1-factor	
sum 8	17, 26, 35	
odd add 4	15, 37	This pair and the next
even nim-add 6	06, 24	combine to make a 1-factor

Here are several drawings of each of the 3-valent graphs. Note that isomorphic graphs can look quite different, and nonisomorphic ones can look very similar. For the 4-valent graphs, see **Hamilton circuits**.

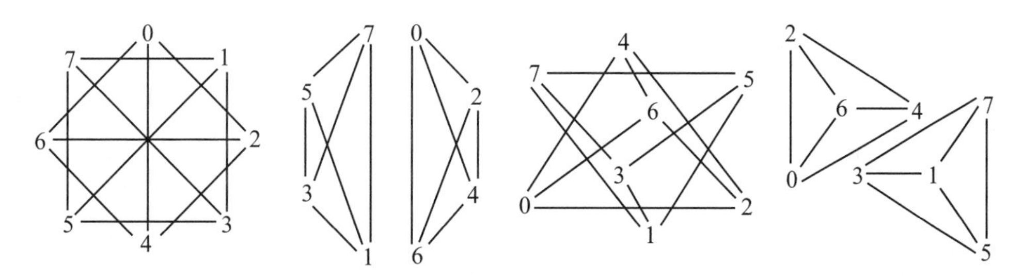

Kepler's stella octangula. Planar. Two components, each the complete graph on four points. The edges may be thought of as the 12 face diagonals of a cube.

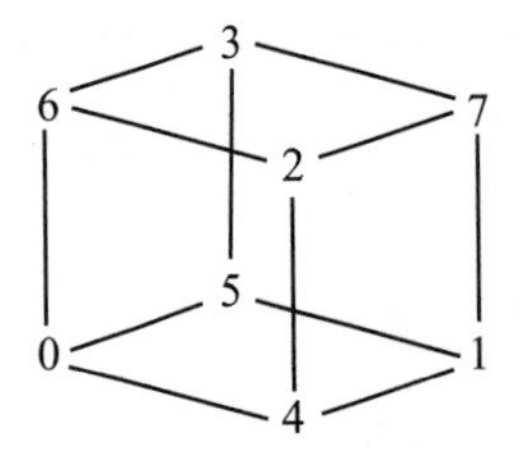

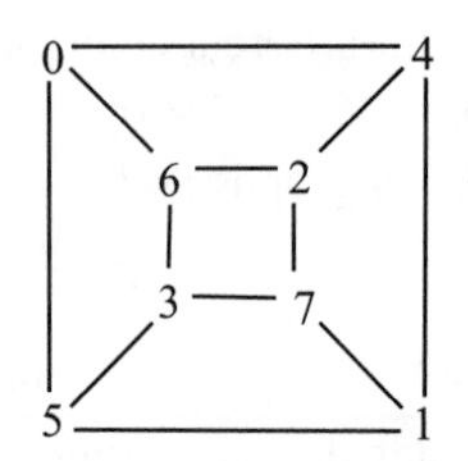

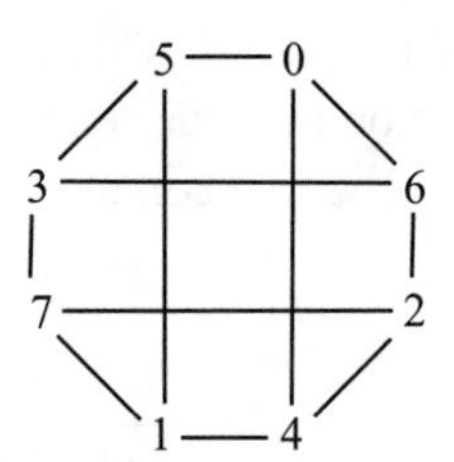

Cube edges. Planar. Bipartite.

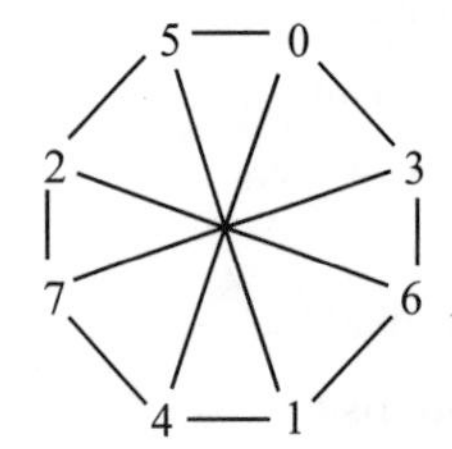

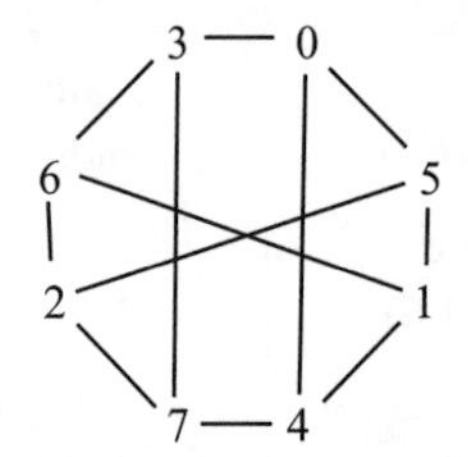

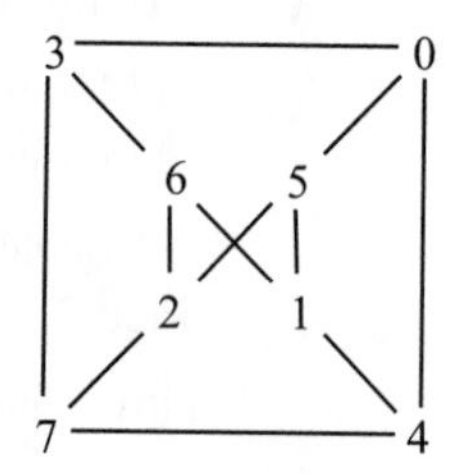

Möbius ladder. Crossing number 1.

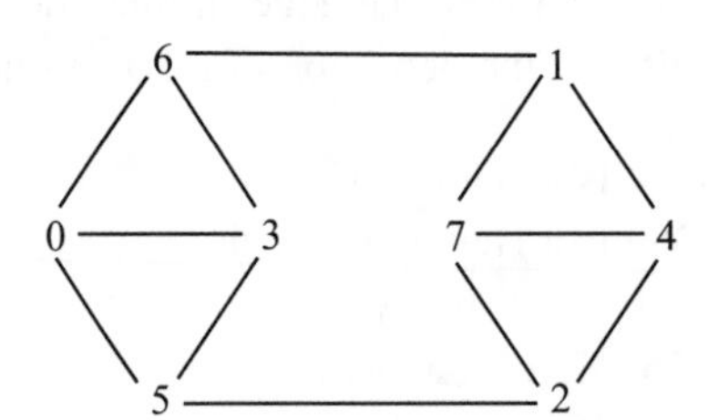

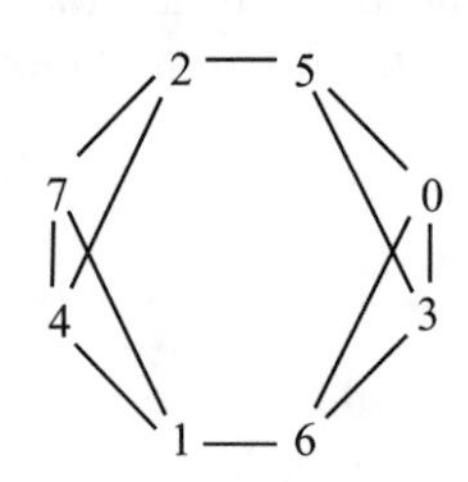

Double spindle. Planar.

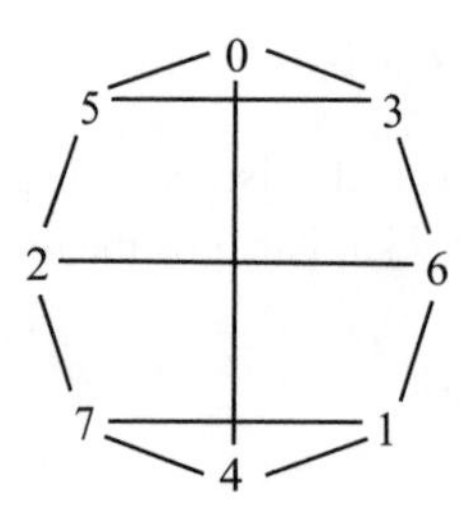

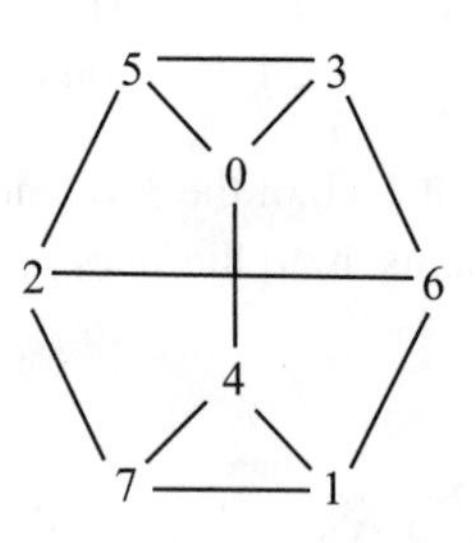

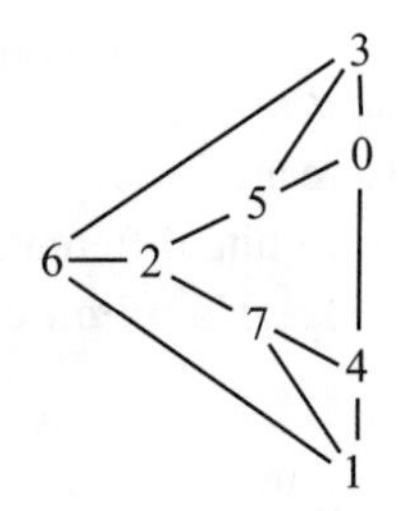

Semi-truncated tetrahedron. Planar.

Hamilton circuits. A Hamilton circuit in a graph is a circuit (see **Euler circuits**) that passes through each vertex of the graph exactly once (the beginning and ending vertices coincide). Although there is a simple condition for a connected graph to have an Euler circuit, there is no such simple check for the existence of a Hamilton circuit.

Here are the problems which can be thought of in terms of Hamilton circuits. We will discuss the first and last of these in some detail.

	Vertices	**Edges**
P32	squares	rook moves
P50	dots in squares	hor. & vert. segments
Q86, P87	squares	ringworms
P88, P89	squares	rook circuits
P113, P114	grids of points	polygons
P158	tiles	pairs of segments
Q205	squares	rook moves
P210	lily pads	frog hops
P223	towns	roads
P243	animals	compatibility

Problem **P32** asks for a Hamilton circuit for (i). Even if it is known that a Hamilton circuit exists, there isn't an "automatic" method for finding it. One approach, which is a little better than simple trial and error, is to do as well as possible in a first attempt. When stymied, begin again on some unvisited vertex and continue until you get stuck. Repeat this until all vertices are part of some short path. For example, we might begin as in (ii).

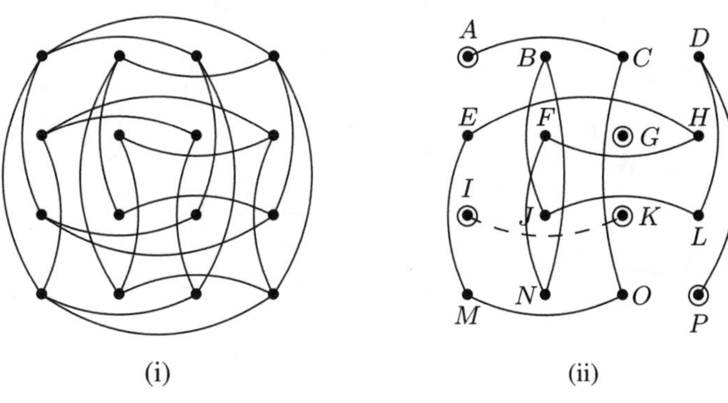

(i) (ii)

We have endpoints at vertices A, G, I, K, and P. To get rid of these, we find "add-delete" paths as follows. Add PM, delete ME, add EG; add GO, delete OC, add CK; finally, add IA. This eliminates the endpoints and creates two circuits shown in (iii). Connect these loops by deleting edges AI and DL and adding edges AD and IL. The final circuit is shown in (iv).

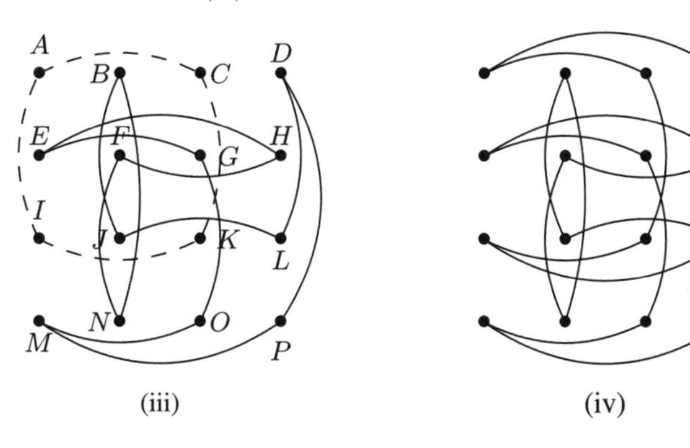

(iii) (iv)

We have seen at **R243** that the problem of caging the animals can be solved by using **Dirac's theorem** in the 'compatibility graph'. Under **Graphs** we listed the five different (that is, no two are isomorphic) regular 4-valent graphs on 8 points. We make several drawings of each, with vertices labeled so that 012345670 is a Hamilton circuit. For an numerical description of the edges, see the table in **Graphs**. Alternate edges of a Hamilton circuit join four pairs of compatible animals, so that pairs of animals with nim-sum 1, that is 01, 23, 45, 67, may occupy the same cage. This is a **1-factor** or **matching**, a regular subgraph of valence 1. Another way of pairing the animals is 12, 34, 56, 70, and other examples are noted in the numerical descriptions given under **Graphs**. Curiously, finding a matching is not a 'hard' problem.

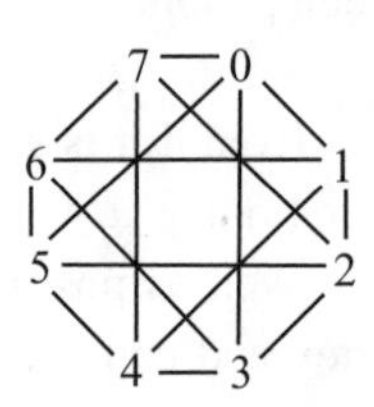

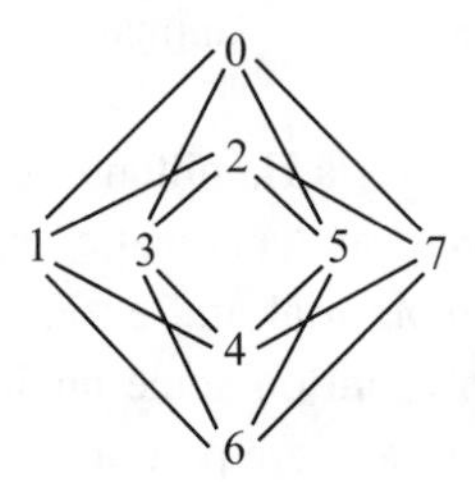

Complete bipartite graph $K_{4,4}$. Crossing number 4.

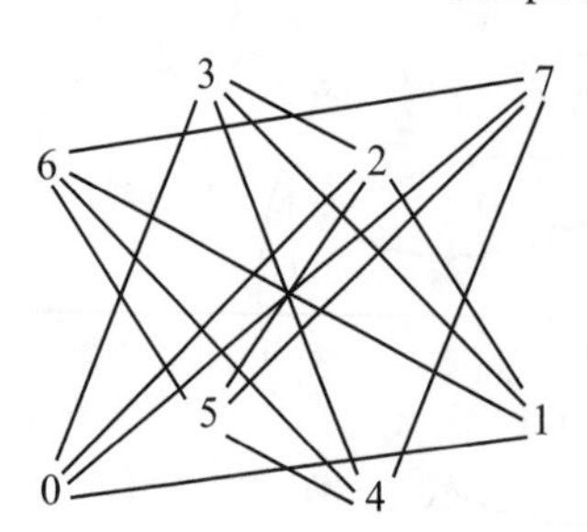

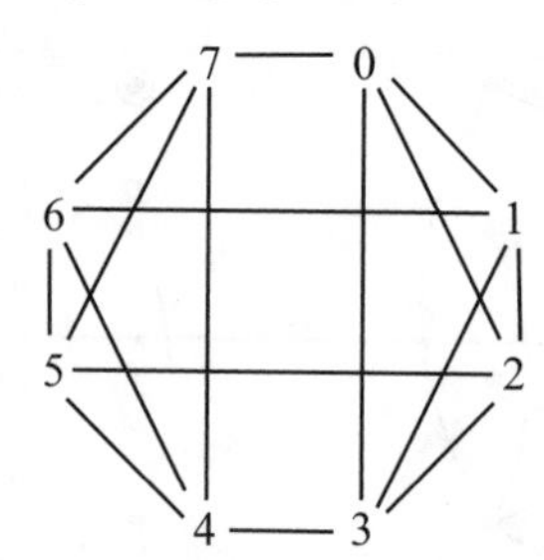

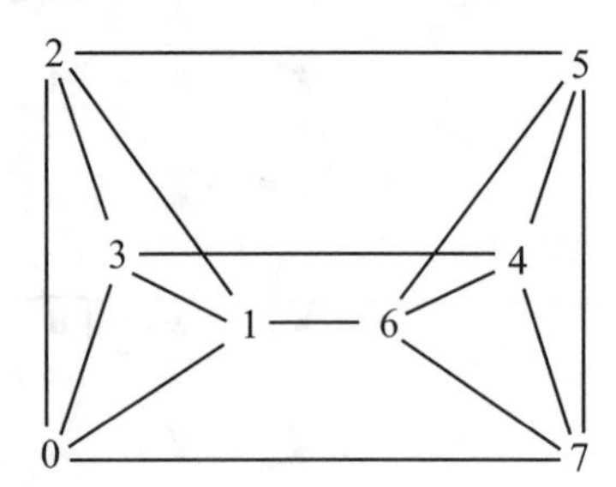

Cube diagonals. Crossing number 2.

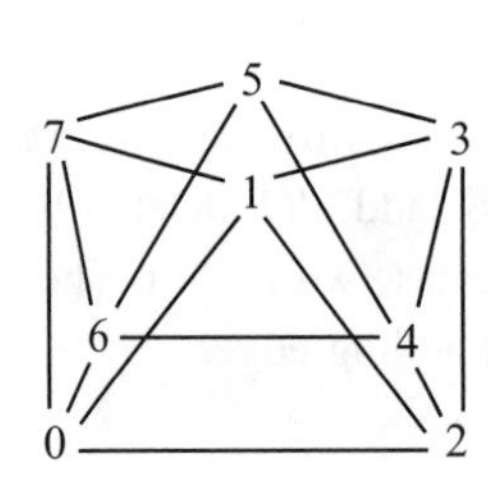

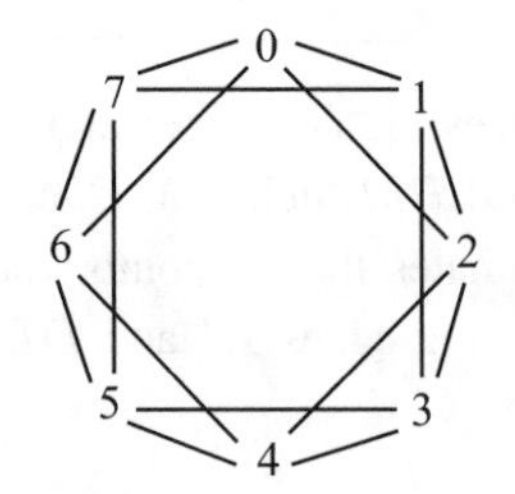

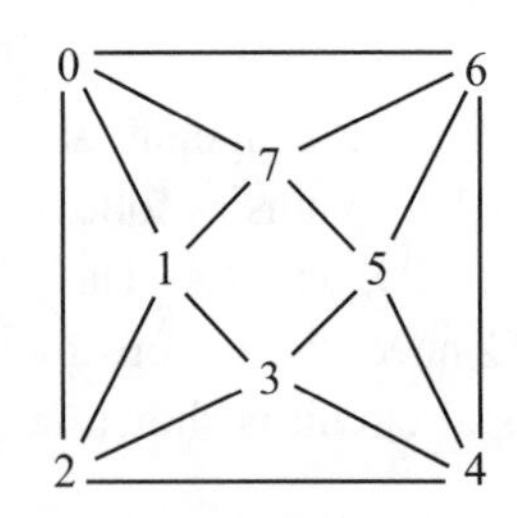

Square Antiprism. Planar.

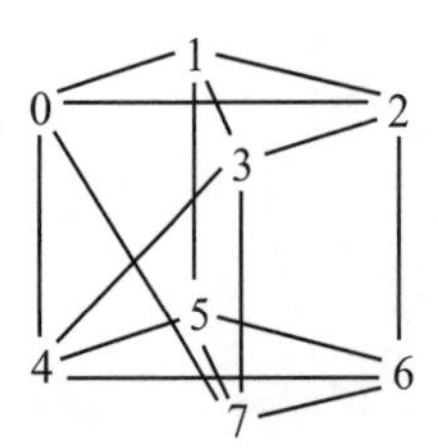

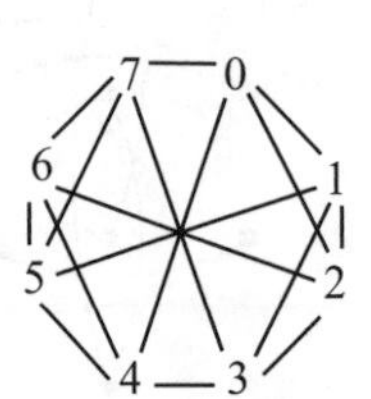

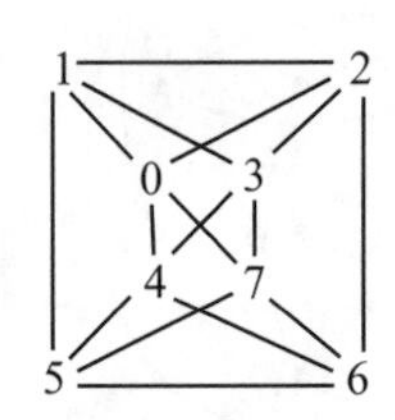

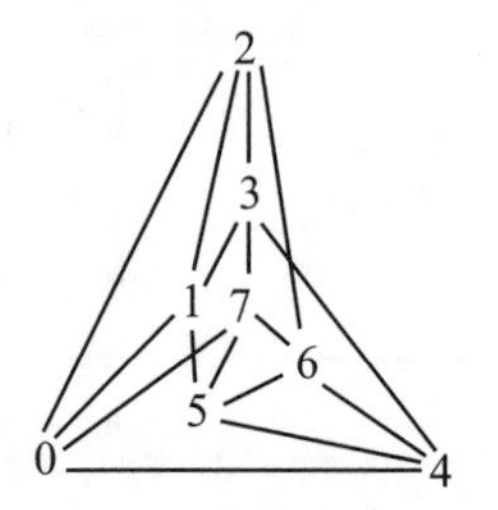

Crossed cube. Crossing number 2.

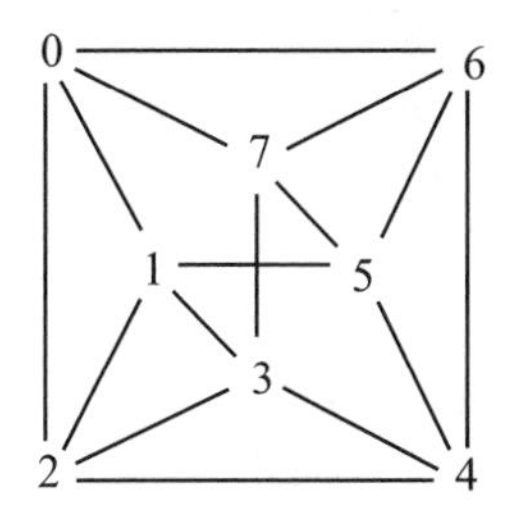

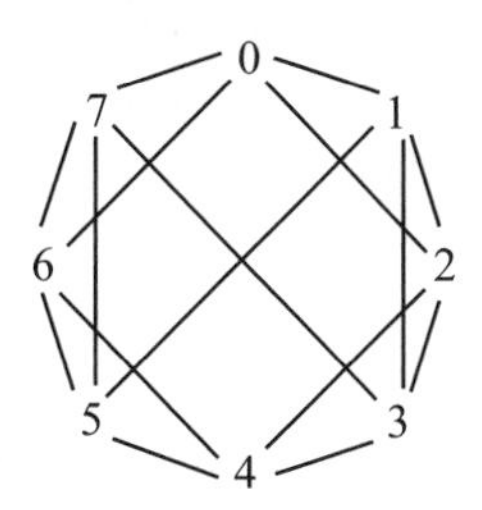

Crossed antiprism. Crossing number 1.

Hamming code. See **Error-correcting codes**.

Handshake lemma. In a graph (see **Graphs**) the number of vertices of odd valence (i.e., having an odd number of edges) is even. This is an immediate consequence of the fact that each edge has two ends. We use the Handshake lemma in **P18**, and in **S197b** where the children are vertices and acquaintance is an edge, and in **S221** with islands and flights.

Impartial games. See **Games**. Games in which the same options are available to the two players, regardless as to whose turn it is to move. (See Richard K. Guy, *Fair Game: how to play impartial combinatorial games*, Comap Mathematical Exploration Series, 1989.)

Indirect proof. See **Proof by contradiction**.

Induction. Logical induction is contrasted with mathematical deduction. Mathematical induction is a method for proving the truth of an infinite sequence of propositions. Prove the first proposition, the *basis* for the induction, then prove that each proposition follows from the previous one, the *inductive step*.

We used induction in **Q113**, **Q114**, and **Q210b**: the inductive step was to supply a border taking one size of grid to a larger one. In **R165** we started from some small stacks of pizza, and the inductive step was to add pairs of pieces. In **R232b** the step was to add 2 coins to neighboring bowls.

Inequalities. These are both similar to and yet different from equalities (equations). We can perform the same elementary arithmetic operation to each side, except that when we multiply or divide, the multiplier must be positive. We can multiply by a negative quantity provided we remember to reverse the inequality at the same time:

$$\begin{array}{ccccc} 5 > 3 & \text{and} & 20 > 12 & \text{but} & -20 < -12 \\ -5 < 3 & \text{and} & -20 < 12 & \text{but} & 20 > -12 \\ -5 < -3 & \text{and} & -20 < -12 & \text{but} & 20 > 12 \end{array}$$

Similar care must be taken when squaring or taking square roots. It's fairly easy to see mistakes when dealing with numbers, but what about algebraic expressions? If $a < b$, then $a - b < 0$, but is $(a-b)^2 < 0$? And what about $\sqrt{a-b}$?

Problems **P22**, **P51**, **P78**, **P85**, **P126**, **P135**, and **P145**, can be formulated in terms of algebraic inequalites, and inequalities figure in the solutions of minimum- and maximum-value problems **P44**, **P48**, **P231**, and existence problems **P45**, **P92**, **P154**, **P227**, and **P235**.

Invariants. One way of analyzing whether or not something can be done according to certain prescribed rules is to find some property that doesn't change (that is, an invariant) when the procedure is carried out. Often this is the **parity**, that is, the oddness or evenness of some quantity: the number of tails in **S90**, the difference of the black and white numbers in **P104**, or of the number of disks on black and white squares in **P128b**, or the number in red places in **P129**. The sum of the numbers in the stack of **P165** is always even, as is the number of queens on white squares in **P217** or the number of flights in **P221**, while in **P128a** the number of disks is always odd, as are the number of odd numbers in **P178** and the number of beans left after Pryor's move in **P206**.

In **R104** the invariant is the difference between the sums of the numbers on the black and white squares. In **S33** it's divisibility by three, and in **S186** and **S208**, divisibility by nine, and in **S242** the remainder on division by 13.

If you attempt to exchange 3 pairs of knights, instead of the 4 pairs of **P187**, then the knight moves form a cycle of eight, and the order of the six knights on the cycle is invariant, with ABC being clockwise, say, and abc counterclockwise. No amount of rotation of the cycle can interchange these.

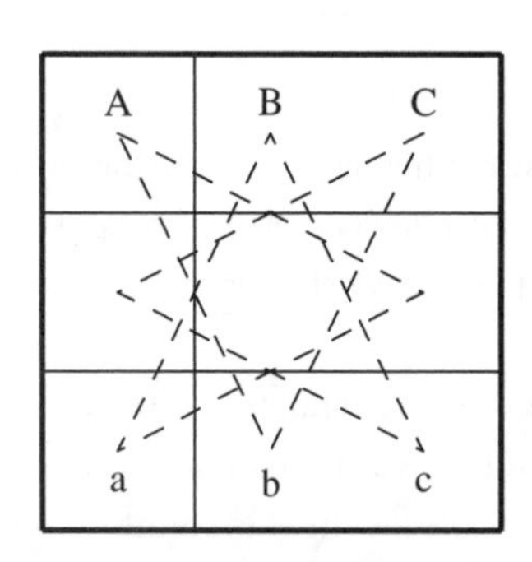

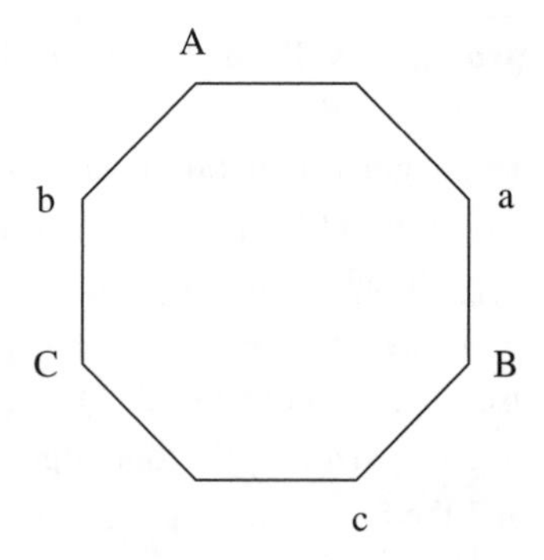

Latin squares. An square array, say $n \times n$, of numbers from 1 to n, inclusive, in which each number appears once in each row of the array and once in each column. See **P116**, **P131**.

Lexicographic order. An ordering on ordered sets of the same size. Arrange the elements of the sets in decreasing order, say $X = \{x_1, x_2, \ldots, x_m\}$ and $Y = \{y_1, y_2, \ldots, y_m\}$, $x_{i+1} > x_i$, and $y_{i+1} > y_i$ for all i. Then X is greater than Y in their lexicographic order if $x_1 > y_1$, or $x_1 = y_1$ and $x_2 > y_2$, or $x_1 = y_1$, $x_2 = y_2$, and $x_3 > y_3$, and so forth.

For example, in **R34**, in the **EMJA**, we were reducing the set $\{12, 9, 5, 1\}$ to three members. One way was to $\{12, 8, 4\}$, where we agreed to take out the common factor, leaving $\{3, 2, 1\}$. There are many other ways of reducing the set. Here they are in lexicographic order:

3 2 1	8 5 1	9 8 1	11 9 5	12 9 1
5 3 1	9 2 1	11 8 4	12 4 1	12 9 4
5 4 1	9 5 1	11 8 5	12 5 1	12 9 5
7 4 1	9 7 1	11 9 4	12 8 5	

Logic problems. These can be of a variety of types, but are usually concerned with the deduction of true statements from other true statements, or examining the consistency of statements which may be true or false.

In **P51** we wish to select exactly three consistent statements from five, each of which may be true or false. In **P55** to **P58** we are similarly concerned with the truth or falsehood of various statements, while **P148** underlines the principle that from a false statement, one can deduce any other, true or false. In **P253** we have seven statements, one of which may be false; see **Error-correcting codes** for a neat solution.

For a useful tool see **Truth Tables**.

An interesting kind of logic problem is that of the 'know/don't know' type. Here is an intriguing one, proposed by D. Sprows as Problem 977 in *Mathematics Magazine*, **49**(1976) p. 96.

Prudence (P) is told the product, Sam (S) is told the sum, of two (not necessarily distinct) integers in the range 2 to 100.

(1) P: I don't know the two numbers.
(2) S: Nor do I, and I know you don't know.
(3) P: Aha! Now I know the numbers!
(4) S: Aha! So do I!

But do you, the reader, know them?

This was discussed by Martin Gardner in his Mathematical Games column in *Scientific American*, **241** No. 6(Dec 1979) and, in more detail, with extension to more than two interchanges of information, by John O. Kiltinen & Peter B. Young in their paper: Goldbach, Lemoine and a know/don't know problem, *Math. Mag*, **58**(1985) pp. 195–203.

Here's the beginning of an analysis sent to us by Andrew Bremner:

Statement (1) tells us that the two numbers are not both primes, otherwise P would be able uniquely to factor the product and deduce the two numbers.

Statement (2) tells us firstly ("Nor do I") that the sum is at least 6, and secondly ("and I know you don't know") that the sum does not allow a partition as the sum of two primes (for example a sum of 14 admits the decomposition $3+11$, and the corresponding product 33 would have allowed P to deduce the numbers). By the Goldbach Conjecture the sum must be odd. Moreover, it is not of the form $p+2$ with p a prime.

Can you complete the solution?

Logical induction. Compare **Logic problems** and **Induction**.

Lucas numbers. See **Fibonacci numbers** and **Recurrence relations**.

Magic squares. An $n \times n$ magic square is an array of n^2 numbers in which the sum of the numbers in each row, column, and diagonal is the same. Magic squares occur in

P60 and **P103**, and also in **P136**, though here we use the product in place of the sum, and we don't include the diagonals. To make it into a (vector) sum write it as

$$\begin{array}{ccc} 200 & 101 & 010 \\ 011 & 000 & 300 \\ 100 & 210 & 001 \end{array}$$

where the components are the exponents on the primes 2, 3, 5. For example, 210 stands for $2^2 3^1 5^0 = 12$.

In **P72** we ask for a rectangle with magic properties and **P202** reveals the connexion between Tic-Tac-Toe and a magic square. Other magic configurations are Barry Cipra's magic line, **P200**, and magic triangle, **P218**. For more, see W. S. Andrews, *Magic Squares and Cubes*, Open Court, Chicago, 1917, reprinted Dover, 1960.

Barry Cipra noticed the following striking connexion between solutions to the limerick puzzle posed in **Configurations** and 4×4 semi-magic squares ("semi-" means that you're only concerned with row and column sums, not with diagonals). Identify each piece of LeWitt's puzzle (see **P127**) with a string of 1's and 0's corresponding to the presence or absence of the directions vertical, horizontal, up diagonal, and down diagonal (in that order). The strings are then clearly readable as binary representations of the numbers 0 through 15. Each solution of the limerick clearly produces a 4×4 semi-magic square with sum 30.

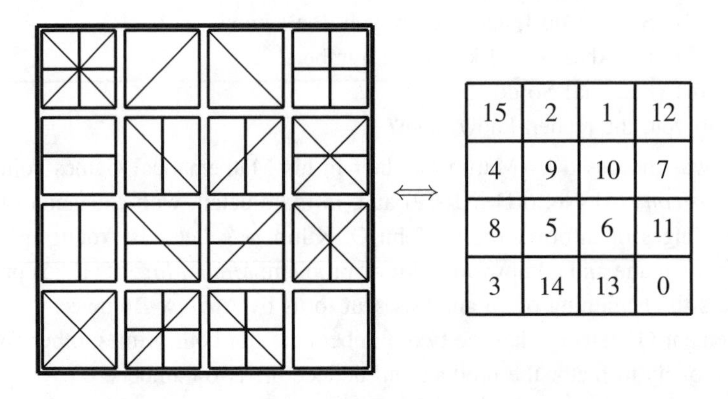

But more importantly, the converse is true: every 4×4 semi-magic square corresponds to a solution of the limerick! And the 880 different 4×4 magic squares (enumerated by Frenicle de Bessey in 1693) yield solutions to the limerick in which the two diagonals also contain two copies of each of the four slopes. And the 48 of these which are **pandiagonal** or **Nasik** squares yield solutions in which the property extends to all diagonals, including the 'broken' ones. For details, and the connexion between pandiagonal squares, **nim-addition**, and the Magic Tesseract, see pages 778–783 of *Winning Ways*.

Mathematical induction. See **Induction**.

Maximum/minimum problems. When a problem has several different solutions, it is natural to ask about the "best" solution. The interpretation of "best" might involve calculating a maximum or a minimum value, or the fewest number of steps needed to carry out an algorithm.

We want the least number of Varmlanders in **Q18**, of employees in **P22**, of visits by the Cookie Monster in **Q34**, of circuit segments in **Q89**, of black squares in **P177** and of knight moves in **Q187**. On the other hand we want the greatest number of polygon vertices in **Q113**, of coins in **R146** and **Q252**, and both the least and greatest number of coins in **R192**.

We want the least length of path in **P31**, **P48**, and **P109**, the least spread of scores in **P44**, the least distance between the terms of arithmetic progressions in **P61**; and we want to minimize the total of the remaining numbers in **Q104b**, the number of black squares in **P177** and the incompatibility in **P166** and **P243**.

Modular arithmetic. One of Gauss's many valuable contributions to mathematics is his introduction of the use of *congruences*: the idea that for many purposes we only need to work with the remainder after division by a fixed number, the *modulus*. Sometimes we need only to know the day of the week, so we work *modulo* 7, or the time of day, when we work modulo 12 (or preferably 24).

If we work modulo 9, we see that the remainder when you divide a number (in decimal notation) by 9 is the same as when you divide the sum of its digits by 9. This is the idea behind "casting out the nines" (see **Divisibility tests**).

If two numbers a and b leave the same remainder when we divide them by m, Gauss said that they were *congruent modulo* m and wrote $a \equiv b \pmod{m}$.

To add (or multiply) a and b modulo m, we compute $a + b$ (or $a \times b$) in the usual way, and then take the sum (or product) modulo n to be the remainder when it is divided by m.

For example, $8 + 7 \equiv 3 \pmod{12}$ and $8 \times 7 \equiv 8 \pmod{12}$; $6 + 9 \equiv 5 \pmod{10}$ and $6 \times 9 = 4 \pmod{10}$; and $3 + 3 \equiv 1 \pmod{5}$ and $3 \times 3 \equiv 4 \pmod{5}$.

In **S191a**, we found that $13333 + 665B$ was a multiple of 16, so to find B we need only look at remainders modulo 16 and $5 + 9B \equiv 0 \pmod{16}$, $9B \equiv -5 \pmod{16}$, then because $9 \times 9 \equiv 1 \pmod{10}$, we have $B \equiv 9 \times 11 \equiv 3 \pmod{16}$, so, because B is a digit, $B = 3$.

In **P193**, to find the last digit, we work modulo 10. For powers of 7, we see that $7^n \equiv 1, 7, 9$ or $3 \pmod{10}$ according as $n \equiv 0, 1, 2$ or $3 \pmod{4}$, so we need to calculate $7^{7^{7^7}} \pmod{4}$. We have $7 \equiv 3 \pmod{4}$, $7^7 \equiv 7^3 \equiv 3 \pmod{4}$, $7^{7^7} \equiv 7^3 \equiv 3 \pmod{4}$ and $7^{7^{7^7}} \equiv 7^3 \equiv 3 \pmod{4}$, so our answer is 3.

In **P248** we can reduce the search by looking at the possibilities in various modular systems.

Modular arithmetic is sometimes referred to as "clock arithmetic" because you can think of the numbers as on a clock. For $m = 4$ we have the clock and the tables as shown.

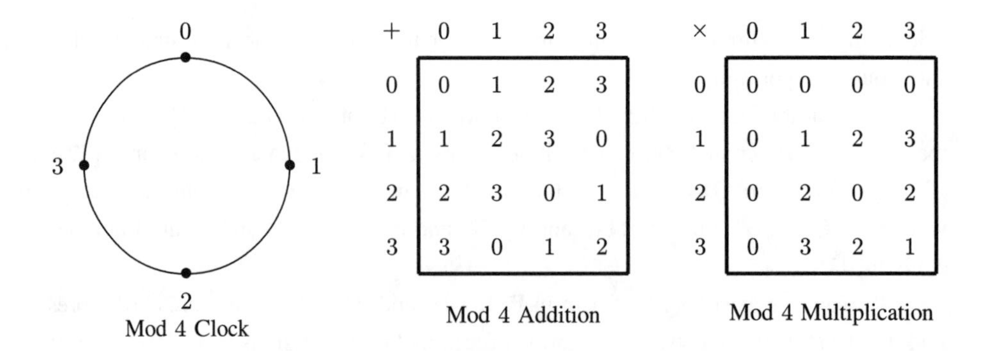

Mod 4 Clock

+	0	1	2	3
0	0	1	2	3
1	1	2	3	0
2	2	3	0	1
3	3	0	1	2

Mod 4 Addition

×	0	1	2	3
0	0	0	0	0
1	0	1	2	3
2	0	2	0	2
3	0	3	2	1

Mod 4 Multiplication

An integer is divisible by m if and only if it is equal to 0 modulo m. In this way, questions about divisibility can be transformed into the language of modular arithmetic. Places where modular arithmetic can be used are mod 19 in **P75**, where $8 \equiv -11$ (eight floors up or 11 floors down); also mod 3 in **P147** and **P182**, mod 9 in **P183** and **P186**, and both 3 and 9 in **P184**; mod 2 and 4 in **P150** and powers of 2 in **Q177**; mod 11 in **P239**, mod 13 in **P242**, mod 25 in **P245**, 37 and 41 in **P216**, mod 65 in **P174**, mod 2010 and mod 2011 in **P247**, and various moduli, especially 3, in **Q182**.

Motion problems. Problems on relative velocities are often most easily dealt with by bringing one of the moving objects to rest; that is, by observing the problem as though you were sitting on that object. For example, you can look at **P95** from the point of view of Juliet, at **P106** from that of Harry, and **P137** by looking through the goggles.

Multiplicand. When you multiply a by b to get c, then a is the **multiplicand**, b is the **multiplier**, and ab is the **product**.

Multiplier. See **Multiplicand**.

Nim. A game between two players with heaps of beans. When it's your turn to move, you choose a heap and take a number of beans, perhaps the entire heap, but at least one. In normal play, the player who takes the last bean wins. In *misére* play, the last player loses. It has a neat mathematical theory, discovered by Charles L. Bouton (Nim, a game with a complete mathematical theory, *Ann. of Math., Princeton* (2), **3**(1901–02), pp. 35–39).

A Nim position is a P-position (previous player winning) exactly if the nim-sum (see **Nim-addition**) of the number of beans in the heaps is zero. The theory was extended by Sprague and by Grundy independently to cover all **impartial games**. But theory and practice are not always the same thing. Consider the game, played with heaps of beans, where the move is to split the heap into two smaller heaps (not necessarily the same size). You'll soon find out the theory behind that. But now look at Grundy's Game, in which the move is to split a heap into two *unequal* heaps (heaps of two can't be split). Dan Hoey and Akim Flammenkamp have analyzed this game for heaps up to 20 billion beans, but they haven't yet found a pattern!

Nim-addition. This is addition without carrying in base 2 or as mathematicians would say, "vector addition over GF(2)." Computer scientists and logicians will recognise it as "exclusive or": $0 \oplus 1 = 1 = 1 \oplus 0$ of course, and $0 \oplus 0 = 0$, but also $1 \oplus 1 = 0$. Here's a nim-addition sum: 53 is 110101 in binary, and 45 is 101101 in binary, and they nim-add to

$$\begin{array}{cccccccc} & 1 & 1 & 0 & 1 & 0 & 1 \\ \oplus & 1 & 0 & 1 & 1 & 0 & 1 \\ \hline & 0 & 1 & 1 & 0 & 0 & 0 \end{array}$$

which is 24, so, if you're playing Nim, and confronted with three heaps of 54, 45 and 24 beans, you'd better hope that it's your opponent's turn to play!

Notice that the nim-sum of the numbers on either side of each of the lines of **R21** is zero. Nim-addition is useful for flipping coins in **R90**.

Nim-addition can be used to solve the following problem from the *Math Intelligencer*, Winter, 2001, Problem 937.

> Charlie has 7 cards numbered 1,2,3,4,5,6,7 and randomly deals 3 of them to Alice and 3 to Bob. All three people look at the cards that they hold.
>
> Can Alice and Bob communicate with each other, in the presence of Charlie, so that after the communication Alice knows which cards Bob has, and Bob knows which cards Alice has, but, for any card except the one he has, Charlie does not know whether Alice or Bob has it?

One way to do it is for Alice and Bob to announce their nim-sums. The nim-sum of those is Charlie's card.

Reason: nim-sum of 1 2 3 4 5 6 7 is 0

Key idea: Charlie will know Alice's and Bob's nim-sums, but cannot know how the remaining two possibilities are distributed.

For example, if A says 3 and B says 5, then C has 6, A has 5, B has 3 and they have 2 4 or 1 7 in some order.

245 or 157 (Alice)
137 or 234 (Bob)

One-to-one correspondence. One-to-one correspondence together with graph theory can be used to solve **P107** in which the squares of a 10×10 grid are numbered consecutively from 0, starting in the upper left corner and moving to the right and then down the rows as on a printed page. Each number is given a $+$ or $-$ sign so that half of the numbers in each row and in each column are positive, and the other half are negative. Construct a graph with the squares of the grid as vertices. Obviously there is a one-to-one correspondence between the positive numbers and the negative numbers in each row; draw horizontal edges showing this correspondence. Similarly, there is a one-to-one correspondence between the positive numbers and the negative numbers in each column; draw vertical edges showing this correspondence. Each vertex in the resulting graph of 100 vertices and 50 horizontal and 50 vertical edges has valence 2, so it consists of a set of closed circuits (as you traverse each circuit, the associated numbers will alternate in sign). The sum of the numbers in each circuit is zero (do you see why?), so the total of all the numbers in the grid is zero.

In **S23b** there was a correspondence between even and odd numbers; and between positive and nonpositive sequences in **S27**, with a single exception in each case. In **S189** we saw a one-to-one correspondence between two kinds of ticket number and between numbers with digital sums $3k + 1$ and $3k + 2$ in **S249**.

Here is an example of a theorem that can be proved using one-to-one correspondence. The number of partitions of a number n into unequal (that is, distinct sized) parts is equal to the number of partitions of n into uneven (that is, odd sized) parts. The one-to-one correspondence was established by Franklin using diagrams such as

```
A A A A A A A A A        R S S S S S S S S
A B B B B B B            R R S U U U U
A B C C C C C C          R T R S U W W W
A B C D D D              R T T R S U
A B C D E                R T V T R
A B C D   F              R T V V R
A B C                    R T V
A   C                    R   V
A                        R
    17.11.11.5.1.1            14.11.7.6.5.3
```

which illustrate the number 46, arranged in a symmetrical pattern about the falling diagonal, and partitioned into the odd parts 17.11.11.5.1.1 by the right-angled gnomons A, B, C, D, E, F and into the distinct parts 14.11.7.6.5.3 by the 45 degree gnomons R, S, T, U, V, W.

For example, there are five partitions of 7 of each kind:

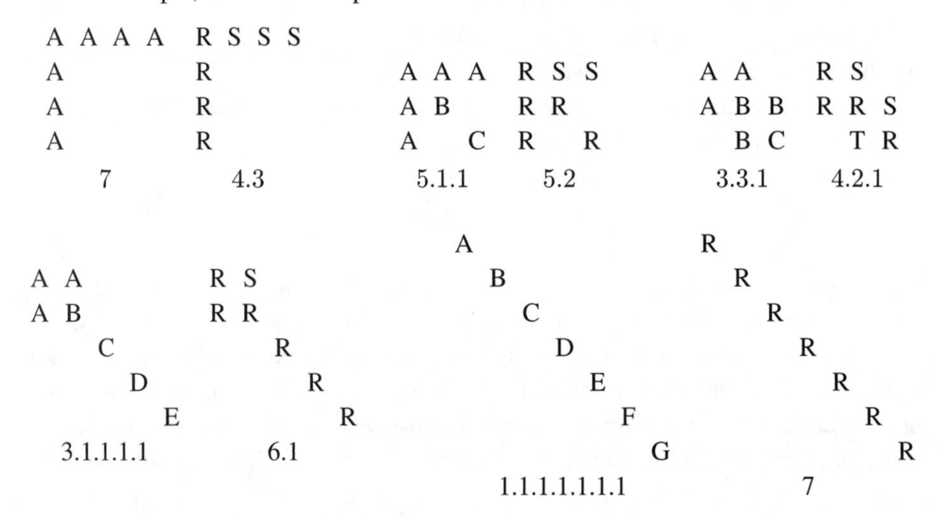

Optimization problems. See **Maximum/minimum problems**.

Packing problems. These may be exact, in which case they may be regarded as **Tilings** or they may simply require that you pack as many objects as possible into a given space, such as the tennis balls in **P211**.

Parity. The parity of a number refers to its property of being even or odd, but it has much wider application to situations in which there are just two states: clockwise or counterclockwise gears in **P5**, equality or difference of gender of adjacent people in **P157** and **P174** and of guests in **P179**, winning or losing service in **P64**, appearance of adjacent coins in **P39** and color of adjacent squares in **P40**, **P50**, and **P121**, horizontal and vertical segments in **P89**, positive and negative products in **P77**, zeros and ones in binary strings in **P13** and **P59b**, and occurrence or non-occurrence of coins in **Q192**.

In many situations the parity of some quantity remains unchanged. Examples are listed under **Invariants**. Other places where parity enters are the digital sum in **P23**, permutations in **P25**, the numbers in **P66**, the number of arm-wrestlers in **P67**, that $a \pm b$ are of the same parity in **P79**, the sums of subsequences in **P115**, the total of the page-numbers in **P150**, the number of 2×2 squares in **P156**, the number of children's acquaintances in **P197**, the sum of consecutive numbers in **P227**, the number of black squares covered by the T-tetromino in **P238**, and the parity of the pairs chosen from coins, cards and candy in **P122**. Note that in **P253** it is useful to remember that the nim-sum, although usually different from their ordinary sum, always has the same parity.

Parity arithmetic. The parity of a sum or product of two integers can be determined by the parity of the integers to be added or multiplied. The rules are given in the following table. Here, "0" represents an even number and "1" an odd number. For example, "an even number plus an odd number is an odd number" is encoded as "$0 + 1 = 1$".

+	0	1
0	0	1
1	1	0

Parity Addition

×	0	1
0	0	0
1	0	1

Parity Multiplication

See **P37**, **Q90**, **Q177**. Also, compare and contrast with **Nim-addition**.

Partitions. The partitions of a number n are ways of writing it as a sum of positive numbers. The order is immaterial. For example, the partitions of 5 are

$$5, \quad 4.1, \quad 3.2, \quad 3.1.1, \quad 2.2.1, \quad 2.1.1.1, \quad 1.1.1.1$$

In **Q9** we required the partitions of 12 into *exactly* 4 parts. We can find these expeditiously by subtracting 1 from each part and finding the partitions of $12 - 4 = 8$ into *at most* 4 parts:

$$\begin{array}{ccccccc} 8, & 7.1, & 6.2, & 6.1.1, & 5.3, & 5.2.1, & 5.1.1.1 \\ 4.4, & 4.3.1, & 4.2.2, & 4.2.1.1, & 3.3.2, & 3.3.1.1, & 2.2.2.2 \end{array}$$

and then adding the ones back in:

$$\begin{array}{ccccccc} 9.1.1.1, & 8.2.1.1, & 7.3.1.1, & 7.2.2.1, & 6.4.1.1, & 6.3.2.1, & 6.2.2.2 \\ 5.5.1.1, & 5.4.2.1, & 5.3.3.1, & 5.3.2.2, & 4.4.3.1, & 4.4.2.2, & 3.3.3.3 \end{array}$$

More generally: the number of partitions of n into *exactly* p parts is the same as the number of partitions of $n - p$ into *at most* p parts.

In **Q21** and **P100** we wanted the partitions of 14 into 3 distinct parts no bigger than 9. Here we can take off 1 from the smallest part, 2 from the next, and 3 from the biggest, and write down the partitions of $14-(1+2+3)=8$ into at most 3 parts of size at most $9-3=6$. We have already seen these:

6.2, 6.1.1, 5.3, 5.2.1, 4.4, 4.3.1, 4.2.2, 3.3.2

and we add 3.2.1 to them, giving

9.4.1, 9.3.2, 8.5.1, 8.4.2, 7.6.1, 7.5.2, 7.4.3, 6.5.3

Our general theorem is now: the number of partitions of n into p *distinct* parts is the same as the number of partitions of $n-p(p+1)/2$ into *at most* p parts.

These theorems are clearly illustrated by means of **Ferrers's diagrams**:

$\odot \times \times \times \times$	
$\odot \times \times \times$	$\odot \odot \odot \times \times \times \times \times$
$\odot \times$	$\odot \odot \times \times \times$
$\odot$	$\odot$
5421 from 431	851 from 53

Another theorem obtained by reading Ferrers's diagrams by rows or by columns;

A A A A A A A A		R S T U V W X Y
B B B B B B		R S T U V W
C C C C C C	=	R S T U V W
D D D D		R S T U
E E E		R S T
8.6.6.4.3		5.5.5.4.3.3.1.1

states that the number of partitions of n into parts not exceeding m is equal to the number of partitions of n into at most m parts.

For example, the partitions of 6 into parts not exceeding 3 are

3.3	3.2.1	3.1.1.1	2.2.2	2.2.1.1	2.1.1.1.1	1.1.1.1.1.1
2.2.2	3.2.1	4.1.1	3.3	4.2	5.1	6

where the second row contains corresponding partitions into at most three parts.

				A A A	R S T		
		A A A	R S R	B	R	A A	R S
A A A	R S T	B B	R S	C	R	B B	R S
B B B	R S T	C	R	D	R	C C	R S
3.3	2.2.2	2.2.2	3.2.1	3.1.1.1	4.1.1	2.2.2	3.3

				A	R
		A A	R S	B	R
A A	R S	B	R	C	R
B B	R S	C	R	D	R
C	R	D	R	E	R
D	R	E	R	F	R
2.2.1	4.2	2.1.1.1	5.1	2.1.1.1	6

For yet another theorem, see **One-to-one correspondence**.

We said at the outset that the order of the parts is immaterial. If the order does matter, then the partitions are called *compositions*. For example, the compositions of 1, 2, 3, 4 are

1
2, 1.1
3, 1.2, 1.2, 1.1.1
4, 3.1, 1.3, 2.2, 2.1.1, 1.2.1, 1.1.2, 1.1.1.1

There's a simple formula for the number of compositions of n. Can you find it, and a simple proof that it's correct?

Problems which involve partitions are

	partition	**into parts of size**
P9	12	exactly four
P35	71 points	25, 20, 10, 5, 3, 2, 1
P38	35	distinct, < 15
Q38	40	distinct, < 16
P42	62 points	35, 25, 18, 13, 11, 10, 9, 1
P54	13	distinct, < 11
P91	8 or 10 points	three distinct
P233	8 or 22	three distinct

Pascal's triangle. See **Binomial coefficients**. Although it is usually called by this name in the West, it was known long ago by the Japanese, by the Chinese before them, and goes back at least to Omar Khayyam, author of the famous Rubaiyat.

If we tilt Pascal's triangle over to one side, then it is easy to see the ones in the zeroth column; the counting numbers in the first column, got by adding up the ones; the **triangular numbers** in the second, got by adding the counting numbers; the **tetrahedral numbers** in the third, got by adding the triangles; and so on, into the fourth, fifth, ..., dimensions, with the d-dimensional pyramid numbers occurring in column d.

And the sums of the rows are the **Fibonacci numbers**!

0	1	2	3	4	5	6	7	8	9	10	11	12	13	14	15	Total
1																1
	1															1
1		1														2
	2		1													3
1		3		1												5
	3		4		1											8
1		6		5		1										13
	4		10		6		1									21
1		10		15		7		1								34
	5		20		21		8		1							55
1		15		35		28		9		1						89
	6		35		56		36		10		1					144
1		21		70		84		45		11		1				233
	7		56		126		120		55		12		1			377
1		28		126		210		165		66		13		1		610
	8		84		252		330		220		78		14		1	987

Permutations. A permutation of n objects is a rearrangement or reordering of those objects. A set of n objects can be permuted in $n!$ different ways (see **Factorial** and **Fundamental counting principle**).

In **P190** we wanted to make the *cyclic permutation* (1 2 3 4 5 6 7 8 9 10 11 12 13 14) in which $1 \to 2,\ 2 \to 3,\ \ldots,\ 12 \to 13,\ 13 \to 14,\ 14 \to 1$. We did this by taking the *product* of the two permutations $(1)(2\ 14)(3\ 13)(4\ 12)(5\ 11)(6\ 10)(7\ 9)(8)$ (let the families in apartments 1 and 8 stay put and interchange families in apartments 2 and 14, 3 and 13, 4 and 12, etc., and $(1\ 2)(14\ 3)(13\ 4)(12\ 5)(11\ 6)(10\ 7)(9\ 8)$ (interchange families in apartments 1 and 2, 14 and 3, 13 and 4, and so on). The first takes family 2 to apartment 14 and the second moves them from apartment 14 to apartment 3, so in the product, the family in apartment 2 ends up in apartment 3: $2 \to 14 \to 3$. Similarly, $3 \to 13 \to 4$, $4 \to 12 \to 5$, and so forth.

Pigeonhole principle. An important method for showing the existence (or non-existence!) of an object satisfying stated conditions. The principle states that if $n+1$ pigeons are placed in n boxes, then some box will contain more than one pigeon. In the context of mathematics, one has to decide what corresponds to the pigeons, what corresponds to the boxes, and directions for placing the pigeons into the boxes. For example,

	Pigeons	**Holes**	**Outcome**
P102a	11 coins	6 countries	not 3 in one
P102b	96 coins	5 countries	20 in one
P146	9 dimes	6 rows	6 dimes in 3 rows
P160	55 tables	3 categories	19 available
P161	6 games	4 teams	a team wins 2
P184a	3 digits	4 classes	enough
P184b	5 digits	4 pairs	digit in 2 pairs
P197	198 children	66 groups	3 in each group
P198	9 points	4 labels	3 of the same label
P225	18 roads to B	35 towns	town in route
	18 roads to A		from A to B

Problem **P146** has a neat solution. As suggested in the hint, the pigeonhole principle ensures that there will be three rows which contain, between them, at least six dimes. Choose these rows and the (not more than three) columns which contain the remaining dimes.

Platonic solids. These are the five polyhedra which are completely regular in the sense that all faces are congruent regular polygons which meet the same number at each vertex. They are the regular tetrahedron, cube, octahedron, dodecahedron and icosahedron. Diagrammatic representations of them can be seen at **R111**.

Positional notation. In the positional system of notation base b ($b > 1$), numbers are written as sums of powers of b. Specifically, for each positive integer n there are unique nonnegative integers $a_0, a_1, \ldots, a_k$ less than b with $a_k \neq 0$ such that

$$n = a_k b^k + a_{k-1} b^{k-1} + \cdots + a_1 b + a_0,$$

and we write

$$n_b = a_k a_{k-1} \dots a_2 a_1 a_0.$$

For example, the number 13 in bases $2, 3, 4, \dots, 13$ is

Base b:	2	3	4	5	6	7	8	9	10	11	12	13
13_b:	1101	111	31	23	21	16	15	14	13	12	11	10

A good way to solve **P108** is to count in base 7. Compare **Binary notation** and **Decimal notation**.

Proof by contradiction. If by starting from a statement you can deduce something which is clearly false, then the original statement must be false, and its opposite (if it has one!) is true. For example, if we assume that $\sqrt{2}$ is a rational number, the ratio of two integers, p/q, then we can also assume that p and q are the smallest such integers. But then $p^2 = 2q^2$ and

$$\frac{p}{q} = \frac{p^2 - pq}{pq - q^2} = \frac{2q^2 - pq}{pq - q^2} = \frac{2q - p}{p - q}$$

and we've found smaller ones, contrary to our assumption. We conclude that $\sqrt{2}$ is not rational.

Puzzles. The word 'puzzle' is often contrasted with the word 'problem' with the implication that the former is just a trick while the latter requires a mathematical solution. In fact it isn't possible to draw a line between the two. Each requires the application of logic. Rubik's Cube is a puzzle, but to understand it completely requires a knowledge (possibly subconscious!) of group theory. It is left to the reader to decide to classify as puzzles or problems those items we have listed in the Treasury under the various headings: Algorithms, Arrangements, Cryptarithms, Euler or Hamilton circuits, Logic problems, Magic squares and other magic configurations, Tilings, Dissections, Coverings, Packings.

Pythagorean triples. Integer solutions to the Diophantine equation $x^2 + y^2 = z^2$; for example, $(3, 4, 5)$, $(5, 12, 13)$, $(8, 15, 17)$, $(20, 21, 29)$. They are called *primitive* if their greatest common divisor is 1. The equation can be solved by writing it in the form

$$y^2 = z^2 - x^2 = (z - x)(z + x)$$

and putting $z + x = r^2$, $z - x = s^2$ so that

$$(x : y : z) = (r^2 - s^2 : 2rs : r^2 + s^2).$$

General triples are found by multiplying primitive ones by a common factor. For example $(3, 4, 5)$ gives $(6, 8, 10)$, $(9, 12, 15)$, and so on.

Quadratic equations. Equations that involve the square of a variable, in addition to the variable itself; for example, $x^2 - 5x + 6 = 0$. This particular equation can be factored as $(x - 2)(x - 3) = 0$, and if the product of two quantities is zero, at least one of them must be zero, $x = 2$ or $x = 3$. That is, quadratic equations have two solutions, or *roots*.

We can solve the general quadratic equation by "completing the square":

$$\begin{aligned} ax^2 + bx + c &= 0 \\ 4a^2 + 4abx + 4ac &= 0 \\ 4a^2 + 4abx + b^2 &= b^2 - 4ac \\ (2ax + b)^2 &= b^2 - 4ac \\ 2ax + b &= \pm\sqrt{b^2 - 4ac} \\ x &= \frac{-b \pm \sqrt{b^2 - 4ac}}{2a} \end{aligned}$$

The quantity $b^2 - 4ac$ is called the *discriminant* of the quadratic equation. If it is zero, the quadratic is already a perfect square and its two roots are equal. If a, b, c are rational, and the discriminant is the square of a rational number, then the roots are rational. If the discriminant is negative, the roots are complex.

Quadratic residues. See **Triangular numbers**. In **S182**, 44 is not a quadratic residue of 3.

Rational numbers. A rational number is one that can be written as a ratio of two integers, a/b.

Recurrence relations. The terms of many sequences can be calculated from earlier terms. For example the geometric sequence 1, 2, 4, 8, 16,... satisfies the relation $a_{n+1} = 2a_n$, and the Fibonacci and Lucas sequences satisfy $u_{n+1} = u_n + u_{n-1}$. If these recurrence relations do not involve any powers or products of the variables, they are called *linear* and there is a way of finding a formula for the general term. As an example of how to do this, consider the Fibonacci recurrence, and suppose we guess the answer to be of the form $u_n = Ax^n$ and substitute in the equation:

$$Ax^{n+1} = Ax^n + Ax^{n-1}.$$

Divide through by Ax^{n-1} to get $x^2 = x + 1$, a quadratic equation whose roots are

$$x = \frac{1+\sqrt{5}}{2} \quad \text{and} \quad x = \frac{1-\sqrt{5}}{2}.$$

Because our recurrence is linear, its general solution has the form

$$A\left(\frac{1+\sqrt{5}}{2}\right)^n + B\left(\frac{1-\sqrt{5}}{2}\right)^n.$$

To calculate the values of A and B in the special cases of the Fibonacci and Lucas sequences we must write down the formula for two values, say $n = 0$ and $n = 1$, and solve the resulting simultaneous equations. We get, for the Fibonacci numbers u_n and Lucas numbers v_n:

$$u_n = \frac{1}{\sqrt{5}}\left(\left(\frac{1+\sqrt{5}}{2}\right)^n - \left(\frac{1-\sqrt{5}}{2}\right)^n\right) \qquad v_n = \left(\frac{1+\sqrt{5}}{2}\right)^n + \left(\frac{1-\sqrt{5}}{2}\right)^n.$$

The first of these is sometimes known as Binet's formula, but it was known to Euler.

Recurrence relations arise in connexion with the Thue-Morse sequence in **Q8**, and with the sequence of Fibonacci numbers in **P246**.

Residues. This is another word for remainders. For example, a **complete set of residues** is the set of all possible remainders when you divide by a given number, the **modulus**. In **S248** we examined the complete set of residues 0, 1, 2, 3, 4, 5, 6, 7, 8, 9 modulo 10; also the **reduced set of residues** 1, 3, 5, 9, which have no common factor with the modulus other than 1. See also, **Triangular numbers** for **quadratic residues** and **triangular residues**.

Round robin tournaments. A round robin tournament is one in which each player meets each of the other players once (for a total of $\binom{n}{2} = n(n-1)/2$ games); such a tournament is usually conducted in rounds. See **P15**, **P67**, **P161**, **P169**, **P231**, **P250**.

Can a round robin tournament be carried out in $n - 1$ rounds? For this to happen, each player has to play a different player in each round. However, if n is odd, it cannot be done, because only an even number of players can be scheduled to play in a single round. So, in this case, at least n rounds are required, and this can happen only if each player is given a bye in one of the rounds and plays in all the other rounds.

The pairings for a round robin tournament can be obtained by sliding blocks in a tray as illustrated below for the case of 9 players.

1	2	3	4	5
	9	8	7	6

For the first round, pair 2 vs 9, 3 vs 8, 4 vs 7, 5 vs 6, and give 1 a bye. For the second round, rotate the blocks clockwise to obtain

9	1	2	3	4
	8	7	6	5

and pair 1 vs 8, 2 vs 7, 3 vs 6, 4 vs 5, and 9 gets a bye. Continue in this way to get the pairings shown in the following schedule.

Rounds \ Players	1	2	3	4	5	6	7	8	9
1	×	9	8	7	6	5	4	3	2
2	8	7	6	5	4	3	2	1	×
3	6	5	4	3	2	1	9	×	7
4	4	3	2	1	9	8	×	6	5
5	2	1	9	8	7	×	5	4	3
6	9	8	7	6	×	4	3	2	1
7	7	6	5	×	3	2	1	9	8
8	5	4	×	2	1	9	8	7	6
9	3	×	1	9	8	7	6	5	4

This procedure works when 9 is replaced by any odd number. In the case of an even number of players, add a block in the blank space and use the same procedure (or set up a schedule for all but one of the players, and then the player in each round with the bye can be paired with this excepted player; e.g., read $\times$ as 'ten').

Sequences. A succession, or ordered list, of numbers or objects. For example, the Fibonacci sequence, $0, 1, 1, 2, 3, 5, 8, 13, 21, 34, 55, \ldots$ which occurs in **P246** and in dozens of places in mathematics. The Thue-Morse sequence occurs in **P8** and the sequences of triangular and square numbers in **P245**. The terms of the sequence in **P37** are the sums or differences of their neighbors, and those in **P43** form bits of a geometric series. The sequence of natural numbers occurs in **P108**, omitting those with certain digits, and in **P105** with signs attached, and their squares with signs. In **P115** we're concerned with a sequence whose subsums have a given parity, while **P132** concerns 'balanced' sequences and **Q176** 'elevator' sequences. You can solve **P247** by looking at a sequence of remainders.

Series. The sum of the terms of a sequence. The sum of the first $n+1$ terms of the Fibonacci sequence, $u_0 + u_1 + u_2 + \cdots + u_n$ is $u_{n+2} - 1$; for example,

$$0 + 1 + 1 + 2 + 3 + 5 + 8 + 13 = 34 - 1.$$

See **Arithmetic series** and **Geometric series**.

Simultaneous equations. If there is more than one unknown, then we need more than one equation in order to find the unknowns. If there are more unknowns than equations, then there are usually an infinite number of solutions. See **Diophantine equations**, for example. If there are more equations than unknowns, then, unless the equations are related, there won't be a solution. But with the same number of equations as unknowns, there are usually a finite number of solutions. In fact there is just one solution if the equations are linear (contain no powers or products of the unknowns). The simplest case is a pair of linear simultaneous equations, such as

$$37x + 13y = 5$$

$$5x + 2y = 25.$$

These are easily solved. For example, multiply the second by 37 and subtract 5 times the first, eliminating x:

$$(37\times 2 - 5\times 13)y = 37\times 25 - 5\times 5$$

$9y = 900$, $y = 100$. We can now either eliminate y by a similar process or substitute the value we've found into one of the equations and find $x = -35$.

If we replace the second equation by $111x+39y = 15$, then when you try to eliminate one variable, you'll find that you eliminate them both! In fact this new equation is only a thin disguise for the original first equation and we really only have one equation for two unknowns, for which we found infinitely many solutions under **Diophantine equations**.

If on the other hand we replace the second equation by $111x + 39y = 17$, then the equations are *inconsistent* and don't have any solutions at all.

Some simple simultaneous equations occur in **P85**, **P142**, **Q165**, **Q192**, and **Q234**.

Sine formula. The sines of the angles A, B, C of a triangle are proportional to their opposite sides a, b, c:

$$2R = \frac{a}{\sin A} = \frac{b}{\sin B} = \frac{c}{\sin C}$$

where R is the radius of the circumcircle.

Strategy. See **Winning strategy** and **Games**.

Strategy stealing. The **Tweedledum and Tweedledee strategy** is a very simple example of strategy stealing. John Nash, co-inventor with Piet Hein of the game of Hex, used a more subtle argument to show that the second player can't have a winning strategy in that game. Because, if the second player had such a strategy, then the first player could make an arbitrary move, which can't do the first player any harm, and then 'steal' whatever strategy the second player might have had. However, this non-constructive argument doesn't help you to play the game! Berlekamp uses a similar argument in his analysis of Dots-and-Boxes:

Elwyn R. Berlekamp, *The Dots-and-Boxes Game*, A. K. Peters, 2000.

Von Neumann used it to show that the first player always had a win in his version of Hackenbush (see E. R. Berlekamp, J. H. Conway, & R. K. Guy, *Winning Ways for your Mathematical Plays*, Academic Press, London, 1982; 2nd edition, A. K. Peters, Natick, MA, 2001, p. 572). Robert Hutchings also used it in John Conway's game of Sylver Coinage (*Winning Ways*, pp. 581–583, or Richard K. Guy, Twenty questions concerning Conway's Sylver Coinage, *Amer. Math. Monthly*, **83**(1976), pp. 634–637.)

Tetrahedral numbers, or **triangular pyramid numbers**, are obtained by adding the triangular numbers $1 + 3 + 6 + 10 + 15 + 21 + 28 + \cdots$:

$$1 \quad 4 \quad 10 \quad 20 \quad 35 \quad 56 \quad 84 \quad \ldots$$

You can see them in column 3 of the figure under **Pascal's triangle**.

Tilings. This is a convenient name for exact packings or exact coverings: problems in which planar or solid figures are to be put together to fill some prescribed area or volume, often subject to certain conditions.

Dominoes are the pieces in **P4**, **P172**, **P173** and **P246**, trominoes in **P155**, tetrominoes in **P119** and **P238**, and pentominoes in **P162** and **P163**. Equilateral triangles do the tiling in **P164** and right-angled triangles in **P76**. Quadrilaterals tile a regular hexagon in **P65**, the whole plane in **P181** and special rectangles (don't always) tile a square in **P199**, **P156** and **P212**, while different squares tile a rectangle in **P167**, and diamonds (sometimes) do the tiling in **P152**. The Sol-LeWitt–Barry Cipra puzzle of **P127** and Loren Larson's modification of it in **P158** are examples of MacMahon-like tilings: see **S158** for a fascinating reference.

In three dimensions, nine cubits form a cube in **P70**, while 13 blocks fail to leave a hole in the middle of a cube in P118. The notched blocks of Loren Larson's dedication to David Klarner are the pieces of a pretty puzzle in **P180**.

There's a beautiful tiling on Jim Propp's home page at

www.math.wisc.edu/~propp/trg.html.

This gives a 'proof without words' of the striking fact:

THEOREM. *If you tile a regular hexagon with calissons* (see **P152**) *there are equal numbers of calissons in each of the three possible orientations.*

Here's a picture showing the idea of the proof, with 25 calissons in each orientation. Focus on one of each of the three visible faces of the stacked cubes, those that are similarly illuminated.

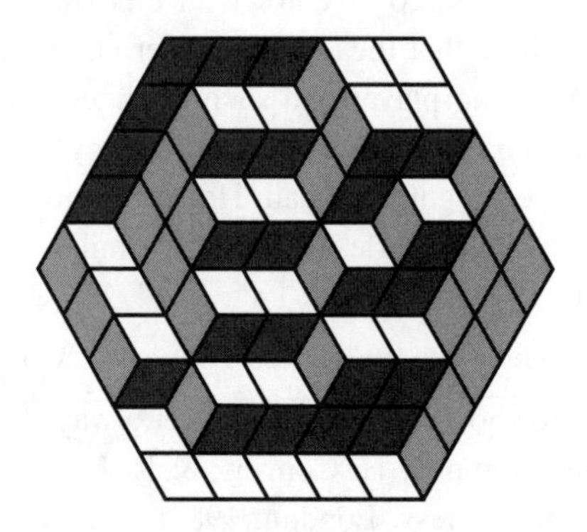

Tournaments. The most common forms are knockout (**P68**) and round-robin tournaments (see **Round-robin tournament**). The latter can be represented by the complete graph (a graph with an edge between every pair of vertices) with the players as vertices, which can be turned into a directed graph by making each edge into an arrow from the winner to the loser. Tournaments in this graph-theoretic sense have many interesting properties: see John W. Moon, *Topics on Tournaments*, Holt, Rinehart & Winston, New York, 1968. The road system of **P223** can be thought of as such a tournament.

Tree diagrams. A tree diagram is a way of visualizing the special cases that need to be considered in a problem (each branch of the tree represents a special case). In **Q6** we have a tree to illustrate the states of the sandglasses, in **P58** the truth or falsity of statements, in **P159** the results of weighings, and in **Q176** to display the possible elevator trips. Even though a tree may not be explicitly drawn, the concept is a useful construct in the mind's eye; for example, see **Backtracking**.

Triangle inequality. The length of a side of a triangle is less than (or equal to, in a degenerate triangle) the sum of the lengths of the other two sides. We use the triangle inequality to position the gazebo in **P48** and to decide where to cut the wires in **P82**. For real numbers a and b, $|a+b| \le |a| + |b|$.

Triangular numbers. The sequence of numbers, T_k, defined by the number of dots in the following triangular patterns.

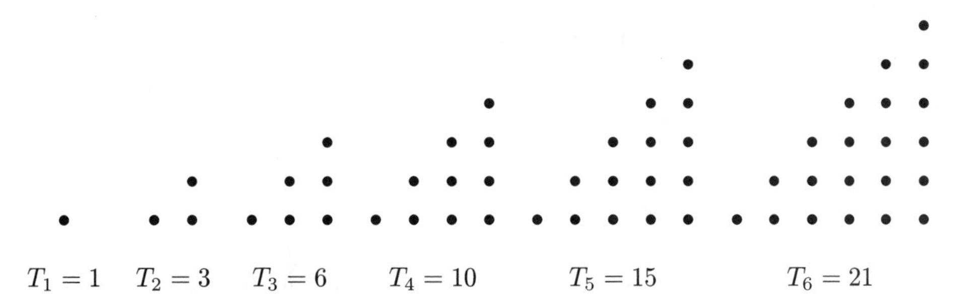

Diagram (i) makes it clear that the kth triangular number is equal to $k(k+1)/2$,

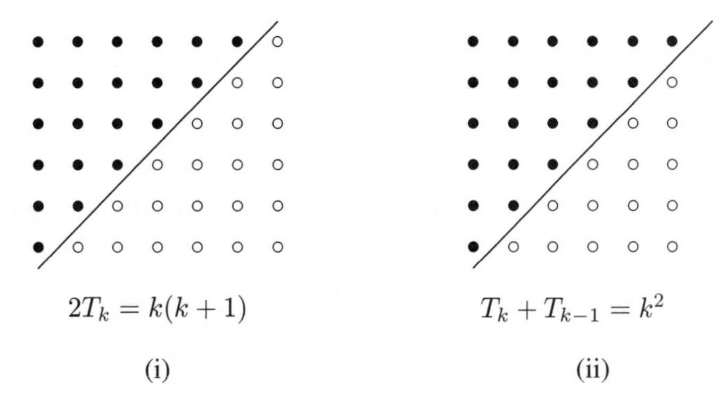

(i) (ii)

and diagram (ii) shows that two successive triangular numbers add to a perfect square,

$$T_k + T_{k-1} = \frac{k(k+1)}{2} + \frac{(k-1)k}{2} = k^2.$$

And eight copies of a triangular number plus one make an odd square:

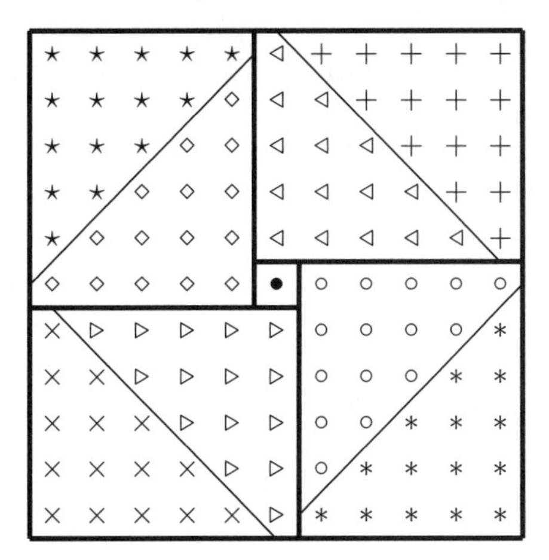

Richard Guy calls this Theorem 0 of number theory:

Odd squares are congruent to 1 *modulo* 8.

In **S245**, Bruce tasted cakes with a triangular number, and in (b) he tasted those with a square number. But in each case he was tasting modulo 25. So we can call the tasted numbers **triangular residues** and **quadratic residues**.

Here are the triangular numbers written modulo 25, the cakes Bruce tasted in part (a). We don't need to go past the middle (the underlined numbers) because the second half is just a reflexion of the first.

$$\begin{array}{ccccccccccccccccc} 0 & 1 & 2 & 3 & 4 & 5 & 6 & 7 & 8 & 9 & 10 & 11 & \underline{12} & 13 & 14 & 15 & \dots \\ 0 & 1 & 3 & 6 & 10 & 15 & 21 & 3 & 11 & 20 & 5 & 16 & \underline{3} & 16 & 5 & 20 & \dots \end{array}$$

Now, suggested by the last diagram, multiply the triangular residues by 8 and add 1, still working modulo 25 you get the squares (the cakes Bruce tasted in part (b)).

$$\begin{array}{ccccccccccccccccc} 1 & 9 & 0 & 24 & 6 & 21 & 19 & 0 & 14 & 11 & 16 & 4 & \underline{0} & 4 & 16 & 11 & \dots \end{array}$$

If you square the odd numbers modulo 25, you'll get the same answers (notice that there's no such thing as an odd number if you're working with an odd modulus!).

$$\begin{array}{ccccccccccccccccc} 1 & 3 & 5 & 7 & 9 & 11 & 13 & 15 & -8 & -6 & -4 & -2 & \underline{0} & 2 & 4 & 6 & \dots \\ 1 & 9 & 0 & 24 & 6 & 21 & 19 & 0 & 14 & 11 & 16 & 4 & \underline{0} & 4 & 16 & 11 & \dots \end{array}$$

You can see the triangular numbers in column 2 of **Pascal's triangle**.

Trilinear coordinates. See **H252**.

Truth tables. A method for determining the truth of a compound statement from the truth of its constituent parts. See **P56** and **R57**.

Tweedledum and Tweedledee Stategy. A way of winning a game by imitating your opponent's strategy. For example, if you are playing Nim with two equal heaps, this is a P-position. Whatever the Next player does (takes b beans from one of the heaps, the Previous player can imitate (take b beans from the other heap) and win. See **P203**.

Unique factorization. See **Fundamental theorem of arithmetic**.

Vectors. These are just sequences of numbers, but, when combined with certain rules, can be very useful. If there are n numbers in the sequence, it is also called an n-tuple. The numbers can be thought of as the coordinates of a point in n-dimensional space. The case $n = 3$ is especially useful in describing physical phenomena: velocity, acceleration, momentum, force (including gravity, electromagnetic force, electrostatic force), moment (turning effect) of a force, couple (a pair of equal and opposite parallel forces), are all examples of (3-dimensional) vectors. But vectors are very important algebraic quantities, too; they were used to solve **S256**.

If $n = 3$, we can write a vector in the form

$$V = \begin{pmatrix} x \\ y \\ z \end{pmatrix},$$

where x, y, z are numbers. Two vectors are equal if their coordinates are equal. The zero vector has each coordinate zero.

The product kV, for a number k, is the vector

$$kV = k \begin{pmatrix} x \\ y \\ z \end{pmatrix} = \begin{pmatrix} kx \\ ky \\ kz \end{pmatrix}.$$

If $W = kV$ for some k, we say that W is a multiple of V. Their coordinates are proportional to each other.

The **sum** of

$$A = \begin{pmatrix} a_1 \\ a_2 \\ a_3 \end{pmatrix} \quad \text{and} \quad B = \begin{pmatrix} b_1 \\ b_2 \\ b_3 \end{pmatrix}$$

is the vector

$$A + B = \begin{pmatrix} a_1 + b_1 \\ a_2 + b_2 \\ a_3 + b_3 \end{pmatrix}.$$

It's easy to see that for any k, $k(A + B) = kA + kB$.

The **scalar product** of vectors A and B is the number $A \cdot B = a_1b_1 + a_2b_2 + a_3b_3$, and you can show that for numbers k and vectors A, B, C, $(kA) \cdot B = k(A \cdot B)$ and $A \cdot (B + C) = A \cdot B + A \cdot C$

Two nonzero vectors A and B are called **orthogonal** if their scalar product, $A \cdot B$, is zero, because in this event the lines from $\begin{pmatrix} 0 \\ 0 \\ 0 \end{pmatrix}$ to $\begin{pmatrix} a_1 \\ a_2 \\ a_3 \end{pmatrix}$ and to $\begin{pmatrix} b_1 \\ b_2 \\ b_3 \end{pmatrix}$ are perpendicular.

Suppose that A and B are nonzero vectors such that neither is a multiple of the other (that is, their coordinates are not proportional). Then we know from high school algebra that the equations

$$a_1x_1 + a_2x_2 + a_3x_3 = 0$$

$$b_1x_1 + b_2x_2 + b_3x_3 = 0$$

have a nontrivial solution. In the language of vectors, these equations may be written $A \cdot X = 0 = B \cdot X$; in other words, there is a nonzero vector E that is orthogonal to both A and B, and all solutions to these equations are multiples of E. In fact E is the **vector product** of A and B,

$$E = A \times B = \begin{pmatrix} a_2b_3 - a_3b_2 \\ a_3b_1 - a_1b_3 \\ a_1b_2 - a_2b_1 \end{pmatrix}$$

[BEWARE! Vector product is not commutative: $B \times A = -A \times B$.]

This is the key idea in Cedric Smith's solution to Herbert Wright's problem, **P256**, and the vectors $U(r)$ in **S256** are (simple multiples of) the vector products $V(r) \times M$. Compare the table of $U(r)$ given there with the following table of $V(r) \times M$:

Coin # r	1	2	3	4	5	6	7	8	9	10	11	12	13
Coin	D	E	F	R	O	C	K	I	N	G	H	A	L
$V(r) \times M$	−6	3	12	12	12	−9	0	−6	3	3	−6	−9	−9
	8	8	8	−1	−10	−9	9	−10	−10	−1	−1	9	0
	−2	−3	−4	−1	2	4	−3	4	3	0	1	−2	1

Now the vector product of a vector with itself is the **zero vector**, $\begin{pmatrix} 0 \\ 0 \\ 0 \end{pmatrix}$, so we can solve the vector equation $W = gM - fV(r)$ that was found in **S256** by vector multiplying it by M:

$$\begin{aligned} W \times M &= gM \times M - fV(r) \times M \\ &= -fU(r) \end{aligned}$$

So to find the fake we calculate the vector

$$W \times M = \begin{pmatrix} 9w_2 - 3w_3 \\ w_3 - 9w_1 \\ 3w_1 - w_2 \end{pmatrix}$$

and find which coin it corresponds to in the above table.

There is a connexion between this solution and the **balanced ternary representations** of the numbers from -13 to 13. Each of these numbers can be uniquely expressed in base 3 using digits -1, 0, and 1. For example, $-5 = (1)1 + (1)3 + (-1)9$, which is the scalar product of the vector of coefficients $\begin{pmatrix} 1 \\ 1 \\ -1 \end{pmatrix}$ with $\begin{pmatrix} 1 \\ 3 \\ 9 \end{pmatrix}$, the vector of powers of 3, that is, $V(\text{Coin O}) \cdot M$. In fact the labels $r = 1$ to 13 that were used in **S256** are arbitrary, and do not enter into the calculations. If we are to number the coins, it would be more meaningful to use the scalar products $-V \cdot M$. Here is the other table from **S256**, with these numbers appended as a last row:

Coin # r	1	2	3	4	5	6	7	8	9	10	11	12	13
Coin	D	E	F	R	O	C	K	I	N	G	H	A	L
Vector $V(r)$	-1	-1	-1	0	1	1	-1	1	1	0	0	-1	0
	-1	0	1	1	1	-1	0	-1	0	0	-1	-1	-1
	-1	-1	-1	-1	-1	0	0	-1	-1	-1	-1	0	0
$-V(r) \cdot M$	13	10	7	6	5	2	1	11	8	9	12	4	3

If you try to attach 13 different letters to the numbers in this order and make easily memorable phrases to give a solution, you will appreciate just how ingenious Cedric Smith has been.

Venn diagram. A figure in the plane whose regions represent the different subsets of a given set. In **S7** the 'tennis' and 'soccer' regions overlap in the set of those who play both games; while the diagram in **H85** depicts all subsets of the set of three possibilities – compare this with our example under **Backtracking**.

Valence. The number of edges at a vertex of a graph; often referred to as the degree of the vertex.

Winning strategy. If, in a two-person game, there's complete information (both players know what is going on), no chance moves (no shuffling of cards, no rolling of dice) and no ties or draws, then one of the players must have a winning strategy in the sense that she has a scheme so that, no matter what moves her opponent makes, she always has a winning reply. That is, the positions of the game can be classified as either P-positions (Previous player winning) or N-positions (Next player winning). Terminal positions (when the game is over) are P-positions, of course. Any position having an option which is a P-position is an N-position. If all its options are N-positions, a position is a P-position.

For example, the game of Nim is played with heaps of beans. In turn, each of two players chooses a heap and removes as many beans as he wishes, perhaps the whole heap, but in any case at least one. The winner is the person who takes the last bean. The P-positions are those for which the nim-sum (see Nim-addition) of the numbers of beans in the heaps is zero. If the nim-sum is not zero, it's an N-position, and the next player has at least one winning move. So here the winning strategy is always to move to a position with nim-sum zero.

About the Authors

Paul Vaderlind was raised in Warsaw, Poland. In 1969, at the age of 21, he emigrated to Sweden. Vaderlind had been studying math at the University of Warsaw. He finished his education in Stockholm and got his PhD in the field of discrete mathematics at the department of mathematics of Stockholm University. He has taught mathematics at Stockholm since 1974. Vaderlind's first contact with problem solving was when he, at age 10, was given this beautiful puzzles book *The Moscow Puzzles* by Boris Kordemsky (now a classic). Since then problem solving and problem constructing have been his main hobbies. He has published (in Swedish) several books on this subject for both adults and children. Besides his work at the university Vaderlind is engaged in mathematical competitions and in the education of gifted youth. For several years he was the leader of the Swedish team to the IMO. He has two children and lives in a suburb of Stockholm.

Richard K. Guy has taught mathematics at all levels from kindergarten to post-graduate, in Britain, Singapore, India, and Canada. He has been involved in the publications program of The Mathematical Association of America (MAA) for quite some time, serving for twenty-five years as editor of the Unsolved Problems Section of the American Mathematical Monthly, as well as serving on several MAA editorial boards. The MAA selected him as the Hedrick Lecturer in 1978, and he won the MAA Lester R. Ford Award for Expository Writing in 1989. He is the author of a dozen books, notably among them *Winning Ways,* with Elwyn Berlekamp and John Conway; *The Book of Numbers,* with John Conway; *Unsolved Problems in Number Theory, and Unsolved Problems in Geometry,* with Hallard Croft and Ken Falconer.

Loren Larson was Professor of Mathematics at St. Olaf College from 1963–1996, and Benedict Distinguished Visiting Professor of Mathematics at Carleton College, 1999–2001. He has been active in the MAA as Governor of the North Central Section, and as Associate Director of the William Lowell Putnam Mathematical Competitions. His publications include: *Algebra and Trigonometry Refresher for Calculus Students, Problem-Solving through Problems,* and *The Wohascum County Problem Book,* with George Gilbert and Mark Krusemeyer published by the MAA.